Modeling for Preparative Chromatography

Modeling for Preparative Chromatography

Georges Guiochon
Department of Chemistry
The University of Tennessee
Knoxville, TN, USA
and
Chemical and Analytical Sciences Division
Oak Ridge National Laboratory
Oak Ridge, TN, USA

Bingchang Lin
Center of Separation Technology
Anshan University of Science and Technology
People's Republic of China

ACADEMIC PRESS

An imprint of Elsevier Science

Amsterdam - Boston - London - New York - Oxford - Paris
San Diego - San Francisco - Singapore - Sydney - Tokyo

TABLE OF CONTENTS

PREFACE

The development of appropriate mathematical models of the phenomena of interest, the study of the analytical or numerical solutions of these models, and the investigation of their properties are at the foundation of research in modern science. This applies particularly well to nonlinear chromatography, a field that is at the border of chemical engineering and physical chemistry. In turn, the theory of nonlinear chromatography is the foundation of preparative chromatography. This separation process has lately become of considerable interest in the pharmaceutical industry. During the 1990's, the development of the life sciences and of their applications have required increasingly sophisticated separations and purifications. Only chromatography is sufficiently flexible and powerful to satisfy the practical requirements encountered in most difficult separations of pharmaceuticals and pharmaceutical intermediates. For this reason, scientists and engineers involved in the separation sciences are paying considerable attention to preparative chromatography.

Obviously, fundamental research is considerably more difficult in nonlinear than in linear chromatography. The main reason for the major difficulties encountered in the understanding of nonlinear chromatography and in the development of its preparative applications is the lack of maturity of the entire field of nonlinear mathematics involved. The optimization of the design and operating conditions of a chromatography unit for the most economical achievement of a certain separation is the optimization of a nonlinear, multivariate system, a problem which cannot be solved in the general case. For several important fundamental problems relevant to preparative chromatography, it is often not even possible to write a correct algorithm after which the computer program required to calculate numerical solutions could be derived. There are no theorems demonstrating the existence of a solution or showing convergence of a numerical method for the calculation of a solution. Although scientists have been studying nonlinear chromatography since the early 1940's, the solution of many problems remains incomplete or not satisfactory.

The complexity of this fundamental problem and its lack of satisfactory ready-made solutions have the nefarious consequence that, practical solutions being required, the only issue that seems to appeal to the separation scientist is the search for empirical solutions. Unfortunately, most such attempts are extremely costly, wasteful, and often do not give a satisfactory answer. The conventional methods of trial-and-error allow the rapid achievement of acceptable results only for linear problems. By contrast, the proper recourse to a heavy dose of theory of nonlinear chromatography and to extensive numerical calculations allows the satisfaction of many of the practical needs of the pharmaceutical industry. Engineers need only the proper road maps of the field rapidly to locate the best approach to solve any new problem. The purpose of this book is to provide a guide among the various mathematical approaches that have brought some clarity to the field.

Since "nonlinear" behavior is strictly a mathematical concept, it is difficult to leave out mathematics from any fundamental study of nonlinear chromatography. Therefore, we will describe in this book the different mathematical models of chromatography, examine the assumptions on which they are based, consider their properties, and discuss their solutions. We will do all this from the point of view of mathematical analysis, paying considerable attention in this study to the influence of the nonlinear behavior of the relevant functions on the results. Still, we should not forget that the fundamental requirement of modeling is to remain germane to the physical processes studied.

ACKNOWLEDGEMENTS

An important part of this book is the result of numerous discussions that the two authors had with their coworkers over many years. We wish to express here our gratitude toward all those who, by their work and their questions, helped us in improving our understanding of the models of nonlinear chromatography. Many ideas conveyed in this work are from Professors Bingcheng Lin, Attila Felinger, Francois James, and Krzysztof Kaczmarski, and from Drs. Eric V. Dose, Anita M. Katti, Michel Martin, and Sadroddin G. Shirazi. The results of many numerical calculations are from Professor Wang Jida. The experimental and calculation results shown in Tables IX-5 and IX-6 and in Figures IX-19 and IX-20 were made by Drs. Dongmei Zhou and Fabrice Gritti. In the typing and editing of the manuscript and the drawing of the graphics, MM. Guangzhen Hu, Feng Song, Guoyong Xiao, Jiayuan Zhang, and Xuanbo Zhang provided great help for which we are grateful. Fruitful discussions of the manuscript with Professor Attila Felinger are gratefully acknowledged. This book could not have been completed and published without the contribution of all these friends and that of numerous others with whom we have interacted.

CHAPTER I
INTRODUCTION

I - THE PRINCIPLE OF CHROMATOGRAPHY

Chromatography was invented and developed in 1902–1906 by the Russian botanist Tswett [1]. It is a separation process based on the difference between the migration velocities of the different components of a mixture when they are carried by a stream of fluid percolating through a bed of solid particles called a column. The fluid and the solid phases constitute the chromatography system. Between the two phases of this system, a phase equilibrium is reached for all the components of the mixture. The separation may be successful only if the equilibrium constants of all these components have "reasonable" values. If they are too small for some components, these compounds travel at a velocity too close to that of the solvent and their complete separation cannot be achieved. If these constants are too large, the corresponding components do not leave the column or leave it so late and in bands that are so dilute that no useful purpose can be achieved. Temperature, the nature of the solid surface (or the bulk solid composition if its entire volume is accessible to the analytes or the feed components, as it is for many polymeric materials), and the nature and the composition of the fluid used as the mobile phase, together control these equilibrium constants. If the solid phase is an adsorbent, its specific surface area and its pore volume are also important. All particles used in preparative applications are porous or penetrable by the molecules of the compounds investigated, in order to maximize the capacity of the corresponding column and to allow the handling and the separation of large samples. The fluid used as the mobile phase can be a liquid, a gas, or a supercritical fluid. There are, thus, three possible forms of chromatography and all three have been used for analytical and preparative applications. They all have numerous applications although the first two forms widely dominate the fields of analytical and preparative separations.

Best results are obtained if the phase equilibrium is reached rapidly. This requires the percolation of the solvent through an homogeneously packed bed of porous particles. Furthermore, these particles should be sufficiently small to ensure a rapid diffusion of the component molecules into these particles and back out to the bulk solvent which conveys along the column the batch of product to be separated. Nevertheless, equilibrium between the two phases is closely approached only in the region of the column where the concentration distribution is at a maximum. Anywhere else, the system is not at actual equilibrium. This resistance of the system to achieve fast equilibrium limits the degree of separation that can be achieved between closely eluted components.

During a chromatographic separation, the mobile phase stream is not homogeneous in the axial direction. In all analytical and most preparative applications, chromatography is implemented as a batch process. A pulse of the mixture to be separated is injected periodically. Therefore, at any given time, the concentrations of the different components of the sample vary along the column. At any location along the column, these concentrations vary with time. Thus, diffusion tends to relax the corresponding concentration gradients, to broaden the concentration profile of each band, and to erode the effects of the separation in progress. The combination of these two types of effects, the resistances to mass transfer delaying equilibration and axial diffusion eroding the band profiles, reduces the quality of the separation achieved below what thermodynamics alone would have allowed.

It is for the separation scientists, whether analytical chemists or chemical engineers, to design and operate columns implementing a selected chromatographic system in order to achieve the fastest or the most sensitive analysis possible, the most economical or the highest production rate possible, respectively. This goal cannot be achieved without a profound understanding of the chromatographic process. Especially in preparative chromatography, such an understanding cannot be achieved without the extensive use of suitable models. To be economical, a preparative separation requires the collection of bands as concentrated as possible, hence the use of large samples of feed material. In turn, this demands that the column be operated under experimental conditions such that the behavior of the phase equilibrium is no longer linear, as it can be for analytical separations. The optimization of a separation performed by preparative chromatography becomes a multiparameter, nonlinear problem. Such problems remain incompletely mastered to this day. Only through the use of proper mathematical models of the process, using numerical solutions calculated with a computer, can we hope to understand the complex interactions between these parameters. Furthermore, the successful development of any new application requires the understanding of the many physicochemical phenomena involved in the chromatographic process. A comprehensive knowledge of the state of the art in methods, materials, equipments, and products, especially the available stationary phases — the list of which increases constantly — is also necessary. These last issues, however, are beyond the scope of this book.

II - THE MODELING OF CHROMATOGRAPHY

Although Tswett had a profound understanding of the physico-chemical phenomena involved in chromatography [2], being a botanist he did not care much for modeling and no attempts were made by him to describe chromatographic separations in mathematical terms. He did not even try to cast his observations in algebraic forms, using the physico-chemical relationships known in his time. Tswett was too far removed from the mainstream of European chemistry to be able to demonstrate the power of chromatography to foreign chemists before he died prematurely, in late 1920. For thirty years, his separation method was forgotten. It was reborn in the mid 1930's when Kuhn and Lederer [3] rediscovered Tswett's work through a German translation of his Dissertation. Following the pioneering work of A. J. P. Martin and his associates [4,5] in liquid, paper, and gas chromatography, this method rapidly became the preferred separation method for all concerned with chemical analysis in the years following World War II. Several dubious claims have been made to independent discoveries of chromatography [2]. The only serious one regards the development of the gas mask at the end of World War I. The cartridge of this device operates under the principle of the breakthrough curve: the mask bearer remains well protected until the wave of toxic gas breaks through the cartridge.

To interpret the results of their work on the cartridges of gas masks, in 1920 Bohart and Adams [6] developed a transport model of the breakthrough curve which can be considered as the first attempt at modeling nonlinear chromatography. The first fundamental investigations in this field, however, and the first theoretical models of chromatography date from the late 1930's and early 1940's. The pioneering work of Wicke [7], Wilson [8], and DeVault [9] on the one hand and of A. J. P. Martin [4,5] and Craig [10] on the other hand, began a two-track approach to developing and improving our understanding of chromatography. This dual perspective still persists today, separating these studies into linear [11] and nonlinear chromatography [12], and also into continuous [7-9] and discrete [4,5,10] models. A third approach, the stochastic theory of chromatography, follows some entirely different principles. It is based on statistics theory, considers the possible behavior of populations of molecules in a chromatographic system, and attempts to derive band profiles as the result of the behavior of a large population of molecules, using the classical results of statistical thermodynamics. The method was brilliantly successful at first and led to the work of Eyring and Giddings [13] modeling adsorption on an heterogeneous surface as the retention mechanism. Many other useful results were obtained in linear chromatography [14,15]. However, enormous difficulties prevented significant progress toward the development of a stochastic theory of nonlinear chromatography and the method was nearly abandoned for thirty years. A recent breakthrough by Dondi and Felinger [16] will rejuvenate it and, combined with the possibility to perform single molecule detection, may trigger the development of molecular chromatography.

In this book, we will consider only the modeling of chromatography as a separation method using two essentially continuous phases, and large enough amounts of sample to neglect statistical fluctuations. Thus, our approach will be based essentially on macroscopic considerations involving the mass balance of the sample compounds, their isotherms in the phase system selected, and the kinetics of their mass transfer between these two phases. We will not discuss the stochastic theory of nonlinear chromatography [13-16], which is too complex, nor the discrete or plate models of chromatography [4,5,10], which are far too simplistic and supply little useful information.

III - THE GOALS OF THIS BOOK

The main purpose of this book is to show how it is now possible: (1) to investigate on a theoretical basis the influence of the different parameters of a chromatographic system on the concentration profiles obtained at the outlet of the column; (2) to explain the characteristics of the separations performed in the most important modes of chromatography; and (3) to illustrate the most important of their features. Because there is an excellent agreement between the experimental results and the theoretical predictions [12], theory allows a rapid and inexpensive prediction of many experiments, thus saving effort, time and money. Admittedly, this agreement is achieved only when the physico-chemical parameters of the thermodynamics and kinetics of the specific separation studied have been accurately measured, which may be long and costly. Theory also permits the optimization of separations for any objective function, the comparison between the performance of different implementation schemes, e.g., recycling or simulated moving bed separations, and the derivation of relevant isotherm data from the chromatographic signals acquired. Finally, theory alone makes possible the complete investigation of the inverse problem of chromatography, hence the derivation of the single-compoent and the competitive equilibrium isotherms from chromatographic data. In this area, earlier solutions lead to simple methods such as frontal analysis [17] and the elution by characteristic points [18]. Recent progress in theory combined with the considerable advances made in computer power allow the direct determination of the parameters of an isotherm model from experimental band profiles [19,20].

All the theoretical discussions presented in this book apply in principle to gas, supercritical, and liquid chromatography. There are profound differences between the solutions of the various forms of chromatography models, however. This difference is caused by the widely different compressibilities of these mobile phases. The fluid compressibility controls the retention. Because the flow of the mobile phase is viscous, the local pressure decreases along the column. Accordingly, the velocity of the mobile phase increases along the column since this fluid decompresses and since its specific volume increases with decreasing pressure. For liquids, the effect is generally negligible [21]. For gases, it is most important but it is also relatively easy to calculate, as long as the carrier gas behavior is close to ideal [22]. For nonideal gases, dense gases, or supercritical fluids, the proper equation of state (the so-called P, V, T equation) must be used, rendering the solution of the chromatographic problem specific to each particular application of these forms of chromatography. Furthermore, the concentration of a solute in a compressible mobile phase decreases when the local pressure decreases, because of the fluid expansion. While in liquid chromatography the elution of a breakthrough curve depends only on one point of the equilibrium isotherm, in gas chromatography it depends on a whole arc of this isotherm. This physical dilution must be included in models of nonlinear gas chromatography. A mass balance equation must be added for the carrier gas and, thus, the investigation of these models is far more complex than that of their liquid chromatography counterparts, as explained elsewhere [23]. In this book, we will assume that the mobile phase is not compressible. So, in effect, many of the conclusions reported here apply only to liquid chromatography. These results can be used as a first step in the investigation of nonlinear gas or supercritical chromatography.

IV - BOOK OUTLINE

To be able to achieve our goals, we need first to cover a rather extensive amount of background material. Chapter II covers the wide range of the physico-chemical phenomena involved in chromatography. It discusses the relationships which all models of chromatography must use. These include the thermodynamics and the kinetics of phase equilibrium, the equilibrium isotherms of pure compounds and the competitive isotherms of mixtures of solutes, the phenomena responsible for axial dispersion in chromatographic columns, the mass transfer resistances, and the important Darcy law that controls the convective flow of the mobile phase [24]. Although the Langmuir isotherm model is used in almost all the theoretical developments, this chapter discusses briefly some fundamental problems related to the derivation of models of equilibrium isotherms from the general principles of statistical thermodynamics and from considerations of molecular thermodynamics. In Chapter III, the derivation of the differential mass balance equations that are used in all models of chromatography is presented, the selection of the proper model of mass transfer kinetics is discussed, and the various boundary conditions that are the mathematical translation of the different modes of operation of a chromatographic column are described. Finally, important and useful results on the theory of partial differential equations are briefly reviewed.

In Chapter IV, we describe and discuss the effects of diffusion and dispersion on the band profiles of pure compounds obtained for the common boundary conditions, with emphasis on elution and frontal analysis, the effects of the mass-transfer kinetics and those of the stochastic nature of a bed of packing material. Most of this material relates to linear chromatography, a simple case in which it is possible to investigate more accurately these kinetic effects. In Chapter V, we analyze the ideal, nonlinear model of chromatography and its properties. This model is the limit case in which diffusion and the mass-transfer resistances approach zero. Under these conditions, the nonlinear behavior of the isotherm of phase equilibrium is the only factor that determines the shape of the elution profiles. The importance of this model arises from it showing the best possible performance of a given chromatographic scheme implemented with a certain phase system. An important feature of this model is that a concentration shock will form during the chromatographic process. The formation of this shock, its migration velocity, and its retention time are analyzed in Chapter V. In order not to limit the discussion to the Langmuir adsorption (or partition) model, we also discuss the more general Lax solution. It satisfies the chromatographic equation whenever the isotherm is a convex function. To permit a better understanding of the meaning of the "discontinuous solution", the concept of weak solution is also analyzed. In Chapter VI, nonlinear reaction chromatography is introduced as an application of ideal, nonlinear chromatography in heterogeneous catalysis.

The solution of the important nonideal, nonlinear model of chromatography is discussed in Chapter VII for the single component case. This model is the most realistic one discussed in this book. It has no closed-form solutions but for a few exceptional cases that are reviewed. Approximate solutions, valid at low or high concentrations and derived from the corresponding solutions of the ideal, nonlinear and the nonideal, linear models, exist. They

provide useful insights and are discussed in depth. We also present and illustrate in this chapter the important concept of the shock layer. Finally,
we show how the transition from linear to nonlinear conditions can be expressed with a simple number. In Chapter VIII, we discuss the application of
the ideal, nonlinear model to the two-component case. Fundamentally, the
purpose of chromatography is to separate substances, so a deep understanding of multicomponent chromatography is the essential goal of theoretical
investigations. The principle of superposition does not hold in nonlinear
chromatography. The profiles of the elution bands of the individual components in multicomponent chromatography cannot be expressed merely after
the results obtained in single component chromatography. Because nonlinear isotherms are competitive, the solution of the chromatographic problem
for each component is coupled with the solutions for all the other components, so a global analysis of the problem is necessary. In this book, only
two-component chromatography is discussed. Although this is the simplest
of the problems arising in multicomponent chromatography, this is also the
most fundamental one. We show how it is possible, through the method
of characteristics, to derive the characteristic direction and the Riemann
invariant of the two-component system and, from these results, the simple
wave solution, the shock solution, and the retention time under condition of
Langmuir isotherm behavior.

Although the solutions of the ideal model are most useful to understand
how separations proceed in multicomponent nonlinear chromatography, the
solutions of more sophisticated models, those taking axial dispersion and
mass transfer resistances into account, are needed for all practical applications. It is not possible, however, to obtain analytical solutions of systems
of nonlinear Partial Differential Equations (PDE) when both diffusion and
the mass-transfer resistances are included. So, numerical analysis plays an
important role in the calculation of solutions of the mathematical models
of nonlinear chromatography, in the study of their properties, and in their
applications. We discuss the search for numerical solutions of these problems
in Chapter IX. In this chapter, we present the general concepts of numerical
analysis and numerical calculations in nonlinear chromatography and discuss
the relationship between the discrete methods and the conventional kinetic
theory of chromatography, often referred to as the Height Equivalent to a
Theoretical Plate (HETP) theory. On the concrete side, the characteristics
scheme, the conservation scheme, and a few other computational schemes are
presented. We show how it is possible to eliminate the errors arising from
the pseudo–oscillations of the numerical solutions in the Total Variation
Diminishing (TVD) scheme, thus providing a serious improvement because
this error drastically affects the calculations of retention times. Finally, the
Galerkin method and the orthogonal collocation methods are introduced to
solve problems containing diffusion terms.

LITERATURE CITED

[1] M. S. Tswett, *Tr. Protok. Varshav. Obshch. Estestvoistpyt., Otd. Biol.*, **14** (1903, publ. 1905) 20. *"On the new Category of Adsorption Phenomena and their Applications in Biochemical Analysis."* Reprinted and Translated in G. Hesse and H. Weil, *"Michael Tswett's erste chromatographische Schrift,"* Woelm, Eschwegen, 1954. As cited in K. Sakodynskii, *J. Chromatogr.*, **49** (1970) 2.

[2] M. Verzele and C. Dewaele, *"Preparative High Performance Liquid Chromatography,"* TEC, Gent, Belgium, 1986, Chapter I.

[3] R. Kuhn and E. Lederer, *Naturwissenschaften*, **19** (1931) 306.

[4] A. J. P. Martin and R. L. M. Synge, *Biochem. J.*, **35** (1941) 1358.

[5] A. J. P. Martin and A. T. James, *Biochem. J.*, **50** (1952) 679.

[6] G. S. Bohart and E. Q. Adams, *J. Amer. Chem. Soc.*, **42** (1920) 523.

[7] E. Wicke, *Kolloid Z.*, **86** (1939) 295.

[8] J. N. Wilson, *J. Amer. Chem. Soc.*, **62** (1940) 1583.

[9] D. DeVault, *J. Amer. Chem. Soc.*, **65** (1943) 532.

[10] L. C. Craig, *J. Biol. Chem.*, **155** (1944) 519.

[11] J. C. Giddings, *"Dynamics of Chromatography,"* M. Dekker, New York, NY, 1965.

[12] G. Guiochon, S. Golshan-Shirazi and A. M. Katti, *"Fundamentals of Preparative and Nonlinear Chromatography,"* Academic Press, Boston, MA, 1994.

[13] J. C. Giddings and H. Eyring, *J. Phys. Chem.*, **59** (1955) 416.

[14] D. A. McQuarrie, *J. Chem. Phys.*, **38** (1963) 437.

[15] F. Dondi and M. Remelli, *J. Phys. Chem.*, **90** (1986) 1885.

[16] F. Dondi, P. Munari, M. Remelli and A. Cavazzini, *Anal. Chem.*, **72** (2000) 4353.

[17] G. Schay and G. Szekely, *Acta Chim. Hung.*, **5** (1954) 167.

[18] E. Cremer and G.H. Huber, *Angew. Chem.*, **73** (1961) 461.

[19] F. James, M. Sépúlvéda, F. Charton, I. Quiñones and G. Guiochon, *Chem. Eng. Sci.*, **54** (1999) 1677.

[20] A. Felinger and G. Guiochon, *J. Chromatogr. A*, In Press.

[21] M. Martin, G. Blu and G. Guiochon, *J. Chromatogr. Sci.*, **11**, (1973) 641.

[22] A. T. James and A. J. P. Martin, *Biochem. J.*, **50** (1952) 679.

[23] P. Rouchon, M. Schonauer, P. Valentin and G. Guiochon, *Separat. Sci.*, **22** (1987) 1793.

[24] J. C. Giddings *"Unified Separation Science,"* Wiley–Interscience, New York, NY, 1991.

CHAPTER II
THE PHYSICOCHEMICAL
BASIS OF CHROMATOGRAPHY

In order successfully to model any process, we must first understand its physical, physico–chemical, and chemical basis, then we must select and understand the main mathematical tools necessary to build and use appropriate models of this process. Like all separation methods, chromatography involves a combination of selective, convective and diffusive transports [1]. Selective transport is the motor for a separation. Convective transport is indifferent but, properly combined with selective transport, it allows for the generation of extremely powerful separation schemes, with large number of effective stages. Dispersive transport has a negative effect, it tends to undo the separation work already performed and must be limited as much as possible.

Chromatography is a separation process based on the distribution equilibrium of the feed components between two different phases, the stationary and the mobile phase; the selective transport results from the tendency of the system to evolve toward a state of equilibrium that is characterized by the equality of the chemical potentials of each component in these two phases. Hence, a thorough knowledge of the thermodynamics of phase equilibrium is central for the understanding of all processes based on chromatography. We will limit the presentations made in this book to liquid–solid equilibria. Most of the theoretical developments discussed, however, are easily extended to the other separation or retention mechanisms used in practice.

Preparative chromatography is an important separation process used in the pharmaceutical industry. For obvious economical reasons, it is most often carried out at relatively high concentrations and the equilibrium isotherms are rarely linear. Most often, several components being simultaneously present, these isotherms are also competitive. So, we will discuss the properties and the modeling of nonlinear isotherms in a first section.

The transport of the feed components along the chromatographic column is caused by the percolation of the mobile phase through the column bed, which has been prepared by the consolidation of an adequate amount of the packing material. We will, therefore, cover briefly the essentials of flow of liquids through porous media. Finally, dispersive transport arises from axial and eddy diffusion in the mobile phase and from the mass transfer resistances, all phenomena that control the kinetics with which phase equilibria are reached. The thermodynamics of phase equilibria, its kinetics, and flow convection are the three essential physicochemical phenomena that must be known in order to understand the behavior of the bands of feed components while they migrate at different velocities along the column, interacting, and separating during their propagation.

The discussion of the physicochemical models of chromatography serves as the conclusion of this chapter. This discussion is essentially a description of the physical and physicochemical phenomena that different models will take into account or neglect, depending on their degree of complexity. This discussion of physical models is devoid of equations. The simplest of these models is the ideal model. It takes exclusively the thermodynamics of phase equilibrium into account. It predicts the best possible results that can be achieved by chromatography. Even when highly efficient HPLC columns are used, corrections are needed to account for the effects of axial dispersion and the mass transfer resistances. Several models accounting for the limited efficiency of actual columns have been suggested in the literature. The more complex of these models can afford only numerical solutions. Under certain conditions, the simplest models lead to useful algebraic solutions. This tradeoff balance between the degree of complexity of the model, the extent of its validity, and the ease of its use is discussed. The translation into mathematical equations of the various features of the physical models described is presented in the next chapter.

For a long time, the analytical applications of chromatography were dominant in this field. Little attention was paid to nonlinear chromatography and to its preparative applications. Recently, however, the traditional separation tools available to the chemical engineer proved to be unable to satisfy the needs of the laboratories and the production plants of the biotechnology companies and even of the traditional pharmaceutical industry. Both require the production of large amounts of highly purified chemicals and biochemicals. Preparative chromatography is now recognized as the most promising general process available for the achievement of sufficiently large production rates of the high purity compounds required in the pharmaceutical industry. This, in turn, has promoted the development of nonlinear chromatography. The study of the properties of the different models of this process, the discussion of the system of equations that model them, and the investigation of the properties of their solutions is not only an important topic of theoretical research, it is also a necessary step in the proper utilization of chromatography. Made possible by modern computer technology, the development of numerical solutions for the mathematical models of nonlinear chromatography allows in most cases the prediction of the optimum design and operation conditions of new separations. It also supplies the basis for the necessary rationalization of the experimental results achieved in preparative chromatography.

We discuss now in detail the different phenomena involved in chromatography to explain the physicochemical basis of the different models used.

I - EQUILIBRIUM ISOTHERMS

Chromatography is based on the differences that exist between the equilibrium constants in a biphasic system of the feed components that one needs to separate. The literature describes a very large number of phase combinations that have been and are used in the different applications of the method. One of these two phases must be a fluid to permit the differential migration needed in chromatography. Gases, liquids, or supercritical fluids can be used. The essential difference between these types of fluids, as far as modeling is concerned, is in their compressibility. Because of the enormous difference between gas and liquid compressibilities, the velocity of a gas phase cannot be considered as constant along the column while that of a liquid phase should be, within an excellent approximation, at least as far as retention is concerned [2-4]. Gas–solid and gas–liquid isotherms depend on the local pressure and, accordingly, the equilibrium constant varies along the column. This introduces additional complexities in the modeling of gas chromatography, which have no equivalent in liquid chromatography. We will not consider here any model of gas chromatography. Some elements on modeling in this field can be found elsewhere [4-6]. We will not consider supercritical fluid chromatography either.

The second phase is most often a solid and is almost always stationary. Accordingly, the words *solid phase* and *stationary phase* will be considered as synonymous in this book. It must be noted, however, that in true moving bed separations, the solid phase is actually moving too, in the direction opposite to that of the liquid phase. By contrast, in simulated moving bed separators, the solid phase is actually stationary but, for modeling purposes, should better be considered as moving in the direction opposite to that of the fluid phase. Fortunately, the same models can be used as in more conventional applications of chromatography, because, with appropriate variables, the relative velocity of migration of the two phases is all that matters. We will assume that such adjustments are not beyond the capacity of the readers when they need to apply a column model to a moving bed implementation. A liquid stationary phase is used in liquid–liquid or countercurrent chromatography (LLC), a highly efficient implementation of liquid–liquid extraction for analytical purposes or for the preparation of small amounts of purified products. However, this process encounters serious difficulties in achieving fast mass transfer between phases in scaled-up implementations and is not much used yet in industrial separations (if at all). All the models discussed later apply to LLC and the transposition requires only the appropriate changes of a few words.

Among the many possible solid–liquid equilibria, a large number are found to be useful for practical applications. The most important of these equilibria are solid–liquid adsorption, ion-exchange, size-exclusion, and liquid–liquid partition. The stationary phases used are, respectively, an adsorbent, a resin the skeleton of which carries ionized groups that can exchange counter-ions with the stream of mobile phase, a solid with pores having sizes of the order of the dimensions of the molecules of the feed components, and a liquid coated on a solid support. Ion exchange and adsorption equilibria are the two types of equilibria most often used in practical applications. Ion exchange is conveniently divided into anion- and cation-exchange chromatography, depending on the nature of the ions (*e.g.*, $-NH_3^+$ or $-SO_3^-$) bonded to the network of the resin used.

Adsorption chromatography itself can be divided into several important subgroups. Carried out with a polar solid adsorbent (*e.g.*, porous alumina, silica, or zirconia) and an apolar solvent (*e.g.*, n-pentane, dichloromethane, or THF), it is known as normal-phase liquid chromatography (NPLC). If the adsorbent is nonpolar (*e.g.*, porous cross-linked polystyrene) and the solvent polar (*e.g.*, water, methanol, or their mixtures), it is reversed-phase liquid chromatography (RPLC). If the surface of the adsorbent is covered with chemically-bonded groups (*e.g.*, octadecyl groups bonded to silica), we have also an RPLC system. However, the liquid–solid interface becomes thicker than a conventional monolayer and the retention mechanism may seem to acquire certain characteristics of partition chromatography. When the adsorbent is nonpolar (*e.g.*, bonded octadecyl groups, alkyl 1,2 diols or polyglycol chains) and the solvent is a polar, dry organic solvent (*e.g.*, diethyl ether), we have NARP (non-aqueous reversed-phase chromatography). In HIC (hydrophobic interaction chromatography), both the adsorbent and the solvent are polar. Finally, if the surface of the adsorbent contains atoms or groups which can form rather strong complexes with a component of the feed or a few of them, and if these complexes can be dissociated in a different mobile phase to recover them, we have bioaffinity chromatography. Note that graphite or graphitized carbon is often used as an adsorbent but carbon-based systems do not enter conveniently in any of these classes. Chromatography with carbon as a stationary phase is not NPLC, nor RPLC, nor NARP. Adsorption on carbon is strong and the mobile phase must be strongly adsorbed on carbon to compete with the feed components and give them the moderate retention required in practical applications. The strong solvents used with these stationary phases have a relatively large molecular weight and a high polarizability, *e.g.*, chlorobenzene, chloroform. Abundant, useful, and relevant information on retention mechanisms can be found in a recent publication [7].

Whatever the actual retention mechanism, however, the models used to account for the experimental results are the same. This illustrates the power and generality of the fundamental approach. To model a separation at high concentrations, it suffices to know the equilibrium isotherms of the feed components between the two phases. We need not be concerned here with the chemistry behind the isotherms nor by molecular thermodynamics. However, we recognize that these fields, because they allow predictions regarding the activity of solutes and adsorbates, may be most useful in the detailed investigation of any practical application, for example in the choice of the phase system. In practical applications, isotherm data are measured, then fitted to the equation of an equilibrium model. Any isotherm equation can be used, whether or not the physical model of adsorption for which it was initially derived applies to the phase system studied. For example, the Langmuir model can be used to account for data collected in RPLC, in ion-exchange chromatography, or in liquid–liquid equilibria. For this reason, in the rest of this section, the word absorption is used often as a synonym for the actual retention mechanism.

Finally, we must stress that, much as for many other experimental issues, isotherm measurements are not within the scope of this book. The interested reader is referred to the comprehensive reviews of Conder and Young for gas–solid isotherms [5], to those of Ruthven [8] and Guiochon *et al.* [9] for liquid–solid isotherms.

1 - Single Component Isotherms

Usually, the equilibrium isotherms of analytes behave linearly under the experimental conditions selected to implement analytical chromatography (eqn. II-1a below). Note, however, that *isotherms are almost always nonlinear and the fact is rather that nonlinear effects in analytical chromatography are merely too small to be measurable.* The isotherms of the feed components are most often nonlinear in the concentration range used in preparative chromatography. Even in analytical chromatography, the isotherms of most mobile phase additives behave nonlinearly because the concentrations of these additives are finite. This may cause a variety of effects (*e.g.*, system peaks [9]). When a solute concentration increases, deviations from linear behavior are observed (*e.g.*, for the main components in some trace analyses). In an intermediate concentration range, low to moderate, the isotherm can usually be expressed as a two-term expansion of the isotherm equation, also known as the parabolic isotherm (eqn. II-1b, below). In the general case, however, the isotherm of a given compound is nonlinear and is often accounted for by a complex model. The Langmuir model (eqn. II-1c, below) is the simplest of the realistic nonlinear models. Thus, we will use for the isotherms of solutes one of the following equations

$$q = aC \tag{II-1a}$$

$$q = aC + AC^2 \tag{II-1b}$$

$$q = \frac{aC}{1 + bC} \tag{II-1c}$$

$$q = f(C) \tag{II-1d}$$

where a, A, and b are numerical coefficients. Note that at moderate concentrations, eqns. II-1b and II-1c are equivalent, with $A = -ab$.

Although the chromatographic process is essentially nonlinear, we can use three different types of models. In the **Linear Model**, which applies to most analytical applications of chromatography because sample sizes are small and the concentrations at injection are dilute, the higher order terms in the isotherm can be ignored and the isotherm may be replaced in practice by a linear approximation (eqn. II-1a). This is why analytical chromatography is usually named linear chromatography. At finite concentrations, for example in most cases of preparative chromatography, the concentrations at injection are large, the retention mechanism (or the column) is said to be overloaded and the whole isotherm must be taken into account (eqns. II-1c or II-1d). The models which assume this feature are the **Nonlinear Models**. So, chromatography at finite concentration is also named nonlinear chromatography. In the intermediate case, the isotherm is too strongly curved for a linear approximation to be satisfactory, yet a second degree expansion may satisfactorily account for the deviation of the isotherm from linear behavior (eqn. II-1b). This simplification allows a detailed study of the onset of column overloading, using *e.g.*, a perturbation method. The results obtained with this simple approximation find some application in analytical chromatography.

2 - Single-Component Isotherm Models

There are many models of isotherms for pure compounds. They have been comprehensively and excellently reviewed in the literature [8]. They have already been discussed in connection with chromatography [9]. A more recent assessment of isotherm equations available in chromatography [10] is also useful. A detailed discussion of isotherm equations is not within the scope of this book, which is limited to the mathematical problems of chromatographic models (Chapter III) and to their analytical (Chapters IV to VIII) and numerical (Chapter IX) solutions. However, a list of single-component models conventionally used in chromatography is reported in Table II-1.

Nevertheless, an isotherm model must be used in any theoretical discussion. For this purpose, we will mainly use the Langmuir isotherm [8-12], as given in eqn. II-1c. This model is as simple as the parabolic isotherm (eqn. II-1b) but it is a more realistic model from a physical viewpoint. It accounts for the most common origin of the saturation of the retention mechanism at high concentrations, and the crowding out of the retained molecules that have limited access to the stationary phase (*e.g.*, in adsorption, the surface area of the solid phase is finite). It otherwise assumes that both the solution and the adsorbed layer are ideal, that there are no adsorbate–adsorbate interactions, and that adsorption takes place in a monolayer. It is logical to take as the unit of concentration in the stationary phase the monolayer

TABLE II-1

EQUILIBRIUM ISOTHERM MODELS[10]

Model	$\theta = q/q_s^\dagger$	Reference
Henry	KC	[8]
Langmuir	$KC/(1 + KC)$	[11,12]
Bilangmuir	$q_1 K_1 C/(1 + K_1 C) + q_2 K_2 C/(1 + K_2 C)$	[9]
Jovanovic	$1 - e^{-KC}$	[13]
Quadratic	$(KC + K^2 hC^2)/(1 + 2KC + K^2 hC^2)$	[14]
Fowler	$KC/(e^{-\chi\theta} + KC)$	[15]
Freundlich	$(aC)^\nu$	[16]
Langmuir–Freundlich	$(aC)^\nu/(1 + (aC)^\nu)$	[17]
Tóth	$KC/[1 + (KC)^\nu]^{1/\nu}$	[18]
Misra	$1 - [1 - (1 - k)KC]^{1/(1-k)}$	[19]
Radke–Prausnitz	$KC/[1 + (KC)^\nu]$	[20]
Dubinin–Radushkevich	$e^{-a[\ln b/C]^2}$	[21]

$\dagger$ In these equations, θ is the surface coverage, q_s, q_1, q_2 are monolayer capacities, C is the mobile phase concentration, K, K_i are equilibrium constants, and a, b, h, k, χ, ν are numerical parameters.

capacity of the adsorbent. Then, for a single component dissolved in a pure solvent, eqn. II-1c can be recast in a more general form

$$\theta = \frac{q}{q_s} = \frac{bC}{1 + bC} = \frac{\Gamma}{1 + \Gamma} \qquad (\text{II-2})$$

where q_s is the saturation capacity. This form allows the reduction of the Langmuir isotherm to a single parameter. In practice, it is most useful because it affords a simple, rapid estimate of the degree of deviation from linear behavior that the isotherm exhibits under a given set of experimental conditions. This degree is given by the value of the product $bC = \Gamma$. If C is small enough that $\Gamma < 0.01$, the isotherm behavior is practically linear; in the concentration range in which $0.1 < \Gamma < 1.0$, it is strongly nonlinear; at high concentrations, when $\Gamma > 1$, the isotherm becomes rapidly close to saturation, hence the amount adsorbed at equilibrium becomes nearly independent of the liquid phase concentration. In many cases, the Langmuir model has proven to afford a good or, at least, a reasonable empirical equation to which experimental adsorption data can be fitted successfully. Obviously, the resulting numerical isotherm equation should not be extrapolated.

Four models of isotherms are illustrated in Figure II-1, the linear, the Langmuir, the bilangmuir, and the Freundlich isotherms. The third model is often found for heterogeneous surfaces (and, notably, for chiral stationary phases). When an adsorbent follows Freundlich adsorption behavior it is not suitable for use in chromatography. Because the ratio q/C increases indefinitely when C tends toward zero, band elution never ends and the peak tails forever.

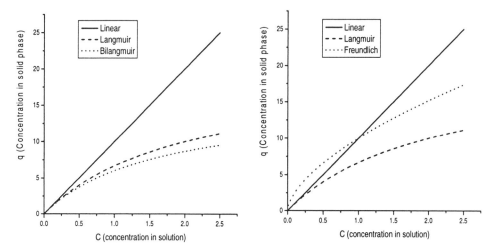

Figure II-1   Illustration of Four Conventional Isotherms. The figure on the left compares the linear isotherm (solid line), the Langmuir isotherm (dashed line), and the bilangmuir isotherm (dotted line). The numerical coefficients were selected so that these three isotherms have the same initial slopes. The figure on the right compares the linear isotherm (solid line), the Langmuir isotherm (dashed line), and the Freundlich isotherm (dotted line). Note the major difference between the initial slopes of these three isotherms. (q and C are in arbitrary units).

3 - Competitive Isotherms

Possibly the most important characteristic of nonlinear isotherms is that they are also competitive. The solid phase concentration of any component of the solution is almost always a function of the concentrations of all the components also present in this solution (additives and all other sample components). So, the equilibrium isotherm of any component in a multi-component system is given by

$$q_i = f_i(C_1, C_2, \cdots, C_i, \cdots, C_n) \tag{II-3}$$

This equation is most general. It provides one of the great sources of complexity in the mathematics of the different models of chromatography. This physical phenomenon of competition explains the interactions observed between high concentration bands in the separation of multicomponent mixtures [9].

As it is necessary in all theoretical discussions to use an isotherm model, in further discussions we will mainly use the competitive Langmuir [8-12] isotherm

$$q_i = \frac{a_i C_i}{1 + \sum_{j=1}^{n} b_j C_j} \tag{II-4}$$

Equation II-4 gives the isotherm of compound i in a solution containing n components. This equation has the great advantage that it is realistic enough to illustrate most features of nonlinear chromatography while leading to simpler mathematical equations than any other nonlinear competitive isotherm. Another important advantage of this equation, when it applies to an actual separation problem, is practical. The competitive form of the Langmuir isotherm (eqn. II-4) is written with the same parameters as the n corresponding single component isotherms (eqn. II- 1c or eqn. II-4 when $n = 1$). In other words, if two components follow competitive Langmuir isotherm behavior, it is sufficient, at least in principle, to measure only their single-component adsorption isotherms. It is not necessary to measure the whole set of competitive isotherm data, which is a great simplification. However, in practice, even if the pure components exactly follow Langmuir isotherm behavior, some competitive data must be acquired. Competition is frequently more complex than anticipated. In numerous cases, each pure component follows Langmuir single-component isotherm behavior but the binary mixture does not follow simple competitive Langmuir isotherm behavior.

To account for a nonlinear isotherm, a minimum of two parameters for each compound is necessary. There are alternative two-parameter, single-component models. Most of these models, however, have no straightforward corresponding competitive form, even for binary systems. For those which have one, it is far more complex. Even for two compounds following Langmuir equilibrium behavior when pure, it was shown [22] that the competitive Langmuir model is thermodynamically consistent only if the saturation capacity of the two compounds involved is the same. Otherwise, a correction is necessary [22]. This correction may be important. It does not require any additional parameters, however, and the practical complication is limited to the need of writing a longer and more complex isotherm equation in a computer program [9]. A recent alternative suggestion has been made to handle the case of the competitive isotherms of two compounds, the adsorption behavior of which follows the Langmuir isotherm when they are pure but which have different saturation capacities, $q_{s,i} < q_{s,j}$ [75]. This approach, which has no general physicochemical justification, consists in assuming that the component that has the lower saturation capacity, $q_{s,i}$, does not have access to the total surface area of the adsorbent. On the fraction of the surface of area $q_{s,i}$, the two components compete. On the rest of the surface (with area $q_{s,j} - q_{s,i}$, only component j has access and there is no competition with component i. In at least one case, this empirical model gave us surprisingly good results [74]. Figure II-2 illustrates the competitive Langmuir isotherm.

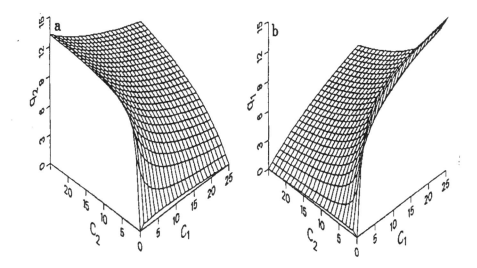

Figure II-2   The Competitive Langmuir Isotherms for the Two Components of a Binary Mixture. Left, isotherm of the second component. Right, isotherm of the first component.

Note that, if the mobile phase is a mixture of several solvents, which is often the case, the concentrations of all the solution components, except that of the weakest solvent, must be included in the isotherm equation. Obviously, the concentrations of the components of the mobile phase (e.g., the organic modifiers in RPLC) influence the isotherms of the feed components, otherwise they would not be used. Two important cases must then be distinguished.

(i) In the general case, the effect of the solvent additives are to reduce the feed-component concentration in the stationary phase at equilibrium with a given mobile phase concentration, hence to reduce retention, by crowding the molecules of feed components out of the surface. This situation is typical of normal phase chromatography, of NARP, or of chiral chromatography. Usually, the concentrations of such additives are low or moderate. They act through the very mechanism of competition and the concentrations of all the corresponding additives must be included in the competitive isotherm equation.

(ii) In other cases, the effect of the solvent additives is merely to increase the solubility of the feed components in the mobile phase. These additives are poorly adsorbed and are used at moderate to high concentrations (*e.g.*, methanol in the water/methanol mobile phases used in RPLC; note that, in RPLC, the retention factor of methanol in pure water, at infinite dilution, is of the order of unity). Then the coefficients of the isotherms of the feed components can simply be considered as functions of the additive(s) concentration(s). As a rule of thumb, a compound can be considered as such an additive if the retention factor of the less retained feed component is more than a few times (at least three times) larger than the additive retention factor in the pure weak solvent [23].

4 - Excess Isotherms

The amount of a compound adsorbed on the surface of an adsorbent is actually not the concentration of the compound at the interface but the excess concentration or difference between the monolayer and the bulk concentrations, provided these concentrations are expressed in consistent units, *e.g.*, in molar fraction. This is why the adsorption of a pure liquid on an adsorbent is rarely discussed, while the gas–solid adsorption of a pure vapor is an important problem (while the adsorption of a permanent gas is most often negligible, at least with the stationary phases used in GSC). A simple model of excess isotherm is

$$n_i^e = \frac{aX_i(1 - X_i)}{1 + bX_i} \tag{II-5}$$

where n_i^e is the excess adsorbed, X_i the mole fraction of the studied compound in the bulk solution, and a and b are numerical coefficients. A correct description of isotherms should be given in terms of excess amounts close to the surface and they should be related to the chemical potentials of each species [24,25]. It is much easier, however, and it has become conventional in applied fields, to substitute adsorbate and fluid phase concentrations for the more correct physicochemical parameters. This practice is commonly accepted in chemical engineering and we will follow it here.

When the mole fraction X_i of the compounds studied is small, eqn. II-5 becomes that of a Langmuir isotherm. Preparative chromatography is carried out at mobile phase concentrations of the feed components that very rarely exceed 10 % w/w. The molecular weight of the solutes is always at least several-fold larger than that of the solvent. Accordingly, the assumption that the mole fractions, X_i, of all the solutes are small is often made in nonlinear chromatography and we will also make it in all the following discussions.

5 - Empirical Isotherms

In practice, experimental data are acquired and fitted to an isotherm model [8,9]. Excellent agreement between the experimental data and the model prediction does not mean that the model is correct, only that it accounts well for the experimental data, within a certain precision, and within a certain concentration range. Accordingly, all isotherm models used in actual experimental investigations are empirical.

It is often observed that the Langmuir model accounts well for the experimental data acquired in HPLC, whether the retention mechanism is adsorption, complexation, or ion-exchange [9]. This is true, for example, in certain cases in which surface diffusion plays a significant role in the mass transfer inside pores (see later in this Chapter, section 4d), although this is not compatible with the Langmuir tenet of localized adsorption. One must observe that the range of concentrations investigated in preparative HPLC is rather narrow (see above). The Langmuir isotherm, like the parabolic isotherm, is often acceptable because it is a satisfactory approximation of any more sophisticated isotherm at low enough concentrations (eqns. II-1b and II-1c are equivalent), for the same reason that, within a narrow range of the variable, a two-term expansion is a reasonable approximation of a function. Not infrequently, the bilangmuir isotherm [9] was found to account better for experimental data. This model is a particular case of the quadratic isotherm which would be the second order expansion of a complex isotherm. Because it makes physical sense, however, it is more suitable for our purpose than a cubic (or third-degree expansion) isotherm. Note that the quadratic isotherm may account for an inflection point in the isotherm data (which the bilangmuir model does not). Because the Langmuir and the quadratic isotherm models both have a saturation limit at high concentrations, these two models are more realistic approaches to a precise description of isotherm data than limited power expansions.

When isotherm data are fitted to an equation, it is important to use programs minimizing the sum of the squares of *the relative differences* between experimental data and best interpolated values, not the sum of the squares of these absolute differences. The latter method would give an isotherm in which the relative weight of the high concentration data is excessive. Because band profiles rise from a concentration equal to zero and return to this value, it is important to take proper account of the low concentration data. Besides, the relative precision of the data is nearly independent of the concentration. Best results are therefore achieved by minimizing the following objective function, using Marquardt's method [26]:

$$\sigma = \sqrt{\frac{1}{N_D - P} \sum_i \left(\frac{q_{ex,i} - q_{th,i}}{q_{ex,i}} \right)^2} \qquad \text{(II-6)}$$

where N_D and P are the number of data points (*i.e.*, of points of the experimental isotherm) and the number of parameters in the isotherm equation, respectively, while $q_{ex,i}$ and $q_{th,i}$ are the experimental value of the solid phase concentration in equilibrium with the mobile phase concentration C_i and the adsorbed concentration calculated for C_i using the equation of the isotherm model, respectively.

The selection of the most adequate model should be performed using the following test which is related to the Fisher's test. The model that best correlates the data is the one that exhibits the highest value of the following parameter [27,28].

$$F = \frac{(n - l) \sum\limits_{j=1}^{n} (q_{ex,j} - \overline{q_{ex}})^2}{(n - 1) \sum\limits_{j=1}^{n} (q_{ex,j} - q_{th,j})^2} \qquad \text{(II-7)}$$

where n is the total number of experimental points, l the number of adjusted parameters of the model (N.B. In eqns. II-6 and II-7, $n = N_D$ and $P = l$), $\overline{q_{ex}}$ the mean value of the vector $\vec{q}_{ex}$, which contains the experimental data as elements $q_{ex,j}$, and $q_{th,j}$ the theoretical value of the amount adsorbed, as calculated from the model equation. The RHS of eqn. II-7 contains the residual sum of squares in the denominator. So, the higher the Fisher factor, F_{calc}, the better the model correlates the data. The first term in the numerator of the RHS of eqn. II-7 decreases with an increasing number of parameters in the model. So, eqn. II-7 has also the advantage of illustrating to which extent the introduction of a new parameter in the model can actually decrease in a significant way the residual sum of squares and improve the quality of the fit to a new model. Thus, it allows the comparison of models having different numbers of parameters [27,28].

6 - Equilibrium Isotherms and Statistical Mechanics

As explained earlier, the equilibrium isotherm of a compound is the relationship between the concentrations of this compound in the two different phases of a system when physicochemical equilibrium is established between these two phases. In chromatography, these phases are the stationary and the mobile phases. The term equilibrium isotherm is generic and applies whatever the physical nature of the retention mechanism (*e.g.*, adsorption, partition, ion-exchange, etc.). As soon as they are nonlinear, equilibrium isotherms in biphasic systems are functions, not only of the concentration of the component considered, but also of the concentrations of all or most other components of the system. This considerably complicates all the aspects of their study, particularly their modeling. A general perspective on the properties of equilibrium isotherms is given by statistical thermodynamics [29-31].

The concentration of a compound is proportional to the number of its molecules in the unit volume of the solution or on the unit surface area of the adsorbent (or rather of the monolayer). The number of molecules in a phase of the system corresponding to a component concentration is the average value of this number measured during an observation period that is short compared to the duration of the macroscopic observations made in chromatography, but is long at the scale of microscopic observations (*i.e.*, at the scale of molecular motions). In a microscopic analysis, adsorption and partition arise from molecular interactions between all the components of the system studied. When feed-component molecules move between mobile and stationary phases, the probability distribution of their presence in each phase and at each instant is determined by the energy of these interactions. The statistical mean values of the local concentrations and of all other macroscopic quantities are also determined by these energies.

The methods of statistical mechanics are required to study the influence of the molecular interactions on the macroscopic thermodynamical properties of systems. The ensemble method allows a statistical analysis of the equilibrium state when the properties of the molecules are known and the study of the influence of molecular interactions on the macroscopic quantities that characterize the equilibrium state. For example, consider one molecule in a box of volume V_B. The probability of finding at a certain time this molecule inside a volume $V_P < V_B$ is $Pr = V_p/V_B$. If there are N molecules, the probability of finding all of them together in the volume V_P at the same instant is $Pr = (V_p/V_B)^N$. This probability is still significant when N is small but becomes negligible if N is commensurate with Avogadro's number. If the interaction energy between the molecules is negligible, this state (all molecules in volume V_P) constitutes one possible example, *i.e.*, a sample or microscopic state of the system. Since, in chromatography we are interested only in molecules, we will consider only the statistics of Maxwell–Boltzmann that assumes that molecules are distinguishable and that there are no limits to the number of molecules that can occupy the same quantum state. Although molecules are really not distinguishable, the results remain valid provided they are properly corrected.

Statistical mechanics considers the collection or ensemble of microscopic states (or instant states) of the system that are all possible physical implementations of the macroscopic system studied and can thus be regarded as samples or examples of this system. Each of these states contains the same number of molecules that possibly interact and it is in one of the many possible quantum states available to the macroscopic system. All these samples are collected and their distribution, *i.e.*, the probability distribution of finding a sample of a given composition, is analyzed. The value of a physical quantity of the system (*e.g.*, its entropy) at a given time is related to the chemical composition of the corresponding sample. We assume that the macroscopic value of the physical quantity studied is equal to the statistical average value of this physical quantity over all the samples in the set. So, the statistical average value of this physical quantity is determined by the probability distribution of the different samples collected. This set of samples is called an *ensemble*, meaning in this case the assembly of all the possible microscopic states consistent with the constraints characterizing the macroscopic system. The ensemble corresponding to even a small speck of matter is enormous, because of the huge value of Avogadro's number. The statistical analysis of an equilibrium state is the analysis of the distribution of the samples in the ensemble.

The distribution of the samples or distribution of the microscopic states is different under different macroscopic conditions, thus the character of the ensemble is also different. In statistical mechanics, the possible macroscopic conditions are divided into three fundamental classes. The first class consists of the *isolated systems* that can exchange neither energy nor matter with the outside; the corresponding ensemble is called the *micro-canonical ensemble*. The second class is that of the *closed systems* that exchange energy but cannot exchange matter with the outside world. This class is important because it includes the isothermal systems. The number of molecules in these systems is constant. The corresponding ensemble is called the *canonical ensemble*. The third class includes the *open systems* that exchange both energy and matter with the outside world. Hence, both the energy of these systems and the number of molecules that they contain are variable. The corresponding ensemble is the *grand canonical ensemble*.

In adsorption or partition chromatography, the set consisting of the molecules of the various components contained in one of the two phases, the mobile or the stationary phase, can be considered as a system. Consider the systems consisting of all the molecules contained in the mobile phase (or of those contained in the stationary phase). The number of molecules contained in each system varies with time. So, both systems are open systems and the corresponding statistical ensemble to consider is their grand canonical ensemble. Thus, the calculation of the average concentration of each component in each phase during equilibration is most directly done using the grand canonical ensemble. When the adsorption or the partition process has reached equilibrium, however, the number of molecules in the mobile phase and that of those adsorbed on the surface or dissolved in the bulk of the stationary phase can both be considered to be constant (or rather, practically constant, the statistical fluctuations being neglected). Thus, the two phases of a system at equilibrium can both be considered as closed systems. In this case, the number of molecules is a macroscopic parameter of the system and the equilibrium isotherm is a relationship between macroscopic parameters of the two phases. This relationship, *i.e.*, the equilibrium isotherm, can be determined from the equilibrium conditions. Since the ensemble that corresponds to a closed system is the canonical ensemble (the system may exchange energy with the outside), adsorption and partition isotherms can be determined by calculating the equilibrium parameters in the canonical ensemble.

7 - Statistical Thermodynamics Isotherms

Applying statistical thermodynamics to the study of adsorption [30], it was shown that the general equilibrium isotherm of one compound should be written as the following ratio of two polynomials of the same degree:

$$\theta = \frac{q}{q_s} = \frac{b_1 C + 2b_2 C^2 + \cdots + nb_n C^n}{1 + b_1 C + b_2 C^2 + \cdots + b_n C^n} \tag{II-8}$$

This equation coincides with the Padé approximation that is known for being able to mimic almost any mathematical function. Thus, it should be used with care. The physical meaning of the coefficients derived from a fit of experimental data to this equation is doubtful, unless few coefficients are used and many data points have been recorded.

The Langmuir isotherm is the first-order approximation of eqn. II-8. The bilangmuir isotherm is a particular case of the quadratic isotherm, the second-order approximation of eqn. II-8, a case in which both roots of the denominator are negative (hence, have no physical meaning). For binary isotherms, the competitive Langmuir isotherm is the first-order approximation of the statistical thermodynamics isotherm. The second-order approximation is written:

$$q_1 = \frac{b_{1,0} C_1 + b_{1,1} C_1 C_2 + 2b_{2,0} C_1^2}{1 + b_{1,0} C_1 + b_{0,1} C_2 + b_{1,1} C_1 C_2 + b_{2,0} C_1^2 + b_{0,2} C_2^2} \tag{II-9a}$$

$$q_2 = \frac{b_{0,1} C_2 + b_{1,1} C_1 C_2 + 2b_{0,2} C_2^2}{1 + b_{1,0} C_1 + b_{0,1} C_2 + b_{1,1} C_1 C_2 + b_{2,0} C_1^2 + b_{0,2} C_2^2} \tag{II-9b}$$

where the coefficients $b_{i,j}$ are numerical coefficients.

II - CONVECTION AND FLOW VELOCITY

The transport of the zones along the column, hence their separation, is controlled by the flow velocity of the mobile phase percolating through the column. This velocity is related to the inlet pressure, ΔP, by the Darcy equation

$$\Delta P = -\frac{\eta\, L\, u}{k_0 d_p^2} \qquad (\text{II-10})$$

where $k_0 d_p^2$ is the column permeability (proportional to the square of the average particle size), η is the mobile phase viscosity, L is the column length, and u is the mobile phase velocity [9]. The coefficient k_0 is usually between 0.5×10^{-3} and 2×10^{-3}, depending on the packing density of the bed (*i.e.*, on its external porosity), on the shape of the particles, on the width of the size distribution and on the smoothness of the external surface of the particles. Information regarding the viscosity of pure liquids and solutions can be found in specialized publications [32-34] or for mobile phases used in HPLC, in [9]. Table II-2 reports the viscosity around room temperature of most solvents used as mobile phases in HPLC. These viscosities are all lower than 1 cP, except for water. The viscosity of most binary mixtures is related to that of their components through an equation due to Arrhenius [9]

$$\ln \eta_m = x_1 \ln \eta_1 + x_2 \ln \eta_2 \qquad (\text{II-11})$$

This equation is not valid when hydrogen bonding between the molecules of the two solvents can be involved. The viscosities of aqueous solutions of methanol and acetonitrile go through maxima of approximately 1.8 and 1.1 cP, for water concentrations of approximately 40 and 20%, respectively [9]. Correlations for the viscosity of mixtures were discussed by Li and Carr [35].

In practice, the values of the different parameters in eqn. II-10 are usually selected on the basis of other considerations than achieving a high column permeability or a low head pressure. The thermodynamics of phase equilibrium and the mass transfer kinetics are of primary concern when selecting the experimental conditions of a separation. Little attention is usually paid to the head pressure, at least as long as the equipment available can supply it (pumps) or accommodate it (valves, tubings, injection device). When this is not possible, compromises must be selected and then eqn. II-10 becomes important.

Note that the flow of the mobile phase percolating through a liquid chromatography column is not only laminar: in almost all cases it has creeping flow behavior. The Reynolds number [34] characterizes the nature of the flow.

$$Re = \frac{u d_p \rho}{\eta} \qquad (\text{II-12})$$

TABLE II-2

VISCOSITY OF SOLVENTS
USED IN MOBILE PHASES[1]

Solvent		η(cP)	$T(^\circ C)$
H_2O	Water	1.0	20
CH_4O	Methanol	0.55	25
C_2H_6O	Ethanol	1.04	25
C_3H_8O	2-Propanol	0.98	52
C_2H_3N	Acetonitrile	0.35	25
C_6H_{14}	n-hexane	0.30	25
C_7H_{16}	n-heptane	0.40	25
C_7H_8	Toluene	0.55	25
$CHCl_3$	Chloroform	0.52	25
CH_2Cl_2	Methylene Chloride	0.41	25
$C_4H_{10}O$	Diethyl ether	0.23	20
C_3H_6O	Acetone	0.32	25

[1] *Data from R. C. Reid, J. M. Prausnitz, and B. E. Poling [32].*

In this equation, d_p is the average particle size, ρ the density of the mobile phase, and η its viscosity. Liquid chromatography is usually carried out at reduced velocities, $\nu = (ud_p)/D_m$, between 3 and 20 (otherwise, the efficiency of the column and the resolution would be too small). D_m is the molecular diffusivity of the feed component of interest, of the order of 0.5×10^{-5} cm^2/s. This gives a value of ud_p which can rarely exceed $20\,D_m$, i.e., is of the order of 1×10^{-4} cm^2/s. The density of the mobile phase being almost always lower than 1.0 and its viscosity rarely lower than 0.5 cP, we obtain a practical maximum value of the Reynolds number of 0.02, a typical value for creeping flow [34]. This means that there can be only very few eddies between particles in packed columns for HPLC.

III - AXIAL DISPERSION

For the sake of brevity and clarity, it is useful to give here the precise definitions of the words diffusion and dispersion which are abundantly used in this book. *Diffusion includes the results of axial diffusion and eddy diffusion, for which the term axial dispersion is sometimes used [9]. Dispersion includes, in addition to these two contributions, the contributions to band spreading originating from the finite character of the kinetics of mass-*

transfer between the two phases of the system.[1] It is often called apparent dispersion in the literature [9]. Dispersion under linear conditions is characterized by the height equivalent to a theoretical plate (HETP) in the equilibrium–dispersive model. The HETP concept is strictly valid only when the rate of the mass transfer kinetics is relatively fast and the profile of the elution peak is Gaussian. It is not valid in nonlinear chromatography.

Diffusion tends to relax all concentration gradients. When a band migrates through a column, the axial concentration profile rises at the front of the band and falls on its tail. In a stream flowing through a porous medium, in the absence of retention and with solid, nonporous particles, diffusion is the sum of only two independent contributions, one related to molecular diffusion in the axial direction, the other, called eddy diffusion, to the heterogeneity of the distribution of the streamlines in a porous medium (when the particles are porous, there is an additional contribution, due to the consequences of the mass transfer resistances across these particles). Both contributions combine to give axial dispersion, characterized by an axial dispersion coefficient, D_a, which plays a role similar to that of molecular diffusivity in the case of axial diffusion in a perfectly still liquid. In a first-order approximation, the two contributions are additive and

$$D_a = \gamma_1 D_m + \gamma_2 d_p u \qquad \text{(II-13)}$$

where D_m is the molecular diffusivity, d_p, the average particle size and u the mobile phase velocity. γ_1 and γ_2 are geometrical constants, usually around 0.7 and 0.5, respectively.

1 - Molecular Diffusivity

The molecular diffusivity of most small molecules in usual solvents can be derived using the Wilke and Chang equation [36]

$$D_m = 7.4 \times 10^{-8} \frac{\sqrt{\psi_B M_B}\, T}{\eta_B \bar{V}_A^{0.6}} \qquad \text{(II- 14)}$$

where the subscripts A and B stand for the solute and the solvent, respectively, ψ_B is an association constant (1 for all nonassociated solvents, 1.5 for ethanol, 1.9 for methanol, and 2.26 for water [36]), M, is the molecular weight, and $\bar{V}$ is a molar volume. The molecular weight and the viscosity of the solvent are usually available from Tables of Constants. The molar volume of the solute can be calculated from structural increments [9,32]. The

---

[1]This issue is discussed in the next section, on mass transfer resistances.

association coefficient is known for pure solvents but is usually unknown for solvent mixtures. In this case, the Perkins–Geankoplis equation [32] can be used to derive the molecular diffusivity in a solution of several solvents from those in the different pure solvents through:

$$D_m \eta_m^{0.8} = \sum x_i D_{m,i} \eta_i^{0.8} \tag{II-15}$$

where i stands for the different components of the mobile phase and $D_{m,i}$ is the molecular diffusivity of the solute in each of these solvents. Other correlations for small molecules were discussed by Li and Carr [35] who recommended the Scheibel correlation as the most accurate estimate of molecular diffusivity, better than the Wilke and Chang equation (eqn. II-14):

$$D_m = \frac{8.2 \times 10^{-8} T}{\eta_B \bar{V}_A^{1/3}} \left[1 + \left(\frac{3 \bar{V}_B}{\bar{V}_A} \right)^{2/3} \right] \tag{II-16}$$

Another useful correlation is available for the molecular diffusivities of proteins in aqueous solutions [9].

2 - Bed Tortuosity

In a packed column, molecular diffusion is slowed down by the presence of the solid particles, which prevent straight trajectories of molecules along any significant distance compared to the particle diameter and force the molecules to diffuse around them. This obstruction explains the tortuosity of the channels open to the molecules. These channels are longer than those of an equivalent bundle of parallel capillaries (same volume and permeability) and their average cross-section is lower. Because the channels do not have a uniform cross-section but are constricted, diffusion through them is further slowed down. This results in a reduction of the apparent dispersion coefficient, although the dispersion effect remains proportional to the molecular diffusivity. The proportionality factor is called the tortuosity of the packed bed. This tortuosity is represented by the coefficient γ_1 in eqn. II-13, a coefficient that is approximately 0.7 for chromatographic columns. Unfortunately, the definition adopted by chromatographers is the reverse of the one used by chemical engineers [37], which might cause some confusion (this explains why the term γ/D_m is often encountered in HETP equations and why the tortuosity values of chromatographic packing materials found in the literature are of the order of 1.4). The band broadening contribution of axial diffusion in linear chromatography is approximately two thirds of the value that would be observed for this contribution if strict piston flow would take place along a cylindrical channel parallel to the column axis and having the same length, L, as that of the column.

3 - Eddy Dispersion

The second term in the RHS of eqn. II-13 accounts for the contribution of eddy dispersion to axial dispersion. Because of the stochastic distribution of particles in the beds of packed columns, there are regions where the packing density is higher, hence the permeability lower, than in other regions. Furthermore, the local velocity of the mobile phase varies widely across a channel, depending on the local position. The anastomosis of these channels contributes actively to the homogenization of the average velocities along each stream-path. Still, these effects combine to contribute to an apparent dispersion of significant magnitude, called the *eddy diffusion*. This phenomenon is not entirely negative, because it also contributes to accelerate radial mass transfers in the mobile phase (see next section). Axial and eddy diffusions have been actively studied in liquid chromatography for over thirty years. They still are. For further information, the reader is referred to an important review by Weber and Carr [38] and to recent publications comparing HPLC and NMR data on axial dispersion in chromatographic columns [39]. Detailed data on molecular diffusivity and on dispersion are in [9].

4 - Column Radial Heterogeneity

Strictly speaking, this contribution is not related to diffusion. In almost all theoretical investigations of chromatography, it is assumed that the column is radially homogeneous. Accordingly, there is no radial component for the local mobile phase velocity in any point of the column. Everywhere across the column, the mobile phase velocity is assumed to be parallel to the column axis and to be constant both in the radial and the axial directions. More refined analyses admit that this is incorrect and that the velocity fluctuates locally but they consider that the above properties remain valid for the velocity averaged over distances of a few particle diameters, d_p. Indeed, all the chromatographic models discussed in this book are defined in a two-dimensional space, the time and the abscissa along the column. A few problems require the consideration of a column which is not radially homogeneous. A simple model of chromatography has been used to study a few such problems, assuming a cylindrical column and two space dimensions, the distance along the column length and the radial position [40]. Models assuming that columns are radially homogeneous are generally acceptable, however, because, in most cases, actual columns are nearly homogeneous.

The consequence of a small degree of radial heterogeneity is to cause another contribution to the apparent axial dispersion. In practice, this means that columns should be carefully packed and that the packing procedure should be designed so as to avoid any systematic radial variation of the packing density. This also means that a significant contribution to the band variance derived from the record of the concentration measured by a detector analyzing the bulk eluent is due to the radial heterogeneity of the column. The interpretation, in terms of mass transfer kinetics through the particles of packing material, of the band variance measured in linear or quasi-linear chromatography must be made cautiously and only on well-packed columns exhibiting a high efficiency for nonretained solutes.

IV - MASS TRANSFER RESISTANCES

When a sudden concentration change takes place in a region of a biphasic system at equilibrium, the equilibrium cannot be restored instantaneously everywhere. This takes a certain time, depending on the rate at which the mass transfer takes place in the system. The passage of a component zone is accompanied by a rapid variation of the local concentration in the mobile phase. Local equilibrium between the stationary phase, which is almost always a porous solid, and the liquid phase percolating through the bed will eventually be restored in a static system but this cannot be immediate. The time needed to achieve equilibrium depends on several characteristics of the compounds studied and of the phase system used for their separation. Actually, however, there is no equilibrium anywhere in a chromatographic column, except around the location of the band maximum. As equilibrium tends to be approached in the stationary phase, the band keeps migrating and the concentration keeps changing in the mobile phase. The equivalence between the contributions of the mass transfer resistances to band broadening and an apparent dispersion was demonstrated by Giddings [1,2]. This analysis was rephrased recently [9]. It is illustrated by the presence of mass transfer resistance terms in the Van Deemter and in the other plate height equations [9].

Several phenomena contribute to mass transfer and the overall kinetics is complex. Most models give priority to one or two of these phenomena and neglect the contributions of the others. The general rate model takes all of them into account (see Chapter III, Sections III to VI). For this reason, it is most complex. The phenomena which contribute significantly to the mass transfer kinetics are called sources of mass transfer resistances. The main ones are now briefly discussed below.

1 - Radial Dispersion in the Mobile Phase Stream

The stream of mobile phase percolating through the column is not well described by piston flow. The distribution of the mobile phase velocities along the channels between particles is extremely complicated. Locally, the velocity increases from the particle surface (against which it is zero) to the center of the channels between the particles (where it is maximum). The solute zone moves faster in the center of the channels than close to the particle surface, thus creating a radial concentration gradient. The irregular distribution of the particles has a profound effect on the radial dispersion. Channels around them split and merge constantly on a length scale equal to a particle diameter. Consider the streamlets, consisting of narrow sets of close streamlines passing between two close particles. These streamlets will be torn apart, part of the streamlines going one side of the next particle, the rest on the other side, when they will hit another particle. This is the origin of Saffman dispersion [39]. The cross-sectional average velocities in two close channels may differ widely. Hydrodynamic problems of this type could, in principle, be solved using the Navier–Stokes equation. However, the distribution of the particles in a packed bed is so extremely complex that a description in mathematical terms, and, accordingly, the boundary conditions for this equation are very complex to write, preventing any attempt to solve it. The power of modern computers is just opening some new avenues to tackle this problem and recent work in this area is offering some new hope [41].

In the meantime, physical chemists must do with more superficial approaches. Only radial dispersion may relax the concentration gradients produced by these wide radial fluctuations of the velocity. Diffusion proceeds swiftly on the scale of a particle diameter, very slowly on the scale of the column diameter. Thus, the contributions of trans-channel and short-range interchannel dispersion to band broadening are relatively small [2]. Unless the column to particle diameter ratio is very small, a few units only, the contributions of long-range inter-channel and trans-column dispersion are important, particularly if the radial distribution of the mobile phase velocity is not flat (see earlier, section III-4). Saffman dispersion leads to two types of contributions. First, it is at the origin of eddy dispersion. It also accelerate radial mass transfer when the mobile phase flow velocity increases [39]. In the same time as they contribute to axial dispersion (see section III), molecular and eddy diffusion contribute both to the relaxation of the radial concentration gradients and to the radial homogenization of the eluent stream. The mechanism of this contribution is similar to that of axial dispersion and it is controlledby much the same phenomena.

Finally, note that the Einstein law of diffusion

$$\sigma^2 = 2D_m t \qquad\qquad\qquad \text{(II-17)}$$

gives a good estimate of the range within which dispersion can relax concentration gradients in a given time, t. Molecular diffusivities in solution are between 1×10^{-5} and 1×10^{-6} cm^2 s^{-1}. Thus, assuming the average value (*i.e.*, $D_m = 5 \times 10^{-6}$ cm^2 s^{-1}, a reasonable estimate of the molecular diffusivity of a compound with a molecular weight of a few hundred Daltons), eqn. II-17 shows that it would take times of the order of 25 ms, 2.5 s, and 250 s to relax a given concentration gradient over distances of 5 μm, 50 μm, and 0.5 mm, respectively (obviously, it takes a longer time, two to three times longer, practically to relax the concentration gradient to a small ripple; the times given here are the units with which the relaxation time should be measured in each case). While the first of these times is of the order of what it takes for the band of a nonretained compound to migrate a distance of about one particle diameter downstream, the third is of the same order of magnitude as the hold-up time. This simple calculation shows why trans-column effects are so pernicious for the column efficiency and why narrow-bore columns have great potential advantages in term of column efficiency.

2 - Film Mass Transfer Resistance

Each particle of the column bed is surrounded by a layer of stagnant fluid. Mass transfer from the mobile phase stream percolating through the bed to the stagnant mobile phase, inside the pores of the particles, can take place only by molecular diffusion through this layer. Its thickness, which controls the rate of this contribution to the mass transfer kinetics, depends on the velocity of the fluid.

The film mass transfer kinetics is usually well accounted for by a linear driving force equation:

$$\frac{\partial \bar{q}}{\partial t} = k_f A(C - C^*) = \frac{6k_f}{d_p}(C - C^*) \qquad \text{(II-18)}$$

where k_f is the effective mass transfer coefficient, A the external surface area of the particles per unit volume ($a = 6/d_p$ for spherical particles), $\bar{q}$ the concentration in the adsorbed phase averaged over the whole particle, C the solute concentration in the mobile phase, and C^* the value of C which would be in equilibrium with $\bar{q}$. The numerical value of k_f is given by an empirical correlation, the Wilson and Geankoplis equation [42]

$$\text{for} \qquad 0.0015 < \text{Re} < 55 \qquad\qquad \text{Sh} = \frac{1.09}{\varepsilon_e} \text{Re}^{0.33} \text{Sc}^{0.33} \quad \text{(II-19)}$$

where ε_e is the interparticle or external porosity, $Sh = (k_f d_p)/D_m$, the Sherwood number, and $Sc = \eta/(\rho D_m)$, the Schmidt number. As we have

shown earlier, the Reynolds number (eqn. II-12) is usually less than 0.02 in
liquid chromatography, so correlations at high values of the Reynolds number
are useless in high performance liquid chromatography.

3 - Intraparticle Pore Diffusion

The intraparticle pore space is immensely more complex than the extra-
particle space. Horváth and Lin [43] gave a rather dramatic illustration
(Figure II-3). We must point out, however, that this figure shows only an illustration of a two-dimensional representation of the three-dimensional pore
structure. It can hardly begin to give a good idea of its intricate 3-D nature.

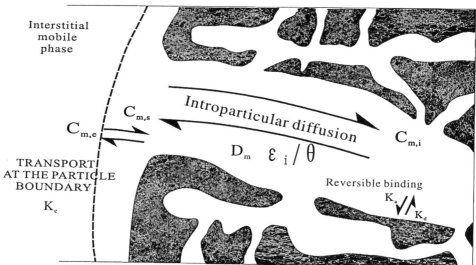

Figure II-3   **Schematic Illustration of the Different Mass Transfer Steps Involved in Liquid Chromatography [43]. k_a and k_d, rate constants for adsorption and desorption, respectively; k_e, mass transfer coefficient at the particle
boundary; θ, tortuosity factor; ϵ_i, internal porosity, $C_{m,e}$, $C_{m,s}$, $C_{m,i}$, concentrations of the solute in the bulk mobile phase, in the mobile phase inside the
particle but at the surface of the particle, and in the pores of the particle,
respectively.** *Reproduced with permission by Elsevier.*

With modern packing materials, the pore size distribution is much wider
than the particle size distribution. Another important characteristic of the
pore network is its connectivity [44] which indicates the number of the different trajectories which, on the average, can lead from one point to another
in the pore space. If this number is low, a few partial obstructions may ren-
der an important part of the intraparticle pore space inaccessible to certain

types of molecules (*e.g.*, proteins). Materials with a low or moderate connectivity are more sensitive to fouling than those with a high connectivity. Finally, while mineral-oxide based adsorbents (*e.g.*, silica, alumina, zirconia) have impermeable walls between their pores, this is not always the same for resin-based packing materials. In some cases, the solutes may diffuse through the walls made of polymeric cross-linked chains. Since intraparticle diffusion depends so much on the structure of the pore network, it is difficult to predict accurately the effective intraparticle dispersion coefficient. A correlation [45] is often used to estimate it:

$$D_p = \left[\frac{\varepsilon_p}{2 - \varepsilon_p} \right]^2 D_m \qquad \text{(II-20)}$$

where ε_p is the intraparticle porosity and D_m the molecular diffusivity. For typical chromatography packings, ε_p is between 0.30 and 0.70, hence D_p may be between 3 and 30 times smaller than D_m. It is often even still lower [51].

Improving the mass transfer kinetics is a constant concern in preparative HPLC because it is an important key to increasing the production rate. This result can be achieved either by reducing the distance over which mass transfer has to proceed, *i.e.*, by using finer particles (at the cost of using a higher inlet pressure, see eqn. II-10), or by increasing the rate of intraparticle diffusion. From eqn. II-20, we derive that the higher the internal porosity of the packing material, the larger the value of D_p. Other approaches have been tried to improve the mass transfer properties of packing materials. Two are worth mentioning.

In the last few years, considerable attention was paid to the preparation of monolithic silica. The column is filled with a block of porous silica that adheres to the column wall. This block has a bimodal pore-size distribution [46]. A network of large pores (average size, *ca* 2 μm) lets the mobile phase stream percolate through the monolith. A second network of pores having a much smaller average pore size (*ca* 150 Å), pores that are well connected to the first pore network, ensures a sufficient specific surface area, giving retention and capacity to the stationary phase. Monolithic columns have most interesting properties for analytical applications [47]. The extrapolation of their synthesis process to the preparation of wide bore columns for preparative HPLC raises serious difficulties and will require considerable effort. Success in this venture would afford columns having a higher permeability than conventional columns and performance otherwise quite comparable to those of conventional packed columns.

A particulate packing material presented as having a similar bimodal porosity was developed ten years ago [48]. The large pores would have had dimensions intermediate between those of the pores found in all adsorbents

(5 to 30 nm) and those of the interparticle volume (dimensions between 0.1 and 0.5 particle diameter). These large pores would have been large enough to be swept by a slow stream of mobile phase, a phenomenon which would considerably accelerate intraparticle diffusion [49], if it would take place. This phenomenon was named the *"perfusion"* effect. Unfortunately, if the principle sounds most attractive [48], its implementation remained controversial [50] and there are still serious doubts whether this secondary flow actually did take place in Poros. No final demonstration by an independent observer of a perfusion mechanism in Poros has been published yet. Other mechanisms may explain the relatively fast mass transfer process observed in Poros [48,49]. Minor changes in the intraparticle porosity may cause important changes in D_p (see eqn. II-20). Also, surface diffusion, which is generally ignored in HPLC, seems to contribute largely to mass transfer, especially in RPLC [51], as we show in the next section. Inadequate handling of the scientific information could explain some errors of interpretations [52].

The pore-size distribution of adsorbents can be measured by mercury porosimetry or may be derived from nitrogen adsorption isotherms [53]. The data obtained by these methods must be interpreted with caution. The accuracy of the data given by the former method relays on the accuracy of the contact angle of mercury for the surface of the adsorbent studied, a parameter which, at least to a degree, depends on the chemistry of the surface and might vary from batch to batch (and certainly does so from brand to brand). Similarly, the amount of nitrogen adsorbed at equilibrium is measured in the gas phase and the structure of a layer of bonded alkyl groups may be quite different under the conditions of this measurement and under those of the actual applications of the packing material.

4 - Surface Diffusion

The mass transfer kinetics through a particle depends on both pore diffusion and surface diffusion. The latter phenomenon takes place only in the case of adsorbents having a relatively low surface energy. It seems more prevalent for nonpolar adsorbents and under experimental conditions such that there is little competition for adsorption between the feed and the mobile phase components (see section I-3). Under certain conditions, sorbed molecules can migrate on the surface while remaining adsorbed. Depending on the chemical nature of the surface, the activation energy required for a molecule to migrate laterally, *i.e.*, to move to a nearby adsorption site while remaining adsorbed, can be less than the energy it needs to desorb from the surface. Because adsorption energies are relatively low in RPLC, surface diffusion is often important in this case, particularly with alkyl-bonded silicas

[51]. Suzuki [54] has shown that the effective intraparticle diffusivity is given by

$$D_e = D_p + \rho_k K D_s \qquad \text{(II-21)}$$

where D_p and D_s are the coefficients of pore and surface diffusion, respectively, ρ_k is the particle density, and K is the Henry constant of adsorption (linear isotherm). Similar to molecular diffusivity, the surface diffusion coefficient is the proportionality coefficient between the surface diffusive flux and the intensity of the concentration gradient along the adsorbed phase. A surface tortuosity may also be defined. However, surface diffusivity is already extremely difficult to measure, so this last concept does not seem useful at this stage.

Depending on the relative value of the isosteric heat of adsorption and the activation energy of surface diffusion, the adsorbed phase can be considered either as a two-dimensional fluid (low activation energy for diffusion) in which molecules individually adsorbed are able to hop from one site to the next, or as immobile adsorbed molecules [54]. The activation energy for surface diffusion is usually taken as a fraction of the isosteric heat of adsorption [55]. Surface diffusion in RPLC was studied in detail by Miyabe and Suzuki. The results of this investigation were reviewed recently [51]. The most interesting feature of surface diffusion is that the surface diffusion coefficient increases with increasing concentration in the adsorbed phase. It was shown that

$$D_s(q) = D_{s,0} \frac{d \ln C}{d \ln q} \qquad \text{(II-22)}$$

where $q(C)$ is the isotherm equation (eqn. II-1d). This might explain why it is often reported that the mass transfer kinetics appears to be faster at high concentrations [56]. However, the concentration dependence of the intraparticle diffusion coefficient could also be attributed to parallel intraparticle diffusion, in which case both pore and surface diffusions are significant. If a parallel diffusion process is modeled with either a pure surface diffusion model or a pure pore diffusion model, the effective intraparticle diffusion coefficient obtained turns out to depend on both the liquid phase concentration of the solute and the isotherm equation. For example, assuming a Langmuir isotherm, it was found that the effective surface diffusion coefficient decreases with increasing liquid phase concentration with a surface diffusion model whereas it increases with a pore diffusion model [57].

5 - Kinetics of Adsorption–Desorption

It takes a finite time for a molecule to get adsorbed from the liquid phase and then to be released back into the mobile phase. As a matter of fact, the net retention time is the integral of all these incremental times spent by molecules in the adsorbed state (averaged over the total sample population). Several kinetic models are available to interpret this kinetics. They were summarized by Ruthven [8]. The simplest of these models are often used in chromatography, sometimes to derive values of the rate constants. Most often, they are treated as empirical models to which the experimental data are fitted to allow further calculations, *e.g.*, those of band profiles. In this last case, the numerical estimates of the "rate" constants obtained may have no real physical meaning (owing to the approximate character of these simple rate models). The kinetic models most commonly used are the Langmuir kinetic model and the linear kinetic models.

a. The Langmuir Kinetic Model

This is the model used in the "kinetic" demonstration of the Langmuir isotherm model [8,11]. The rate of variation of the adsorbate concentration is the sum of the rate of adsorption, proportional to the mobile phase concentration and to the unoccupied fraction of the monolayer, and of the rate of desorption, proportional to the surface coverage, hence:

$$\frac{\partial C_{s,i}}{\partial t} = k_a(q_{s,i} - q_i)C_i - k_d q_i \qquad \text{(II-23)}$$

where $q_{s,i}$ is the column saturation capacity (in the same units as q_i) and k_a and k_d are the rate constants of adsorption and desorption of the component i considered. This model was used by Thomas [58], Goldstein [59], and Wade *et al.* [60] and will be discussed later (Chapter IV, Section III-2).

b. The First-order Kinetic Model

This very simple model was used by Lapidus and Amundson [61]. It assumes simple, first-order kinetics for both adsorption and desorption. Hence,

$$\frac{\partial q_i}{\partial t} = k_a C_i - k_d q_i \qquad \text{(II-24)}$$

where k_a and k_d are the rate constants of adsorption and desorption, respectively.

c. The Linear Driving Force Models

In these models, it is assumed that the driving force for adsorption or desorption is the deviation of the local concentrations from their equilibrium value, *i.e.*, that the rate of adsorption is proportional to the difference between the actual concentration and the concentration which would ensure equilibrium. This concentration is the one given by the equilibrium isotherm (eqn. II-1d). Hence, there are two such models, depending on whether we consider the liquid or the solid phase of the chromatographic system as the more important.

The *solid film linear driving force model* is written:

$$\frac{\partial q_i}{\partial t} = k_f(q_i^* - q_i) \tag{II-25}$$

where q_i^* is the equilibrium value of q_i for a mobile phase concentration equal to C_i and k_f is the lumped mass transfer coefficient. This model was used by Glueckauf and Coates [62], Hiester and Vermeulen [63], Lin *et al.* [64], Golshan–Shirazi *et al.* [65], and Phillips *et al.* [66].

The equation of the *liquid film linear driving force model* is:

$$\frac{\partial q_i}{\partial t} = k_m'(C_i - C_i^*) \tag{II-26}$$

where C_i^* is the solute concentration in the mobile phase which is in equilibrium with the solid phase concentration q_i. Thus, C_i^* is the root of the isotherm equation solved for the mobile phase concentration in equilibrium with the stationary phase concentration $q_i = q_i^*$. Under linear conditions, $C_i^* = q_i^*/a = q_i/a$ where a is the slope of the linear isotherm ($a = k_0'/F$, F, phase ratio), and k_m' is the apparent (lumped) mass transfer coefficient. This model was also used by Lapidus and Amundson [61].

Under linear chromatography, these two models are particular cases of the first-order kinetic model. This is no longer true in nonlinear chromatography and these different models give different results. These models are most often used as empirical models, to which experimental data are fitted. This explains why the rate coefficient, k_f or k_m, may appear to increase with increasing concentration [56]. If surface diffusion is an important contribution to the mass transfer kinetics, this last result is a direct consequence of the properties of surface diffusion [51], see eqn. II-22. However, this dependence may also arise from another model error [67].

The Langmuir kinetic model, the first-order kinetic model, and the linear driving force models are called lumped kinetic models because the contributions of all the mass transfer resistances are lumped together into a single empirical constant.

d. The Equilibrium–Dispersive Model

In this model, the mass transfer resistances are assumed to be small enough to be negligible. This situation is typically encountered in HPLC, when modern, highly efficient columns are used with small molecules of moderate polarity. In this case, as demonstrated by Giddings [2], the contribution of the mass transfer kinetics is equivalent to an increase in the axial dispersion coefficient. The apparent axial dispersion coefficient is then the sum of the contributions of the axial and eddy diffusions and of the contribution corresponding to the mass transfer resistances. In most cases, this coefficient is considered to be independent of the solute concentration. Its value is derived from the column HETP (height equivalent to a theoretical plate) measured under linear conditions (*i.e.*, with a very small sample) through the equation

$$D_L = \frac{Hu}{2} = \frac{Lu}{2N} \tag{II-27}$$

where H is the column HETP and N the column efficiency [9].

V - PHYSICAL MODELS OF CHROMATOGRAPHY

Mathematical models are sets of mathematical expressions that describe through mathematical relationships, the features of the physical model of the chromatographic process that was selected for the study and the particular experimental conditions under which the process is operated. In other words, mathematical models describe an ideal physical experiment from which part of the extreme complexity of nature has been eliminated. The physical experiment has always to be simplified to design a tractable mathematical model. The art of the model designer is in eliminating the unimportant or irrelevant complexities while keeping those which are important and relevant, and, in the process, still obtaining a model for which solutions can be calculated, one way or another, *i.e.*, through analytical or numerical methods.

Thus, our essential problem now is to derive simple yet relevant models of chromatography. In this quest, we can start by combining the previous two main characters of models which have been described earlier in this part, the linear or nonlinear character, the ideal or nonideal character. This approach gives us four types of models which are unevenly fruitful.

1 - The Four Basic Combinations of Models

These four basic families of models have widely different degrees of complexity. The simplest ones contain only a single model, the more complex ones a whole variety.

a. The Linear, Ideal Model

This model predicts that all the feed components give bands whose profiles are identical to the injection profile. Each one of these bands moves at its own velocity that is proportional to the mobile phase velocity and is simply related to the slope of the linear isotherm. Separation takes place rapidly since the bands never broaden. This model accounts only for the effect of a linear thermodynamics of equilibrium in an infinitely efficient column. It is generally useless. It proved actually useful only in simulated moving bed separations, with a linear isotherm [68].

b. The Linear, Nonideal Models

These are realistic models for analytical chromatography. Regarding the relative migration velocity of the component bands and their retention times, they give the same results as the previous model. However, they can also account fairly well for the broadening of these bands under the influence of the different factors involved. There are as many such models as there are useful models of mass transfer kinetics (see previous section). They may even account for the tailing of these bands, provided a suitable kinetic model is used to account for the nonideal behavior [69,70].

c. The Nonlinear, Ideal Model

Like the first model in this series, this third model accounts only for the effects of the hydrodynamics (band migration) and the thermodynamics of equilibrium (now, with a nonlinear isotherm). It cannot be entirely realistic either since all actual columns have a finite efficiency. However, this model is most useful because the effects of a nonlinear isotherm are always important at high concentrations, sometimes so important that actual band profiles may be closely approximated by this model when the column efficiency is high. Also, the nonlinear effects are often sufficiently complex that a degree of simplification is often worthwhile.

This model has the great advantages of simplifying the interpretation of the experimental results and of allowing separate discussions of the problems of (i) band migration and change in concentration profiles due to the nonlinear thermodynamics of equilibrium and of (ii) band broadening and asymmetry originating from the axial dispersion and the mass transfer resistances. In many practical cases, by contrast with the familiar situation observed in linear chromatography, the contributions of the nonideal effects to the elution band profiles are minor and appear as necessary corrections, not as critical contributions [9].

d. The Nonlinear, Nonideal Models

This group of models include the most realistic models and also the most complex ones. Models of this group aim at accounting fully for most of the phenomena contributing significantly to the band profiles in a specific experiment. They require the selection of a proper model of mass transfer kinetics, among the ample selection illustrated in the previous section. There are numerous models in this group. They differ by their degree of sophistication.

2 - The Models Discussed in this Work

It is impossible to consider in this book all the possible models of chromatography, especially all those belonging to the fourth family described above. It is not possible even to consider all the models that have been described and studied in the literature. We will limit our discussions to the following models.

• **The linear, equilibrium–dispersive model.** This is the realistic and practical model used to account for analytical chromatography. The equilibrium isotherm is linear. Because the mass transfer kinetics is usually fast, there are no explicit contributions of this kinetics. The model includes an apparent axial dispersion coefficient, however, that accounts for the contributions of the mass transfer resistances to band broadening.

• **The parabolic, equilibrium–dispersive model.** This model is similar to the previous one, except that it uses a parabolic isotherm (eqn. II-1b). Its main interest is in allowing the investigation of the onset of column overloading.

• **The nonlinear, ideal model.** This model allows the study of the influence of a nonlinear isotherm on the band profile. Exact analytical solutions can be derived, which greatly facilitates general discussions. In most cases of practical importance, these solutions are algebraic. The solutions of this model are close to the corresponding experimental profiles when the column efficiency is high and the column is strongly overloaded.

• **The nonlinear, equilibrium–dispersive model.** This is the most practical model for preparative chromatography when the column efficiency is high. In many practical cases, the mass transfer resistances are sufficiently low to be accounted for by an additional contribution to the apparent axial dispersion coefficient. This model may not be valid for chiral separations nor for the separation of proteins, cases in which the mass transfer kinetics is often too slow.

• **The nonlinear, transport–dispersive model.** This model combines a nonlinear isotherm with a finite axial diffusion and a lumped kinetic model of the mass transfer kinetics. It can be used when the band broadening contribution due to axial and eddy diffusion is small compared to that of a slow mass transfer kinetics.

Although more realistic, the general rate model is not discussed here. This model is too complex to have an analytical solution. Its mathematical investigation does not lead to intermediate results of interest nor does it bring any new concept of general interest. Its numerical solutions may be calculated with sufficiently powerful computers [71-73].

EXERCISES

1) Show that the bilangmuir isotherm is a particular case of the quadratic isotherm.

2) Commercial packages of mathematical tools for computers have good nonlinear fitting programs (*e.g.*, Mathcad, Mathematica, Matlab). Use such a program for the following series of exercises. (i) use a spreadsheet to generate 50 points of a Langmuir isotherm in a concentration range such that θ varies between 0 and l (between 0.10 and 0.50). (ii) Multiply each value of θ by a random number of average value 1.0 and standard deviation 1.05. (iii) Fit the "experimental data" so obtained to a Langmuir isotherm model. (iv) Compare the best values obtained for the numerical coefficients of the isotherm to those used initially. (v) Repeat the process with different values of l.

3) Repeat exercise 2 by changing the standard deviation of the random number (*i.e.*, the error of measurement) or by trying now to fit the "experimental data" to another isotherm model.

4) Operated each under a constant pressure drop, a packed column and a tube have flow rates $F_{v,pc}$ and $F_{v,t}$, respectively. If the diameters of both tubes are doubled, nothing else (pressure drop, lengths of the tubes, particle size, liquid viscosity) being changed, what are the new flow rates?

5) Water flows through a 20 cm long column packed with particles having a 10 μm average size. The inlet pressure is 100 bar (10 MPa or 1×10^7 Pascal). The flow resistance parameter is $k_0 = 1 \times 10^{-3}$ (see eqn. II-10). What are (1) the linear velocity of the mobile phase; (2) the residence time of an unretained tracer (*i.e.*, the hold-up time)? (3) the Reynolds number (eqn. II-12)?

(viscosity of water at 20°C, $\eta = 1.0$ cP or 1×10^{-2} Poise; $\rho = 1$). Be extremely careful with the homogeneity of the units. Care is always needed when pressure data are handled.

6) Using eqns. II-14 and II-16, calculate the diffusion coefficient of glucose ($C_6H_{12}O_6$) in water ($V_A = 18$ mL).

LITERATURE CITED

[1] J. C. Giddings *"Unified Separation Science "*, Wiley–Interscience, New York, NY, 1991.

[2] J. C. Giddings, *"Dynamics of Chromatography "*, M. Dekker, New York, NY, 1965.

[3] M. Martin, G. Blu and G. Guiochon, *J. Chromatogr. Sci.*, **11** (1973) 641.

[4] P. Rouchon, M. Schonauer, P. Valentin and G. Guiochon, *Separat. Sci. Technol.*, **22** (1987) 1793.

[5] J. Conder and J. Young, J., *"PhysicoChemical Measurements by Gas Chromatography"*, Wiley, New York, NY, 1979.

[6] P. Rouchon, P. Valentin, M. Schönauer, C. Vidal-Madjar and G. Guiochon, *J. Phys. Chem.*, **88** (1985) 2709.

[7] P. W. Carr, D. E. Martire, L. R. Snyder Eds., *"The Retention Process in Reversed-phase Liquid Chromatography,"* *J. Chromatogr. A*, **656** (1993).

[8] D. M. Ruthven, *"Principles of Adsorption and Adsorption Processes,"* Wiley, New York, NY, 1984.

[9] G. Guiochon, S. Golshan-Shirazi and A. M. Katti, *"Fundamentals of Preparative and Nonlinear Chromatography "*, Academic Press, Boston, MA, 1994.

[10] I. Quiñones and G. Guiochon, *J. Colloid Interface Sci.*, **183** (1996) 57.

[11] I. Langmuir, *J. Am. Chem. Soc.*, **38** (1916) 2221.

[12] G. M. Schwab, *"Ergebnisse der exacten Naturwissenschaften,"* Vol. 7, Springer, Berlin 1928.

[13] D. S. Jovanovic, *Kolloid-Z. Z. Polym.*, **235** (1969) 1203.

[14] M. Moreau, P. Valentin, C. Vidal-Madjar, B. C. Lin and G. Guiochon, *J. Colloid Interface Sci.*, **141** (1991) 127.

[15] R. Fowler and E. A. Guggenheim, *"Statistical Thermodynamics"*, Cambridge Univ. Press, Cambridge 1965.

[16] Freundlich, *"Kapillarchemie"*, Akademische Verlagsgesellschaft m.b.h., Leipzig, 1922.

[17] R. Sips, *J. Chem. Phys.*, **16** (1948) 490.

[18] J. Tóth, *Acta Chim. Hung.*, **32** (1962) 31.

[19] D. N. Misra, *J. Coll. Interface Sci.*, **79** (1980) 543.

[20] C. J. Radke and J. M. Prausnitz, *Ind. Eng. Chem. (Fundam.)*, **11** (1972) 445.

[21] M. M. Dubinin and L. V. Radushkevich, *Dokl. Akad. Nauk SSR, Ser. Khim.*, **55** (1947) 331.

[22] M. D. LeVan and T. Vermeulen, *J. Phys. Chem.*, **85**, (1981) 3247.

[23] S. Golshan–Shirazi and G. Guiochon, *J. Chromatogr.*, **461**, (1989) 1.

[24] G. Schay, *J. Colloid Interface Sci.*, **42** (1973) 478.

[25] F. Riedo and E. sz. Kováts, *J. Chromatogr.*, **239** (1982) 1.

[26] D. W. Marquardt, *J. Soc. Appl. Math.*, **11** (1963) 431.

[27] S. L. Ajnazarova and V. V. Kafarov, *"Methods for Experimental Optimization in Chemical Technology"*, Vishaia Shkola, Moscow 1985.

[28] I. Quiñones, J. C. Ford and G. Guiochon, *Chem. Eng. Sci.*, **55** (2000) 909.

[29] R.H. Fowler, *"Statistical Mechanics,"* Cambridge University Press, London, UK, 1955.

[30] T. L. Hill, *"An Introduction to Statistical Thermodynamics,"* Addison-Wesley, Reading, MA, 1962.

[31] D. A. McQuarrie, *"Statistical Mechanics"*, University Science Books, Sausalito, CA, 2000.

[32] R. C. Reid, J. M. Prausnitz, and B. E. Poling, *"The Properties of Gases and Liquids,"* McGraw-Hill, New York, 4th ed., 1987.

[33] *"CRC Handbook of Chemistry and Physics"*, D. R. Lidde Ed., Boca Raton, FL, 79st ed., 1998.

[34] R. B. Bird, W. E. Stewart and E. N. Lightfoot, *"Transport Phenomena"*, Wiley, New York, NY, 1962.

[35] J. Li and P. W. Carr, *Anal. Chem.*, **69** (1997) 2530.

[36] C. R. Wilke and P. Chang, *AIChE J.*, **1** (1955) 264.

[37] U. Tallarek, D. van Dusschoten, H. Van As, E. Bayer and G. Guiochon, *J. Phys. Chem. B*, **102** (1998) 3486.

[38] S. G. Weber and P. W. Carr, in *"High Performance Liquid Chromatography"*, Brown, P. R. and Hartwick, R. A., Eds., Chapter 1, Wiley, New York, NY, 1989.

[39] U. Tallarek, E. Bayer and G. Guiochon, *J. Am. Chem. Soc.*, **120** (1998) 1494.

[40] T. Yun, M. S. Smith and G. Guiochon, *J. Chromatogr. A*, **828** (1998) 19.

[41] M. R. Schure, R. S. Maier, D. M. Kroll, H. T. Davis, Communication to PREP– 2001, Washington, D.C., May 2001.

[42] E. J. Wilson and C. J. Geankoplis, *Ind. Eng. Chem. (Fundam.)*, **5** (1966) 9.

[43] Cs. Horváth and H. J. Lin, *J. Chromatogr.*, **149** (1978) 43.

[44] A. I. Liapis, *Math. Modeling Sci. Computing*, **1** (1994) 397.

[45] J. S. Mackie and P. Meares, *Proc. Roy. Soc. (London)*, **A232** (1955) 498.

[46] H. Minakuchi, K. Nakanishi, N. Soga, N. Ishizuka, N. Tanaka, *Anal. Chem..*, **68** (1996) 3498.

[47] M. Kele and G. Guiochon, *J. Chromatogr. A*, **960** (2001) 19.

[48] N. B. Afeyan, S. P. Fulton, N. F. Gordon, T. L. Loucks, I. Maszaroff, L. Varady and F. E. Regnier, *J. Chromatogr.*, **519** (1990) 1.

[49] A. I. Liapis and M. A. McCoy, *J. Chromatogr.*, **660** (1994) 83.

[50] D. D. Frey, E. Schweinheim and Cs. Horváth, *Biotechnol. Progr.*, **9** (1993) 273.

[51] K. Miyabe and G. Guiochon, *Advances in Chromatography*, **40** (2000) 1.

[52] Perceptive Biosystems Inc. *versus* Pharmacia Biotech Inc., Sepracor Inc., Pharmacia LKB Biotechnology AB, Pharmacia Bioprocess Technology AB, and Procordia AB, United District Court, District of Massachussetts, Boston, MA, Civil Action No. 93-12237-PBS, Memorandum and Order, March 31, 1997.

[53] H. Guan, G. Guiochon, E. Davis, K. Gulakowski and D. W. Smith, *J. Chromatogr. A*, **773** (1997) 33.

[54] M. Suzuki, *"Adsorption Engineering"*, Kodansha, Tokyo, Japan, 1990.

[55] K. Miyabe and G. Guiochon, *Anal. Chem.*, **72** (2000) 5162.

[56] P. Sajonz, H. Guan–Sajonz, G. Zhong and G. Guiochon, *Biotechnol. Progr.*, **13** (1997) 170.

[57] K. Miyabe and G. Guiochon, *J. Phys. Chem. B*, **106** (2002) 8898.

[58] H. C. Thomas, *J. Am. Chem. Soc.*, **66** (1949) 1664.

[59] S. Goldstein, J. D. Murry, *Proc. Roy. Soc. (London)*, **A252** (1959) 360.

[60] J. L. Wade, A. F. Bergold and P. W. Carr, *Anal. Chem.*, **59** (1987) 1286.

[61] L. Lapidus and N. R. Amundson, *J. Phys. Chem.*, **56** (1952) 984.

[62] E. Glueckauf, J. I. Coates, *J. Chem. Soc.*, **1947**, 1315.

[63] N. K. Hiester and T. Vermeulen, *Chem. Eng. Progr.*, **48** (1952) 505.

[64] B. Lin, S. Golshan–Shirazi and G. Guiochon, *J. Phys. Chem.*, **93** (1989) 3363.

[65] S. Golshan–Shirazi, B. Lin and G. Guiochon, *J. Phys. Chem.*, **93** (1989) 6871.

[66] M. W. Phillips, G. Subramanian and S. M. Cramer, *J. Chromatogr.*, **454** (1988) 1.

[67] K. Kaczmarski, D. Antos, H. Sajonz, P. Sajonz and G. Guiochon, *J. Chromatogr. A*, **925** (2001) 1.

[68] G. Zhong and G. Guiochon, *Chem. Eng. Sci.*, **51** (1996) 4307.

[69] J. C. Giddings and H. Eyring, *J. Phys. Chem.*, **59** (1955) 416.

[70] G. Götmar, T. Fornstedt and G. Guiochon, *J. Chromatogr. A*, **831** (1999) 17.

[71] Q. Yu and N.-H.L. Wang, *Comput. Chem. Eng.*, **13** (1989) 915.

[72] T. Gu, G. Tsai and G.T. Tsao, *AIChE J.*, **36** (1990) 784.

[73] K. Kaczmarski, G. Storti, M. Mazzotti, M. Morbidelli, *Comp. Chem. Eng.* **21** (1997) 641.

[74] R. Bai and R. T. Yang, *J. Coloid. Interface Sci.*, **239** (2001) 296.

[75] K. Kaczmarski, D. Zhou, M. Gubernak, G. Guiochon, *Biotechnol. Progr.*, Submitted.

CHAPTER III
MATHEMATICAL BASIS
OF CHROMATOGRAPHY

Each physical model of chromatography can be translated into a system of equations and conditions that expresses its different features. This set of equations is the mathematical model of chromatography. The degree of correctness of the translation of the physical model into a mathematical model is important. Neglecting or simplifying certain features of the physical model is often necessary but this must be clearly acknowledged, so that it is possible to understand the limits of the validity of the solutions obtained by this mathematical model. Finally, we must remember that the use of a complex model is not very helpful if there are no independent physical methods that allow a reasonably accurate estimate of the parameters involved in the model. From this point of view, the identification of the parameters of complex models is more an act of faith than a scientifically justified procedure, unless the mathematical model and the identification procedure followed have been previously validated. Numerical agreement between a set of experimental data and their best fit to a model can never be taken as a validation of this model.

The equations in a mathematical model include algebraic equations (*e.g.*, the equilibrium isotherm) and partial differential equations stating the mass conservation of each feed or mobile phase component involved, and expressing the mass transfer kinetics of these compounds. The model also includes the boundary conditions of these partial differential equations. These conditions translate the physical description of the experiment actually performed (*e.g.*, elution, frontal analysis, simulated moving bed separations) into mathematical terms. These equations are derived, justified and explained in this chapter. A brief summary of the properties of partial differential equations is also presented to facilitate the understanding of certain mathematical derivations in further chapters.

I - MATHEMATICAL MODELS OF CHROMATOGRAPHY

All mathematical models of chromatography always consist of a differential mass balance equation for each compound involved in the experiment and of the set of equations describing the mass transfer kinetics. It also includes initial and boundary conditions, and the equilibrium isotherms of the relevant compounds. The search for the solution of a mathematical model of chromatography is always a well-posed problem. We describe now the most important of these equations.

1 - The Fundamental Equations Used in Chromatography

The possible isotherm equations were discussed earlier (Chapter II, section 1). There is nothing more to add here.

a. *The Differential Mass Balance Equation*

This equation expresses mass conservation in a physico-chemical process [1]. It states that the difference between the amounts of a compound that enters and leaves a slice of column of thickness dx during the time dt accumulates in this slice. This equation is written

$$\frac{\partial C}{\partial t} + F\frac{\partial q}{\partial t} + u\frac{\partial C}{\partial x} = D_L\frac{\partial^2 C}{\partial x^2} \tag{III-1}$$

In this equation, C and q are the concentrations of the compound in the mobile and the stationary phase, respectively, $F = (1 - \varepsilon)/\varepsilon$ is the phase ratio (with ε the total column porosity), u is the mobile phase velocity, D_L the axial dispersion coefficient, x the space coordinate along the column axis and t the time. The column is supposed to be radially homogeneous. Equation III-1 is a parabolic partial differential equation (PDE) of the second order (see later a discussion of these mathematical terms, in section II).

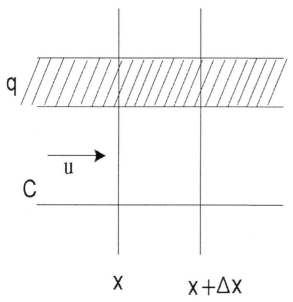

Figure III-1. Derivation of the Differential Mass Balance Equation of a Compound.

A simple derivation of this equation can be given briefly in the case of the ideal model (axial dispersion coefficient $D_L = 0$, infinite rate coefficient of mass transfer, infinite column efficiency). Figure III-1 shows a sketch of a column. Consider a slice of this column having a very small length, dx. The difference between the amount of solute entering the slice at its upstream end, at position x, and the amount of that same solute leaving it at its downstream end, at position $x + \Delta x$, during time Δt is

$$\Delta M = C(x,t)\varepsilon u S\Delta t - C(x + \Delta x, t)\varepsilon u S\Delta t$$

The amount of compound accumulated in the slice during the same time Δt is therefore:

$$\Delta M = [C(x, t + \Delta t)\varepsilon S\Delta x + q(x, t + \Delta t)(1 - \varepsilon)S\Delta x] -$$
$$[C(x, t)\varepsilon S\Delta x + q(x, t)(1 - \varepsilon)S\Delta x]$$

Simple algebraic manipulations, including equating these two values of ΔM, dividing both sides of the equation by $\Delta x \Delta t$, and taking the limit for $\Delta x \to 0$ and $\Delta t \to 0$ give

$$-\varepsilon u \frac{\partial C}{\partial x} = \varepsilon \frac{\partial C}{\partial t} + (1 - \varepsilon)\frac{\partial q}{\partial t}$$

Introducing the phase ratio, we obtain

$$\frac{\partial C}{\partial t} + F\frac{\partial q}{\partial t} + u\frac{\partial C}{\partial x} = 0 \qquad\qquad \text{(III-2a)}$$

This equation is the limit form toward which eqn. III-1 tends when the dispersion coefficient approaches zero. It is a hyperbolic partial differential equation of the first order. Since, under these conditions (infinite rate of mass transfer), q tends toward $q^* = f(C)$, the solid phase concentration is given by the equilibrium isotherm when the column efficiency becomes infinite, and the final equation is

$$\frac{1 + F\frac{dq}{dC}}{u}\frac{\partial C}{\partial t} + \frac{\partial C}{\partial x} = 0 \qquad\qquad \text{(III-2b)}$$

This equation can also be written

$$\frac{\partial Q}{\partial t} + \frac{\partial C}{\partial x} = 0 \qquad\qquad \text{(III-2c)}$$

with $Q(C) = [C + Ff(C)]/u$. This last equation is called the hyperbolic conservation law.

The important properties of the mass balance equation arise from the influence of the isotherm equation on the solution, through the dependence of q on C. For a linear isotherm, all the coefficients of the derivatives in eqns. III-2a and III-2b are constant. For a nonlinear isotherm, the coefficient of the first term of the LHS of eqn. III-2b depends on the concentration. As we shall show later (see Chapter V), eqn. III-2 can propagate concentration discontinuities or shocks. This is an important property of hyperbolic equations that parabolic equations such as eqn. III-1 do not share. Whenever eqn. III-1 applies, as it does in all experimental cases, the influence of the nonlinear isotherm and that of the axial dispersion act in opposite directions on at least certain parts of the concentration profiles (usually the front) and a balance between their effects takes place. We show in Chapter VII that a shock layer is propagated by eqn. III-1, instead of the shock that is propagated by eqn. III-2. This situation is similar to the one encountered in nonlinear, compressible hydrodynamics. In this latter case, the dispersive effect arises from the viscosity of the compressible fluid which prevents it from propagating infinitely fast. When the viscosity is very small and the nonlinear behavior is the dominant factor, the compressible hydrodynamical process produces the "shock layer" known, e.g., as the sound wall effect in aviation or the "shock" wave of an explosion. In chromatography, the dispersive effect arises from axial dispersion (i.e., from axial and eddy diffusion and from the mass transfer resistances). When the dispersive effect is small, as it is in the highly efficient columns of HPLC, and when the nonlinear effect is dominant, the chromatographic process produces a shock, or rather a shock layer, on one side of the profile.

In the compressible hydrodynamic process, there are problems of mass, momentum, and energy transfers. In the chromatographic process, there is little energy or momentum involved in an experiment. Hence, there are important physical differences between the shocks or shock layers encountered in hydrodynamics and in chromatography. But, from a mathematical viewpoint, they are the same because both correspond to the discontinuous solution of a quasi-linear hyperbolic differential equation. When the dispersion is different from zero but tends toward zero, the solution approaches the discontinuous solution (through the formation of a steeper and steeper shock layer). Another common character of these two problems is the simple wave solution, which has an important significance in multicomponent chromatography. From a physical point of view, although mass transfers, differential and integral, are essential to the chromatographic process, there is another aspect of importance which is less obvious: chromatography is also a process of signal propagation. Chromatographic processes may be considered as the migration of a concentration wave along the column. However, it is most important to understand that the transport of matter and signal, as they are described by eqns. III-1 or III-2, are quite different. As we explain in the first section of Chapter V, in nonlinear chromatography, concentrations propagate at a certain velocity but an actual amount of a labeled sample propagates at a velocity that is different from the velocity associated with the corresponding concentration.

We note that eqn. III-2b can be represented by the use of an operator. The operator in this equation is a factor of the operator in the following equation

$$\left[\left(\frac{1 + Fq'}{u} \right)^2 \frac{\partial^2}{\partial t^2} - \frac{\partial^2}{\partial x^2} \right] C = 0 \qquad \text{(III-3)}$$

This result shows that eqn. III-2b can be regarded as a branch of eqn. III-3. It corresponds only to the propagation of the component mass (mass conservation). This type of wave is named a kinetic wave, so the theory of chromatography includes the study of concentration waves. Nonlinear chromatography corresponds to nonlinear concentration waves. As explained above, these waves can produce a shock or discontinuity under certain conditions and this is probably the most important character of nonlinear chromatography.

The solution of a PDE such as those discussed above (eqn. III-1 is a second-order PDE) requires two boundary conditions and an initial condition. These conditions are the values of the function solution of the PDE (or of its derivative) in certain points or along certain curves in the definition space. Conventional initial and boundary conditions used in chromatography are discussed in subsection 1c and 1d of this section, respectively.

b. Mass Balance Equation of the Pore Model

Complex kinetic problems, problems involving a slow mass transfer kinetics with several steps, and problems requiring an accurate solution must be studied using a more complex model. This model recognizes that there are actually two different fractions of the mobile phase and that these two fractions play a markedly different role in the separation process [2]. The first fraction is the extra-particulate fraction of mobile phase, the fraction that percolates across the bed of packing material, ensuring the convective transport of the bands of solute. The second fraction of the mobile phase is the stagnant solution that is located inside the pores of the particles. The sample components have access to this second fraction of the mobile phase by diffusion only. These components must transit through this second fraction of the mobile phase in order to have access to the surface of the stationary phase and to experience retention. So, in the different forms of the general model of chromatography, a mass balance equation is written for each one of these two fractions. These two equations are accompanied by a kinetic equation to account for the mass transfer between these two fractions and a set of appropriate equations accounting for the mass transfer kinetics inside the pores contained in the particles of packing material and between the stagnant mobile phase and the stationary phase.

These two mass balance equations are written as follows

$$\frac{\partial C_{mi}}{\partial t} + u_0 \frac{\partial C_{mi}}{\partial x} = D_a \frac{\partial^2 C_{mi}}{\partial x^2} + Bk_{pm}(C_{pi}|_{r=0}^{R_1} - C_{mi}) \qquad \text{(III- 4a)}$$

$$\varepsilon_p \frac{\partial C_{pi}}{\partial t} = \varepsilon_p D_p \nabla_p^2 C_{pi} + (1 - \varepsilon_p) k_{ps} \left[C_{si} - f(C_{p1}, C_{p2}, \cdots) \right] \qquad \text{(III-4b)}$$

$$\frac{\partial C_{si}}{\partial t} = -k_{ps} \left[C_{si} - f(C_{p1}, C_{p2}, \cdots) \right] \qquad \text{(III-4c)}$$

where $B = \frac{1-\varepsilon_e}{\varepsilon_e}$, with ε_e, the interparticle or external porosity of the bed, ε_p is the intraparticle porosity (or porosity of these particles), D_a is the axial dispersion coefficient, D_p is the intraparticle pore diffusion coefficient, u_0 is the interstitial velocity, r is the radial position in the pores (all the pores are assumed to have a spherical shape and a radius equal to R_1), C_{mi}, C_{pi}, and C_{si} are the concentrations of component i in the percolating mobile phase, the stagnant mobile phase, and the stationary phase, respectively, k_{pm} and k_{ps} are the mass transfer rate coefficients between the two mobile phases and between the stagnant mobile phase and the stationary phase, respectively, and

$$\nabla_p^2 = \frac{\partial^2}{\partial r^2} + \frac{2}{r}\frac{\partial}{\partial r} = \frac{1}{r^2}\frac{\partial}{\partial r}\left(r^2\frac{\partial}{\partial r}\right) \tag{III-5}$$

Substituting eqn. III-4c into eqn. III-4b and adding the resulting equation multiplied by B to eqn. III-4a gives

$$\frac{\partial(C_{mi} + B\varepsilon_p C_{pi} + B(1-\varepsilon_p)C_{si})}{\partial t} + u_0\frac{\partial C_{mi}}{\partial x} =$$
$$D_a\frac{\partial^2 C_{mi}}{\partial x^2} - Bk_{pm}(C_{pi}|_{r=0}^{R_1} - C_{mi}) + B\varepsilon_p D_p \nabla_p^2 C_{pi} \tag{III-6}$$

From the general principle of continuity, the chromatographic equation can be written

$$\frac{\partial(C_{mi} + B\varepsilon_p C_{pi} + B(1-\varepsilon_p)C_{si})}{\partial t} + u_0\frac{\partial C_{mi}}{\partial x} = D_a\frac{\partial^2 C_{mi}}{\partial x^2} \tag{III-7}$$

Comparing eqns. III-6 and III-7 gives

$$k_{pm}(C_{pi}|_{r=0}^{R_1} - C_{mi}) = \varepsilon_p D_p \nabla_p^2 C_{pi} \tag{III-8}$$

Taking the average of eqn. III-8 over the entire pore volume gives

$$\overline{k_{pm}(C_{pi}|_{r=0}^{R_1} - C_{mi})} = \frac{3}{4\pi R_1^3}\int_0^{R_1}\varepsilon_p D_p \nabla_p^2 C_{pi} 4\pi r^2 dr$$
$$= \frac{3D_p\varepsilon_p}{R_1}\frac{\partial C_{pi}}{\partial r}\Big|_{R_1} \tag{III-9}$$

Note that $F = (1-\varepsilon_t)/\varepsilon_t$ is the ratio of the total solid volume to the total volume of mobile phase in the column. Here, ε_t is the total porosity, which is related to ε_e and ε_p as follows

$$\varepsilon_t = \varepsilon_e + (1-\varepsilon_e)\varepsilon_p$$

and B is the ratio of the particle volume to the volume of the liquid phase.

c. The Initial Condition

This condition indicates the status of the column at the beginning of the experiment, *i.e.*, when the injection of the sample or feed into the column begins or, more generally, when the phase equilibrium in the column is perturbed by some change of the concentration of at least one component, somewhere [1]. The initial state of the system is often equilibrium of the stationary phase with the pure mobile phase, which is the general case of elution chromatography, hence

$$C(x,0) = q(x,0) = 0 \qquad \text{(III-10)}$$

More complex cases are encountered. In frontal analysis or in perturbation methods, the column is in equilibrium with a mobile phase solution of the feed at time $t = 0$. The composition of this solution must be clearly stated. In simulated moving bed chromatography or SMB, the process involves an indefinite number of cycles. The initial condition for a column, at the beginning of a new cycle, is the concentration distribution of the feed components in this physical column at the end of the previous cycle. The boundary condition (see below) will then be updated accordingly.

Finally, note that the initial condition is sometimes used in theoretical studies to perform the injection instead of the more natural boundary condition. The instantaneous creation of a Dirac concentration distribution along the column at $t = 0$ is assumed but no suggestions are ever made as to how this could be achieved in practice. The initial condition is then expressed by

$$C(x,0) = A\delta(x) \qquad \text{for} \quad -\infty < x < \infty$$

$$\text{(III-11a)}$$

$$C(0,t) = 0$$

Since $\int_{-\infty}^{\infty} \delta(x)dx = 1$, the dimension of $\delta(x)$ is $[\delta(x)] = [L]^{-1}$. So, the dimension of A is $[A] = [C][L]$, where $[C]$ and $[L]$ are the physical dimensions of a mobile-phase concentration and a length, respectively. In this case, the boundary condition actually consists in pumping the pure mobile phase into the column. Obviously, this model is not realistic. It is merely convenient to derive the analytical solution of some models of chromatography.

d. The Boundary Conditions

These conditions describe exactly the change or changes that are made in the composition of the system during the experiment that is modeled. For example, in elution chromatography, the boundary condition states what is injected into the column, at the beginning and during the rest of the experiment [1]. The main possible cases are listed here.

(i) Elution Chromatography

This method of running chromatography is the most widely used, both for analytical and for preparative applications. A pulse of sample or feed is injected into the column. This injection lasts for a finite period of time, a period that is usually short compared to the elution time of the sample components. Several types of boundary conditions have been imagined, mirroring different common applications.

1 - The Conventional Dirac Injection. The most conventional boundary condition in analytical chromatography is the Dirac injection. The concentration profile injected has a finite area (corresponding to a finite sample size), an infinite concentration, and a zero bandwidth. Although this condition seems to be quite unrealistic, it is often applied successfully in analytical chromatography. Actually, if the injected profile is close to a rectangle and its width is narrow compared to that of the eluted band, this injection model is excellent. There are, however, two different possibilities in modeling this injection. The first injection model consists in actually "injecting" nothing into the column but in assuming the instantaneous creation of a Dirac concentration distribution along the column at $t = 0$, as explained above. In this case, the injection is performed using the initial condition while the boundary condition consists in pumping pure solvent all the time.

In analytical chromatography, the boundary condition consists more often in the injection of a narrow pulse of sample or feed, after which a stream of pure mobile phase is again pumped through the column. The concentration of the sample components in this pulse is given by a Dirac function, $\delta(t)$, in theoretical studies and by a narrow rectangle in numerical calculations. We discuss first the case of a rectangular pulse injection.

2 - Rectangular Pulse Injection. A very general representation of the injection process in chromatography consists in using a rectangular pulse of height C_0 and duration t_p. The boundary condition becomes

$$C(0, t) = 0 \quad \text{for} \quad t < 0$$
$$C(0, t) = C_0 \quad \text{for} \quad 0 < t \leq t_p \qquad \text{(III-11b)}$$
$$C(0, t) = 0 \quad \text{for} \quad t_p < t$$

where C_0 is the concentration of the pulse and t_p its duration. Usually, t_p is finite but small compared to the hold-up time, $t_0 = L/u$. In the extreme case, this pulse is so narrow that it can be represented by the δ function of Dirac. We must note, however, that in this case the amount injected is finite, so the maximum concentration of the Dirac pulse has to be infinite. In practical applications, and particularly in preparative chromatography, where the column has a finite length and a finite efficiency, a more realistic injection profile than the Dirac pulse or the rectangular injection should be used (see next two sections).

If t_p is finite, the solution is obtained as the overlay of the solutions of two successive breakthrough curves of opposite signs, having the same amplitude, C_0, and injected at times $t = 0$ and $t = t_p$, respectively. Obviously, the solution for the Dirac-injection boundary condition is the limit of this last solution when t_p tends toward zero and C_0 toward infinity while the pulse area remains constant.

3 - The Danckwerts Condition. In practice, axial diffusion would proceed at an infinite rate if vertical boundary conditions were considered. Such conditions are not realistic and their use raises serious fundamental difficulties. In practice, in chromatography as well as in chemical engineering, the boundary conditions of Danckwerts are most often used [1].

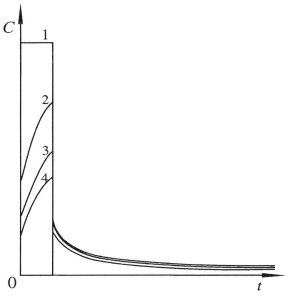

Figure III-2.   Illustration of the Danckwerts Boundary Condition. 1: Rectangular pulse injection. 2, 3, 4: Danckwerts condition; 2: $D = 0.04$ cm^2/s; **3:** $D = 0.08$ cm^2/s; **4:** $D = 0.12$ cm^2/s.

$$\left[uC - D\frac{\partial C}{\partial x} \right]_{x=0} = uC_0$$

$$\frac{\partial C}{\partial x}\Big|_{x=L} = 0 \qquad\qquad \text{(III-12)}$$

From a physical viewpoint, this condition states the conservation of the mass flux in the boundary layer. More specifically, it states that the mass flux in the column at the injection point, $uC(0,t) - D\frac{\partial C}{\partial x}\big|_{x=0}$, is equal to the injection flux in the equivalent pipe, $uC_0(t)$. If we ignore the term $D\frac{\partial C}{\partial x}\big|_{x=0}$, we have $C(0,t) = C_0(t)$. This assumption constitutes the approximation of the ideal model. The Danckwerts boundary condition is illustrated in Figure III-2 for three different values of the coefficient of apparent axial dispersion, D.

Actually, D is finite in the column and $D\frac{\partial C}{\partial x}\big|_{x=0}$ is different from zero. It should not be neglected. For the sake of convenience in numerical calculations, however, the relationship $C(0,t) = C_0(t)$ is often used to replace the Danckwerts condition. This is approximate but remains reasonable when the column has a high efficiency, a few thousand plates or more. If the efficiency is low, e.g., a hundred theoretical plates or fewer, significant errors may occur and the Danckwerts condition should be used. This situation is most typically encountered in numerical calculations carried out in Simulated Moving Bed separations, for example. While the Danckwerts boundary condition must always be considered for numerical calculations, it is rarely considered in fundamental studies. It is a mixed condition, involving values of both the function and its derivative in certain points (at the column inlet and exit). It is complex to handle in algebraic calculations and makes most integrations really difficult if not altogether impossible.

4 - Practical Boundary Conditions. In practical applications, it is an even better approach to use as the boundary condition the concentration profile obtained by recording the signal of a detector located between the injection device and the column (using a detector cell that can hold the pressure). This procedure allows more accurate calculations of band profiles, particularly when column switching procedures are used [1,3].

(ii) Frontal analysis

In frontal analysis, the boundary condition consists merely of substituting a stream of a solution of constant concentration of the feed in the mobile phase for the stream of pure mobile phase. A discontinuity in the concentration of the incoming stream corresponds to the Riemann problem. More realistic conditions are obtained by following the same method as above, under Subsection i-4, Practical Boundary Conditions.

(iii) Displacement chromatography

In this case, the boundary condition consists of a sample injection, as in the elution case. It is no longer followed by the resumption of the injection of a stream of eluent free of sample but by that of a stream of a solution of displacer in the mobile phase. The displacer is a compound which is more strongly adsorbed than any compound of interest in the sample. Because of isotherm competition and because a sufficiently high concentration is selected for the displacer, all the feed components are flushed out of the column ahead of the displacer front, in increasing order of their equilibrium constants [1]. The displacer has to be eliminated from the column before a new experiment can be carried out.

(iv) Gradient elution chromatography

This boundary condition consists of the injection of a feed sample, as in elution chromatography, followed by the injection of a concentration ramp of a strong solvent in the initial mobile phase.

(v) Simulated moving bed chromatography (SMB)

The boundary condition in SMB depends on each column and is updated at the beginning of each new cycle. The boundary condition for each column is derived from the concentration profiles of the two feed components in the previous column (in the direction of the liquid stream), at the end of the previous cycle.

e. Mass Transfer Kinetics and Axial Dispersion

In the equilibrium or ideal model, the mass balance equation(s) and the initial and boundary conditions are merely completed by the isotherm equation(s), the two phases of the chromatographic system being constantly at equilibrium. In the equilibrium–dispersive model, the nonequilibrium due to the mass transfer resistances is assumed to be equivalent to an additional contribution to axial dispersion [1,4]. Accordingly, the diffusion coefficient in the mass balance equation is replaced by an apparent axial dispersion coefficient, D_L (see eqn. II-27), related to the column efficiency under linear conditions, with

$$\sigma_x^2 = 2D_L t_0 = HL \tag{III-13}$$

where σ_x is the standard deviation of the Gaussian profile obtained under analytical conditions, H is the column HETP, L the column length, and t_0 the column hold-up time. This model is valid provided that the mass transfer kinetics is fast. Note that D_L includes a contribution to axial dispersion due to the finite rate of the mass transfer kinetics (see Chapter II, part IV) [1,4].

Under these conditions (axial dispersion different from zero, infinite mass transfer rate coefficient), the mass balance equation becomes

$$\left[1 + F\frac{df(C)}{dC}\right]\frac{\partial C}{\partial t} + u\frac{\partial C}{\partial x} = D_L\frac{\partial^2 C}{\partial x^2} \tag{III-14}$$

This equation is called the equilibrium chromatography equation. In all these cases, no kinetic equations are used.

In simple kinetic models, the most appropriate kinetic equation among eqns. II-20 to II-23, is used. These models may require also the use of an axial dispersion coefficient, D_a, accounting for the effects of axial and eddy diffusion. If the dispersion coefficient in the mass balance equation is set equal to 0, the band broadening effect of diffusion may be lumped with the effects of the finite rate of the mass transfer kinetics. This option is legitimate when diffusion is reasonably fast while the mass transfer kinetics is slow and constitutes the essential source of band dispersion.

To give more generality to the formulation of chromatographic models, dimensionless coordinates are often used. As an example, the combination of the mass balance equation with axial dispersion (eqn. III-1) and the solid film linear driving force model (eqn. II-22) can be written

$$\frac{\partial C}{\partial z} + \frac{\partial C}{\partial \tau} + F\frac{\partial q}{\partial \tau} = \frac{1}{Pe}\frac{\partial^2 C}{\partial z^2}$$

$$\frac{\partial q}{\partial \tau} = St(q^* - q) \qquad\qquad \text{(III-15a)}$$

with

$$Pe = \frac{uL}{D_a} \qquad\qquad \text{(III-15b)}$$

$$St = \frac{k_f L}{u} \qquad\qquad \text{(III-15c)}$$

$$z = \frac{x}{L} \qquad\qquad \text{(III-15d)}$$

$$\tau = \frac{ut}{L} \qquad\qquad \text{(III-15e)}$$

where z and τ are the reduced space and time coordinates, respectively. The dimensionless numbers Pe and St are the column Peclet number and Stanton number, respectively.

f. Multicomponent Case

In any multicomponent case, we must write as many mass balance and kinetic equations as there are active components in the system. Active components include all the retained components of the sample or feed. They include also the components of the mobile phase (except the weak solvent) that interact with the stationary phase and have retention factors at infinite dilution in the weak solvent which are not much lower than those of the feed components [1,5]. Thus, we obtain the following system of equations

$$\frac{\partial C_i}{\partial t} + F\frac{\partial q_i}{\partial t} + u\frac{\partial C_i}{\partial x} = D_a\frac{\partial^2 C_i}{\partial x^2} \qquad\qquad \text{(III-16a)}$$

$$\frac{\partial q_i}{\partial t} = -k_i(q_i - q_i^*) \qquad\qquad \text{(III-16b)}$$

for $i = 1, 2, \cdots, n$, for a n-component system.

In nonlinear chromatography, the isotherms are competitive, *i.e.*, the equilibrium concentration of any component in the stationary phase depends, in addition to its own mobile phase concentration, on the mobile phase concentrations of all the active components present (see eqns. II-3 to II-5). Because of the competitive character of multicomponent isotherms, the equations in the system of PDE's of eqns. III-16a and III-16b are coupled. It is not possible to consider the multicomponent chromatography problem as the superimposition of n single component problems. The solution of multicomponent, nonlinear problems is considerably more complex than that of single component nonlinear or of multicomponent linear problems.

2 - The Quest for Solutions

a. Retrospective

In linear chromatography, the PDE stating mass conservation of a component (eqn. III-16a) is linear. Its analytical solution in the general case was derived in the early 1950s by Lapidus and Amundson [6]. These authors used the Laplace transform and moment analysis, which are important tools to solve linear PDEs (see later, Section II, subsection 3). Further simplifications were introduced shortly afterwards by van Deemter *et al.* [7] who discussed the important particular case in which both axial dispersion and the mass transfer resistances are reasonably small, as they are in most current applications of chromatography, especially with the modern HPLC columns used in analytical and preparative chromatography. These issues will be discussed further in Chapter IV, devoted to linear chromatography.

The Laplace transform is an integral transform (see later, Section II, subsection 3). In general, the solution algorithm cannot be adapted to the case of nonlinear PDEs. Unfortunately, there are no general methods for investigating nonlinear problems, at least until now. As a matter of fact, there are not even theorems demonstrating the existence of solutions for the systems of equations discussed here (except for the ideal model). Only in a few particular cases, are some nonlinear transform available that permits a linearization of the problem. The only general approach for the solution of hyperbolic PDEs is the use of the characteristics method [8-10]. This method gives many important, general results in the case of the nonlinear, ideal model of chromatography [1]. Unfortunately, as explained earlier, this simple model is important from a theoretical viewpoint but its solutions are not directly applicable. For the determination of solutions of practical interest in any specific case, numerical solutions are most often used [1]. Many improvements have been designed to achieve accurate calculations, particularly for the correct determination of the position and the profile of concentration shock layers [1]. Obviously, the application of numerical analysis at the scale practiced currently would be utterly impractical without the use of modern computers.

In their pioneering work in the early 1940s, Wicke [11], Wilson [12], De-Vault [13], and Weiss [14] correctly pointed out that the basic character of nonlinear chromatography is that a migration velocity must be associated to any concentration and that this velocity depends on the concentration itself. Since the velocities associated with high and low concentrations are different, band profiles change during their migration. Later, Thomas [15] solved the equations of the ideal, nonlinear, nonequilibrium chromatography model. Goldstein and Murry [16] used the singular perturbation method to discuss the approximate solution of a general problem that includes, as a particular case, the Thomas model under the condition that the mass-transfer resistances are small (*i.e.*, under conditions of near-equilibrium). Houghton [17] was the first to discuss the effect of axial dispersion in nonlinear chromatography, using a parabolic isotherm, eqn. II-1b. He transformed the mass balance equation into a Burgers equation that he solved using the Cole–Hopf transform. All this work dealt with the single-component problem. Although essential to understand the mechanism of band migration and the evolution of band profiles in chromatography, the solutions of this problem are of limited help in understanding the separation of multicomponent mixtures.

Glueckauf [18], Helfferich [19], Seinfeld and Lapidus [14] and Aris, Amundson and Rhee [9,10,20] discussed the analytical solution of the ideal model for multicomponent, nonlinear chromatography. Of particular fundamental and seminal importance is the systematic work of Aris, Amundson and Rhee [10,20] on the shock theory and the simple wave solution in nonlinear, single and multicomponent chromatography and that of Lax [21], Whitham [22], and Temple [23] on the solutions of hyperbolic PDEs. Another theoretical watershed was the work of Rhee and Amundson [24,25] on the concept and calculations of the shock layer in chromatography. The search for numerical solutions began in 1962, with the work of Ting Chin-Chun and Chu Pao-Lin [26] who used an analog electrical circuit, *i.e.*, a primitive analog computer, the only practical calculation method at that time for this kind of problems. Guiochon *et al.* [1] devoted considerable work to the development of appropriate methods for the numerical calculation of solutions of various models of nonlinear chromatography. Now numerical solutions are available for all models of chromatography, from the simple linear or ideal models [1] to the most complex forms of the general rate models [27]. Solutions based on fast algorithms are available for the rapid calculations of numerical solutions in simple cases or for systematic investigations of general problems. More accurate solutions requiring far longer CPU times are also available. Solutions of either type are available for all kinds of initial and boundary conditions.

In the 1980s, Knox [28], Guiochon *et al.* [1,29-31], and Cox and Snyder [32] began systematic investigations on the optimization of the separation conditions in preparative chromatography based on the theory of nonlinear chromatography. Considerable investigations have been carried out in this area. Comparisons of the performance achieved with different implementations or methods have also been published. Yeroshenkova *et al.* [33] were the first to discuss the effects of the stochastic nature of the column packing.

b. Models and Problems Discussed in this Book

We summarize here the essential equations of the different models and problems discussed in this work. They include most of the classical models of chromatography. A few fundamental problems which have appeared or progressed more recently are also presented.

i. Linear Equilibrium–Dispersive Single-Component Model (see Chapter IV, section 1)

In this model the equilibrium isotherm is linear. There is a finite axial dispersion coefficient which may include a contribution of the mass transfer resistances. The mass balance and isotherm equations are:

$$\frac{\partial C}{\partial t} + F\frac{\partial q}{\partial t} + u\frac{\partial C}{\partial x} = D_L\frac{\partial^2 C}{\partial x^2}$$
$$q = aC \tag{III-17}$$

This model is the classical model used for analytical chromatography [1,4]. It has an analytical solution.

ii. Nonlinear Ideal Single-Component Model (see Chapter V, sections I to III)

In this model the equilibrium isotherm is nonlinear. There is no axial dispersion and no mass transfer resistances. This is the classical ideal, nonlinear model. Its essential equations are

$$\frac{\partial C}{\partial t} + F\frac{\partial q}{\partial t} + u\frac{\partial C}{\partial x} = 0$$
$$q = f(C) \tag{III-18}$$

This model is the simplest one for nonlinear chromatography. It deals only with the influence of the thermodynamics of phase equilibrium on high concentration band profiles. It has an analytical solution.

iii. Nonlinear Equilibrium–Dispersive Single-Component Model (see Chapter VII, section I)

This model is similar to the previous one but considerably more complex because it is a nonideal or nonequilibrium model in addition to being nonlinear. The essential equations are

$$\frac{\partial C}{\partial t} + F\frac{\partial q}{\partial t} + u\frac{\partial C}{\partial x} = D_L\frac{\partial^2 C}{\partial x^2}$$
$$q = f(C) \tag{III-19}$$

An intermediate between this model and the previous two models is provided by using a parabolic isotherm, $q = a_1 C + a_2 C^2$, with $a_2 C < a_1$ in . In this case, the influence of an increasing sample size on the transition between linear and nonlinear behavior is studied. This last model has an approximate analytical solution in the case of a parabolic isotherm (see Chapter VII, section I).

iv. Linear Transport Model (see Chapter IV, section II)

This model is also similar to the first one. In this case, however, if the isotherm remains linear, there is no diffusion and the nonideal behavior arises from a finite rate of mass-transfer kinetics. The essential equations are

$$\frac{\partial C}{\partial t} + F\frac{\partial q}{\partial t} + u\frac{\partial C}{\partial x} = 0 \tag{III-20a}$$

$$\frac{\partial q}{\partial t} = k_1 C - k_2 q = -k_f(q - GC) \tag{III-20b}$$

where k_1, k_2, and k_f are rate coefficients which, in this model, are constant (with $k_f = k_2$ and $G = k_1/k_2$). The second equation, eqn. III-20b is valid only because the isotherm is linear. This model has an analytical solution [6].

v. Nonlinear Transport Model (see Chapter VII, section II)

This model combines the features of the second and third models of this group. There is no axial dispersion but a finite rate of mass transfer and a nonlinear isotherm. The essential equations are

$$\frac{\partial C}{\partial t} + F\frac{\partial q}{\partial t} + u\frac{\partial C}{\partial x} = 0 \tag{III-21a}$$

$$\frac{\partial q}{\partial t} = k_a(q_s - q)C - k_d q \tag{III-21b}$$

where k_a and k_d are the adsorption and desorption rate constants, respectively, and q_s is the saturation capacity (see eqn. II-2). When the equilibrium is established, $\frac{\partial q}{\partial t} = 0$, hence

$$q = \frac{k_a q_s C}{k_d + k_a C} \tag{III-22}$$

The adsorption constant is $b = k_a/k_d$. Obviously, this is the equation of the Langmuir isotherm model. The kinetic equation used above (eqn. III-21b) is that of the Langmuir kinetics. Alternately, the kinetic equation in this model can be written

$$\frac{\partial q}{\partial t} = -k_f(q - f(C)) \tag{III-23}$$

This model is known as Thomas model [15]. It has an analytical solution [15,16,34].

vi. Linear Transport–Dispersive Model (see Chapter IV, section III)

This model combines the first and fourth models above. Its essential equations are

$$\frac{\partial C}{\partial t} + F\frac{\partial q}{\partial t} + u\frac{\partial C}{\partial x} = D_a\frac{\partial^2 C}{\partial x^2} \tag{III-24a}$$

$$\frac{\partial q}{\partial t} = -k_f(q - GC) \tag{III-24b}$$

vii. Nonlinear Transport–Dispersive Model (see Chapter VII, section III)

This model combines the third and fifth models above. The essential equations of this model are

$$\frac{\partial C}{\partial t} + F\frac{\partial q}{\partial t} + u\frac{\partial C}{\partial x} = D_a\frac{\partial^2 C}{\partial x^2}$$

$$\frac{\partial q}{\partial t} = -k_f(q - f(C)) \tag{III-25}$$

This model (which has no analytical solutions) is often used to account for the behavior of compounds that have a relatively slow mass transfer kinetics (*e.g.*, peptides, proteins, or the more retained enantiomer on a chiral phase). It may lead to model errors when the mass transfer kinetics is too slow, the most common one being an apparent dependence of the rate coefficient, k_f, on the concentration [35].

viii. Nonlinear, Ideal Model for Two Components (see Chapter VIII)
   The essential equations of this model are

$$\frac{\partial C_1}{\partial t} + F\frac{\partial q_1}{\partial t} + u\frac{\partial C_1}{\partial x} = 0$$

$$\frac{\partial C_2}{\partial t} + F\frac{\partial q_2}{\partial t} + u\frac{\partial C_2}{\partial x} = 0$$

$$q_1 = f_1(C_1, C_2)$$

$$q_2 = f_2(C_1, C_2) \tag{III-26}$$

This is the simplest model for two components. The two partial differential equations are coupled by the competitive character of the equilibrium isotherms. This model has an analytical solution in the cases in which the adsorption behaviors of the two feed components are well accounted for by the Langmuir competitive isotherm model [36].

ix. Nonlinear Reaction Chromatography (see Chapter VI)
   Reaction chromatography is a complex phenomenon. In Chapter VI, we discuss the behavior of the two components of a binary mixture when, during their elution, the first reacts to give the second one, this reaction product being separated from the reagents under the experimental conditions selected. The essential equations of this model are

$$\frac{\partial C_1}{\partial t} + F\frac{\partial q_1}{\partial t} + u\frac{\partial C_1}{\partial z} = 0 \tag{III-27a}$$

$$\frac{\partial C_2}{\partial t} + F\frac{\partial q_2}{\partial t} + u\frac{\partial C_2}{\partial z} = 0 \tag{III-27b}$$

$$\frac{\partial q_1}{\partial t} = \frac{\partial f_1(C_1, C_2)}{\partial t} + k_r f_1(C_1, C_2) \tag{III-28a}$$

$$\frac{\partial q_2}{\partial t} = \frac{\partial f_2(C_1, C_2)}{\partial t} - k_r f_1(C_1, C_2) \tag{III-28b}$$

where C_1 and C_2 are the concentrations of the reactant and the reaction product in the mobile phase, respectively. We will assume moderate deviation of the reagent isotherm from linear behavior and a low rate of reaction, hence a low extent of the reaction and the linear adsorption behavior of the solute.

x. Linear and Nonlinear Ideal Model for a Single Component, with Random Packing Distribution(see Chapter IV, section IV)

This model considers that the distribution of the particles in the column bed is random. Its essential equations are

$$\varepsilon_\omega(z)\frac{\partial C_\omega}{\partial t} + \mu_\omega(z)\frac{\partial q_\omega}{\partial t} + u_\omega(z)\frac{\partial C_\omega}{\partial z} = 0$$

$$q_\omega = f(C_\omega)$$

(III-29)

where C_ω, q_ω, u_ω, $\varepsilon_\omega(z)$, and $u_\omega(z)$ are the values of C, q, u, ε, and u, respectively, when the packing distribution along the column is stochastic. Depending on the isotherm equation, the model can be either linear or nonlinear.

c. Equivalence between the Equilibrium Dispersive Model and the Transport Model

The set of PDE of the transport model, *i.e.*, the system of eqns. III-21a and III-21b written for a nonlinear isotherm, with a linear driving force kinetics instead of a Langmuir kinetics, is:

$$\frac{\partial C}{\partial t} + F\frac{\partial q}{\partial t} + u\frac{\partial C}{\partial x} = 0 \qquad \text{(III-30a)}$$

$$\frac{\partial q}{\partial t} = k_f(q^* - q) \qquad \text{(III-30b)}$$

with $q^* = f(C)$, where $f(C)$ is the isotherm equation (eqn. II-1d). The kinetic equation is now written with a solid film linear driving force model. This equation can be rewritten:

$$q = f(C) - \frac{1}{k_f}\frac{\partial q}{\partial t} \qquad \text{(III-30c)}$$

When k_f tends toward infinity, q becomes equal to $f(C)$, and we obtain the equation of the ideal model. When k_f is large, we may, as a first-order approximation, combine eqns. III-30a and III-30c, which gives:

$$\frac{\partial C}{\partial t} + F\frac{\partial q^*}{\partial t} + u\frac{\partial C}{\partial x} = \frac{1}{k_f}\frac{\partial q}{\partial t}$$

In the linear case, we have $q^* = a_1 C$ and $x = ut/(1 + Fa_1)$, so we obtain

$$\frac{\partial C}{\partial t} + F\frac{\partial q}{\partial t} + u\frac{\partial C}{\partial x} = \frac{Fa_1 u^2}{(1 + Fa_1)^2 k_f}\frac{\partial^2 C}{\partial x^2}$$

This latter equation is the mass balance equation of the equilibrium dispersive model with an apparent dispersion coefficient equal to $Fa_1 u^2/k_f(1 + Fa_1)^2$. If we use the transport dispersive model instead, we obtain for the apparent dispersion coefficient $D_L = D_a + Fa_1 u^2/k_f(1 + Fa_1)^2$, which is equivalent to the classical van Deemter equation [7].

II - GENERAL PROPERTIES OF
PARTIAL DIFFERENTIAL EQUATIONS

As mentioned earlier, the main mathematical equations involved in the description of the chromatographic process are partial differential equations (PDE). Although this book does not attempt to be a treatise on partial differential equation theory, understanding its substance requires a certain familiarity with the properties and the use of these equations. Accordingly, it is useful, in order to simplify further discussions, to summarize a few basic concepts here.

1 - Definitions

An equation which contains an unknown function (for example, $C(x,t)$) and at least one of its partial derivatives (for example, $\frac{\partial C}{\partial x}$ or $\frac{\partial C}{\partial t}$) is named a partial differential equation.

a. Order of a PDE

The order of the highest partial derivative involved in a PDE is defined as the order of this PDE. For example the diffusion equation and eqn. III-1 are second order PDE, eqn. II-25 is a first order PDE. The mass balance equation of the ideal model of chromatography (eqn. III-2) contains only first order differentials since the axial dispersion coefficient is assumed to be zero. Thus, this equation is a first order PDE.

b. Homogeneous and Non-homogeneous PDE

In a PDE, the term which does not contain either the unknown function or one of its partial derivatives is called the free term. If the free term is equal to zero, then the PDE is said to be homogeneous. If the free term is different from zero, the PDE is said to be nonhomogeneous. For example, if in the following equation

$$a(x,t,C)\frac{\partial C}{\partial x} + b(x,t,C)\frac{\partial C}{\partial t} + d(x,t,C) = f(x,t) \qquad \text{(III-31)}$$

If $f(x,t) \neq 0$, this equation is nonhomogeneous. By contrast, if $f(x,t) = 0$, then it is an homogeneous equation.

c. Linear, Quasi-linear, and Nonlinear PDEs

If, in a PDE, the relationship involving the unknown function and its partial derivatives is linear and if the coefficients of the derivatives do not contain the unknown function, the equation is a *linear PDE*. For example, the equation

$$a(x,t)\frac{\partial C}{\partial x} + b(x,t)\frac{\partial C}{\partial t} + d(x,t)C = f(x,t) \tag{III-32}$$

is a *linear PDE*.

If the coefficient of at least one of the highest order partial derivatives contains the unknown function, the PDE is called a *quasi-linear PDE*. For example, the equation

$$a(x,t,C)\frac{\partial^2 C}{\partial x^2} + b(x,t,C)\frac{\partial^2 C}{\partial t^2} + r(x,t,C)\frac{\partial C}{\partial x} + d(x,t,C)\frac{\partial C}{\partial t} = 0 \tag{III-33}$$

is quasi-linear.

If none of the coefficients of the highest order partial derivatives contain the unknown function but the coefficient of at least one of the lower order partial derivatives contains it, the PDE is called a *semi-linear PDE*. For example the function

$$a(x,t)\frac{\partial^2 C}{\partial x^2} + b(x,t)\frac{\partial^2 C}{\partial t^2} + r(x,t,C)\frac{\partial C}{\partial x} + d(x,t,C)\frac{\partial C}{\partial t} = 0 \tag{III-34}$$

is a semi-linear PDE.

If the relationship defining the PDE is nonlinear with respect to one of the derivatives, the PDE is called a *nonlinear PDE*. For example, the equation

$$\left(\frac{\partial C}{\partial x}\right)^2 + a(x,t)\left(\frac{\partial C}{\partial t}\right)^2 = 0 \tag{III-35}$$

is a nonlinear PDE.

According to these definitions, the mass balance equation of chromatography mentioned earlier (eqn. II-25), which can be written in the case of negligible mass transfer resistances

$$(1 + F\frac{df}{dC})\frac{\partial C}{\partial t} + u\frac{\partial C}{\partial x} = D\frac{\partial^2 C}{\partial x^2} \tag{III-36}$$

is actually a semi-linear PDE in the general case in which the isotherm, $f(C)$, is nonlinear. Moreover, according to these definitions, the equation of the ideal model

$$(1 + F\frac{df}{dC})\frac{\partial C}{\partial t} + u\frac{\partial C}{\partial x} = 0 \tag{III-37}$$

is actually a quasi-linear PDE. However, in chromatography and in chemical engineering, both equations are usually called nonlinear PDEs. In the linear case, however, both eqns. III-36 and III-37 are linear PDEs.

2 - Solution of a PDE

a. *Well-Posed Problems and the Conditions to Solve a PDE*

The PDE itself can only be the mathematical representation of a general principle, such as the conservation of mass in adsorption or partition processes in the case of the fundamental equation of chromatography (eqn. II-25). Alone, it cannot determine the concrete form of the concentration wave. In order to be able to do that, we need to add sufficient, appropriate conditions, for example, the initial and the boundary conditions discussed earlier. The actual solution can then be obtained. This type of condition is called a condition to solve. The search for solutions of a PDE completed with appropriate initial and boundary conditions is a *well-posed problem.* Well-posed problems are those for which there is a single solution. These problems can be divided into three classes:

(1) Problems for which the initial value is known.
(2) Problems for which the boundary conditions are known.
(3) Problems for which initial and boundary conditions are known.

All chromatography problems belong to the third type.

b. *The Stability of the Solution*

If the solution changes to only a small extent when a small change is made in the condition to solve, the dependence of the solution on the condition to solve is continuous. A well-posed problem which also possesses this property is called a *stable problem.* This is an important property, particularly when one needs to calculate numerical solutions. Using numerical analysis, the partial differential equation is then replaced by a difference equation. The stabilities of the calculation of the numerical solutions of a difference equation and of the corresponding differential equation are similar. When the difference calculation is stable, the variations of the solution arising from small perturbations in the calculation are small. Thus, the numerical solution obtained is close to the true solution of the PDE and can be used safely.

c. *The Characteristics of First-order, Quasi-linear PDEs*

The general form of a system of first-order, quasi-linear equations is

$$a_{1,1}\frac{\partial C_1}{\partial t} + a_{1,2}\frac{\partial C_2}{\partial t} + \cdots + a_{1,n}\frac{\partial C_n}{\partial t} + b_{1,1}\frac{\partial C_1}{\partial x} + b_{1,2}\frac{\partial C_2}{\partial x} + \cdots + d_1 = 0$$
$$\text{(III-38a)}$$

$$a_{2,1}\frac{\partial C_1}{\partial t} + a_{2,2}\frac{\partial C_2}{\partial t} + \cdots + a_{2,n}\frac{\partial C_n}{\partial t} + b_{2,1}\frac{\partial C_1}{\partial x} + b_{2,2}\frac{\partial C_2}{\partial x} + \cdots + d_2 = 0$$
$$\text{(III-38b)}$$

$$\cdots \qquad \cdots \qquad \cdots \qquad \cdots$$

$$a_{n,1}\frac{\partial C_1}{\partial t} + a_{n,2}\frac{\partial C_2}{\partial t} + \cdots + a_{n,n}\frac{\partial C_n}{\partial t} + b_{n,1}\frac{\partial C_1}{\partial x} + b_{n,2}\frac{\partial C_2}{\partial x} + \cdots + d_n = 0$$
$$\text{(III-38n)}$$

These equations also be rewritten as

$$\sum_j a_{i,j} \frac{\partial C_j}{\partial t} + \sum_j b_{i,j} \frac{\partial C_j}{\partial x} + d_i = 0 \qquad\qquad i,j = 1, 2, \cdots, n \quad \text{(III-39)}$$

where $a_{i,j} = a_{i,j}(x, t, C_1, C_2, \cdots, C_n)$, $b_{i,j} = b_{i,j}(x, t, C_1, C_2, \cdots, C_n)$, and $d_i = d_i(x, t, C_1, C_2, \cdots, C_n)$. If there is a curve, $x = x(\eta), t = t(\eta)$ where η is the parameter which is such that along this curve we have

$$|b_{i,j} - a_{i,j} \frac{dx}{dt}| = 0 \qquad\qquad \text{(III-40)}$$

this curve is called a *characteristic*. Generally, either $a_{i,j}$ or $b_{i,j}$ is equal to $\delta_{i,j}$ with

$$\delta_{i,j} = 1, \qquad\qquad i = j$$
$$\delta_{i,j} = 0, \qquad\qquad i \neq j$$

Equation III-40 is the eigen-equation of the PDE in eqn. III-38 and dx/dt in eqn. III-39 is the characteristic direction.

Applying this definition to the mass balance equation of ideal chromatography for a single component (eqn. III-2)

$$(1 + F\frac{df}{dC}) \frac{\partial C}{\partial t} + u \frac{\partial C}{\partial x} = 0 \qquad\qquad \text{(III-2)}$$

or

$$\frac{(1 + F\frac{df}{dC})}{u} \frac{\partial C}{\partial t} + \frac{\partial C}{\partial x} = 0 \qquad\qquad \text{(III-41)}$$

we have obviously

$$a_{i,j} = \frac{(1 + F\frac{df}{dC})}{u}, \qquad\qquad b_{i,j} = 1, \qquad\qquad d_i = 0$$

$$\frac{dt}{dx} = \frac{(1 + F\frac{df}{dC})}{u} \qquad\qquad \text{(III-42)}$$

An important example of the use of the characteristics and of their mathematical properties will be provided in Chapter V, where the mass balance of the ideal model, its properties, and its solution are discussed. It is possible to rewrite eqn. III-2 along its characteristics as follows

$$\frac{dC}{dx} = 0$$

$$\frac{dt}{dx} = \frac{(1 + F\frac{df}{dC})}{u}$$

$$\text{or} \qquad \frac{dx}{dt} = \frac{u}{(1 + F\frac{df}{dC})} \qquad\qquad \text{(III-43)}$$

d. The Different Types of PDEs

From the characteristics theory, it follows that if the eigen-equation of a second-order PDE has two real roots (*i.e.*, two eigen-values or two eigen-directions), this PDE is hyperbolic. If there is only one real root, it is parabolic. If there are no real roots, it is elliptic. Obviously, the wave equation is hyperbolic, the diffusion equation is parabolic, and the Laplace equation ($\frac{\partial^2 u}{\partial x^2} + \frac{\partial^2 u}{\partial y^2} = 0$) is elliptic. Generally, if there are n different real eigen-directions in an nth order PDE, this PDE is hyperbolic. In single-component chromatography, the equation of the ideal model is a first-order PDE, its eigen-equation has one eigen-direction, and the PDE is hyperbolic. For n-component chromatography, the set of equations of the ideal model has an eigen-equation with n real roots and n eigen-directions. These equations are hyperbolic.

e. The Discontinuous Solution of a First-order Quasi-linear Partial Differential Equation and its Stability

The general form of a first-order quasi-linear partial differential equation (*e.g.*, eqn. III-18) is:

$$\frac{\partial C}{\partial t} + \frac{\partial \Phi(C)}{\partial x} = 0 \tag{III-44}$$

The discontinuous solution of these equations is the limit of the continuous solution of the corresponding parabolic equation (*e.g.*, eqn. III-1)

$$\frac{\partial C}{\partial t} + \frac{\partial \Phi(C)}{\partial x} = D_L \frac{\partial^2 C}{\partial x^2} \tag{III-45}$$

when D_L tends toward zero. This discontinuous solution satisfies the jump condition

$$u_s = \left(\frac{dx}{dt}\right)_s = \frac{[\Phi]}{[C]} \tag{III-46}$$

which can be derived from the integral of the conservation equation (*i.e.*, of eqn. III-44), where

$$[C] = C_+ - C_-$$
$$[\Phi] = \Phi_+ - \Phi_- \tag{III-47}$$

and $\Phi_\pm = \Phi(C_\pm)$. The subscripts $+$ and $-$ represent the front and the rear sides of the concentration jump in the band profile, respectively.

A stable discontinuous solution of a parabolic equation when $D_L \to 0$ needs to satisfy another condition:

$$\left(\frac{d\Phi}{dC}\right)_+ < \left(\frac{d\Phi}{dC}\right)_s < \left(\frac{d\Phi}{dC}\right)_- \tag{III-48a}$$

or

$$\left(\frac{dx}{dt}\right)_{z+} < \left(\frac{dx}{dt}\right)_s < \left(\frac{dx}{dt}\right)_{z-} \tag{III-48b}$$

where

$$\left(\frac{dx}{dt}\right)_z = \frac{d\Phi}{dC} \tag{III-49}$$

is the characteristic velocity.

The stable discontinuous solution corresponds to a concentration shock. So, the condition mentioned above is the condition for the formation and the stability of a shock. This condition means that the velocity of the discontinuity is larger than the characteristic velocity of the band on its rear side and less than the characteristic velocity of the band on its front side.

f. The Reducible Hyperbolic Equations

Let assume that we have the following two homogeneous partial differential equations

$$\alpha_1 \frac{\partial C_1}{\partial x} + \beta_1 \frac{\partial C_1}{\partial t} + \sigma_1 \frac{\partial C_2}{\partial x} + \delta_1 \frac{\partial C_2}{\partial t} = 0 \tag{III- 50a}$$

$$\alpha_2 \frac{\partial C_1}{\partial x} + \beta_2 \frac{\partial C_1}{\partial t} + \sigma_2 \frac{\partial C_2}{\partial x} + \delta_2 \frac{\partial C_2}{\partial t} = 0 \tag{III- 50b}$$

If the coefficients $\alpha_i, \beta_i, \sigma_i$ and δ_i $(i = 1, 2)$ are continuous functions of C_1 and C_2 which can be differentiated and if the ratios

$$\frac{\alpha_1}{\alpha_2}, \qquad \frac{\beta_1}{\beta_2}, \qquad \frac{\sigma_1}{\sigma_2}, \qquad \text{and} \qquad \frac{\delta_1}{\delta_2}$$

are not simultaneously all equal, the above equations are said to be *reducible*.

If the determinant of the two reducible equations is such that

$$J = \frac{\partial(C_1, C_2)}{\partial(x, t)} = \begin{vmatrix} \frac{\partial C_1}{\partial x} & \frac{\partial C_1}{\partial t} \\ \frac{\partial C_2}{\partial x} & \frac{\partial C_2}{\partial t} \end{vmatrix} \neq 0 \tag{III-51}$$

we can apply the Hodograph transform

$$x = x(C_1, C_2) \tag{III-52a}$$

$$t = t(C_1, C_2) \tag{III-52b}$$

and from the identification relationships

$$C_1 = C_1(x, t) = C_1[x(C_1, C_2), t(C_1, C_2)] \tag{III-53a}$$

$$C_2 = C_2(x, t) = C_2[x(C_1, C_2), t(C_1, C_2)] \tag{III-53b}$$

we derive that

$$\alpha_1 \frac{\partial t}{\partial C_2} - \beta_1 \frac{\partial x}{\partial C_2} - \sigma_1 \frac{\partial t}{\partial C_1} + \delta_1 \frac{\partial x}{\partial C_1} = 0$$

(III- 54a)

$$\alpha_2 \frac{\partial t}{\partial C_2} - \beta_2 \frac{\partial x}{\partial C_2} - \sigma_2 \frac{\partial t}{\partial C_1} + \delta_2 \frac{\partial x}{\partial C_1} = 0$$

(III- 54b)

These two linear equations define the characteristics of $x(C_1, C_2), t(C_1, C_2)$ in the plane (C_1, C_2). These characteristics are the mapping of the characteristics of $C_1(x, t), C_2(x, t)$ in the x, t plane.

g. Simple Wave

If in the above transformation $J = 0$, there are two possibilities. The first one is that the rank of the Jacobi matrix J is equal to zero, *i.e.*

$$\frac{\partial C_1}{\partial x} = \frac{\partial C_2}{\partial x} = 0$$

(III-55a)

$$\frac{\partial C_1}{\partial t} = \frac{\partial C_2}{\partial t} = 0$$

(III-55b)

This means that both C_1 and C_2 are constants. In this case, the mapping of curves in the x, t plane degenerates into a single point of the C_1, C_2 plane.

The other possibility is that the rank of the Jacobi matrix is equal to 1. This means that

$$C_1 = C_1(C_2) = f(C_2)$$

In this case, the mapping of a family of characteristic curves in the x, t plane degenerates into a single curve in the C_1, C_2 plane. In this case, the solution is called a *Simple Wave*.

3 - The Laplace Transform and the Moments of Chromatographic Bands

The Laplace transform is a convenient method to solve complex partial differential equations of the type found in linear nonideal chromatography [10]. Among a wide variety of other applications, it has been used by many authors to derive the solution of numerous models of chromatography. We use it in several places in Chapter IV. Some indications on this method are given below, followed by the definition of the moments of a chromatographic band (*i.e.*, in statistics, the concentration distribution of the band).

a. *The Laplace Transform*

By definition, the Laplace transform of the function $f(t)$ is the following function:

$$F(p) = \int_0^\infty e^{-pt} f(t) dt$$

It is usually denoted as

$$F(p) = L\{f(t)\}$$

The inversion of the Laplace transform is necessarily used to obtain the solution of the problem in the time domain. It is denoted $f(t) = L^{-1}\{F(p)\}$.

This transform has a number of interesting properties that make it most useful in the study of PDE's. Some of these properties that are most relevant to the use of the Laplace transform in the study of models of linear chromatography are listed below. These properties can be demonstrated easily by simple algebraic manipulations.

1. The Laplace transform is linear and:

$$L\{C_1 f(t) + C_2 g(t)\} = C_1 L\{f(t)\} + C_2 L\{g(t)\} = C_1 F(p) + C_2 G(p)$$

2. The Laplace transform is scaled:

$$L\{f(at)\} = \frac{1}{a} F\left(\frac{p}{a}\right)$$

with $a > 0$.

3. A shift in the time domain is easily accounted for:

$$L\{f(t - t_0)\} = e^{-pt_0} F(p)$$

4. A shift in the Laplace domain is also easily accounted for

$$L\{e^{p_0 t} f(t)\} = F(p - p_0)$$

5. If, for a finite integer n we have

$$f(0) = \frac{df}{dt}\Big|_0 = \frac{d^{n-1} f}{dt^{n-1}}\Big|_0 = 0$$

then we have $L\{\frac{d^n f}{dt}\} = p^n f(p)$. This property is particularly important if the function $f(t) = 0$ and $df/dt = 0$. Then, $L\{\frac{d^2 f}{dt^2}\} = p^2 L\{f(t)\}$.

6. Differentiation in the Laplace domain is simple, which is one of the reasons of the great interest of the Laplace transform

$$L\{t^n f(t)\} = (-1)^n \frac{d^n F(p)}{dp^n}$$

7. Integration in the time domain gives a simple result

$$L\{\int_0^t f(\tau)d\tau\} = \frac{1}{p}F(p)$$

8. Integration in the Laplace domain is also simple.

$$L\{\frac{f(t)}{t}\} = \int_p^\infty F(p)dp$$

These properties make the solution of the Laplace transform of a PDE almost as easy to derive in the space of the Laplace variable, p, as those of algebraic equations in the t space. The major difficulty of this approach, however, resides in finding the solution in the real domain, $i.e.$, in deriving the reversed transform, $f(t)$, of the solution, $F(p)$, obtained in the Laplace domain. The only practical general result that can be obtained is the derivation of the moments of $f(t)$. In other words, in most cases the actual solution, $i.e.$, the concentration profile, cannot be calculated but all its moments are easily derived. This provides useful information regarding the solution but, unfortunately, this is insufficient in practice to provide $f(t)$. This major obstacle limits the practical importance of the Laplace transform. This explains, however, the popularity of the moments.

b. Definition and Calculation of Band Moments

Let $\varphi(x,t)$ be a normalized concentration distribution, function of the column length and the position

$$\varphi(x,t) = \frac{C(x,t)}{\int_0^\infty C(x,t)dt} \qquad (\text{III-56})$$

The nth moment of this concentration distribution , $\mathcal{M}_n$, is given by

$$\mathcal{M}_n = \int_0^\infty t^n \varphi(L,t)dt \qquad (\text{III-57})$$

The nth centered moment, μ'_n is defined as

$$\mu'_n = \int_0^\infty (t - \mu)^n \varphi(L, t)dt \tag{III-58}$$

Obviously, the first moment,

$$\mathcal{M}_1 = \int_0^\infty t\varphi(L, t)dt \tag{III-59a}$$

is the average elution time of the band or elution time of its mass center, $\bar{t}$. Note that $\mathcal{M}_1$ is sometimes, by extension but erroneously, named μ_1 or even μ. For a highly symmetrical band profile, we have

$$\mu = \bar{t} = t_R \tag{III-59b}$$

This last relationship is obviously not valid for unsymmetrical peaks. It was shown that the first moment of tailing peaks obtained under linear conditions, as solutions of a model using the lumped kinetic model to account for the mass transfer resistance, is a constant, independent of the mass transfer coefficient. In this case, it is the first moment that is related to the thermodynamic equilibrium constant while the retention time of the peak maximum depends on the rate coefficient.

The second-order centered moment or variance of the band, μ'_2, is related to the width of the elution profile

$$\mu'_2 = \int_0^\infty (t - \bar{t})^2 \varphi(L, t)dt = \mathcal{M}_2 - \mathcal{M}_1^2 \tag{III-60}$$

The column HETP is related to the first and second moments through $H = L\mu'_2/\mathcal{M}_1^2 = L\mu'_2/\mu_1^2$ (see Chapter IV, section III).

The third-order centered moment, μ'_3 is related to the peak asymmetry. It is

$$\mu'_3 = \int_0^\infty (t - \mu)^3 \varphi(L, t)dt = \mu_2 - 3\mathcal{M}_1\mathcal{M}_2 + 2\mathcal{M}_1^3 \tag{III-61}$$

A positive value of μ'_3 corresponds to a tailing peak; a negative value, to a leading peak. The parameter $\mu'_3/\mu'^{3/2}_2$, called the *skew* or the *asymmetry*, is an important characteristic of a chromatographic peak. This parameter is sometimes used in analytical chromatography. It is rarely useful in preparative applications.

In most cases, the solution of the chromatography equation in the Laplace domain takes the following form:

$$\overline{C}(p) = \int_0^\infty C(x,t) \ e^{-pt} dt \tag{III-62}$$

From the properties of the Laplace transform, the definition of the function $\overline{C}(p)$, and the definitions of the moments, we have:

$$\mathcal{M}_n = (-1)^n \lim_{p \to 0} \left(\frac{1}{\overline{C}(p)} \frac{d^n \overline{C}(p)}{dp^n} \right) \tag{III-63}$$

Accordingly, the first moment, $i.e.$, the retention time, is given by

$$\bar{t} = \mathcal{M}_1 = - \lim_{p \to 0} \left(\frac{1}{\overline{C}(p)} \frac{d\overline{C}(p)}{dp} \right) = - \lim_{p \to 0} \frac{d\ln \overline{C}(p)}{dp} \tag{III-64}$$

the variance is given by

$$\mu'_2 = \mathcal{M}_2 - \mathcal{M}_1^2 = \lim_{p \to 0} \left(-\frac{1}{\overline{C}(p)} \frac{d^2 \overline{C}(p)}{dp^2} - \left(\frac{1}{\overline{C}(p)} \frac{d\overline{C}(p)}{dp} \right)^2 \right)$$

$$= \lim_{p \to 0} \frac{d^2 \ln \overline{C}(p)}{dp^2} \tag{III-65}$$

and the asymmetry of the peak is given by:

$$\mu'_3 = \mathcal{M}_3 - 3\mathcal{M}_2\mathcal{M}_1 + \mathcal{M}_1^3 = - \lim_{p \to 0} \frac{d^3 \ln \overline{C}(p)}{dp^3} \tag{III-66}$$

These equations permit an easy derivation of the moments of the peak knowing the solution of the PDE in the Laplace domain, even if the solution cannot be obtained in the time domain.

c. Practical Applications of the Band Moments

Moments of chromatographic bands are easy to calculate, especially now that, with the widespread distribution of data stations and the systematic use of computers as signal recorders, band profiles are recorded as tables of numbers acquired at a constant frequency. Since the period of data acquisition is constant, integration of the profiles requires the mere additions of the corresponding products

$$\mu'_n = \frac{\sum_1^k y_i (t_i - \bar{t})^n}{\sum_1^k y_i}$$

where y_i is the value of the signal acquired at time t_i. So, it is highly tempting to use moments to investigate chromatographic bands. Unfortunately, it turns out that the results of these calculations have often a poor precision, particularly for the third and fourth moments. The reason is illustrated in Figure III-3 which shows the distribution of the contributions to the first four moments of a simple Gaussian band profile (solid line in the figure). When the order of the moment increases, the relative contribution of the parts of the profile that are farther removed from its center increases while the contribution of the central part of the profile decreases.

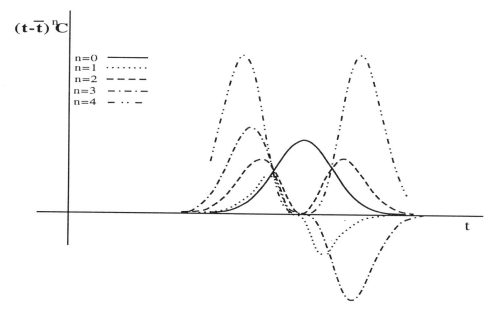

Figure III-3.   Illustration of the Determination of the Moments of a Chromatographic Band. Plots of the contributions of the signal to the moments versus the time.

As a consequence, the low concentration regions of the band, those where the signal is noisy and imprecisely known, weigh more than its center on the final result. The precision of the measurements of the peak area, its retention time, its second, third and fourth moments decreases rapidly in this order. It may easily be between two and three orders of magnitude less good for the fourth than for the first moment. For this reason, the practical use of the moments is limited to those of orders zero to two. Only under exceptional conditions is the third moment a reasonable parameter to characterize band asymmetry and peak tailing.

EXERCISES

1) Derive from the mass conservation law the PDE equation for the non-ideal, nonequilibrium model of chromatography, for a single component. Write the set of equations for two components.

2) Derive the Laplace transforms of the chromatographic equations above and calculate the first and second moments of an elution peak.

3) Give examples of chromatographic (i.e., physical) problems in which either $C(0,t)$ or $C(0,x)$ are different from 0.

4) Analyze the characteristic lines and characteristic velocities of the PDE of the linear, ideal model of chromatography (see Chapter II, section II-2-c). Explain their meaning.

5) Assume that the stationary phase moves with velocity u_s in the direction opposite to the that of the migration of the liquid phase (true moving bed or countercurrent chromatography). Write the equation of the equilibrium–dispersive model of chromatography.

6) Give a practical example explaining the difference between the following two Dirac pulse injections:

$$(i) \qquad C(x,0) = A\delta(x)$$
$$(ii) \qquad C(0,t) = B\delta(t)$$

where $\delta(x)$ and $\delta(t)$ are Dirac δ pulses, the ideal limit of a narrow injection pulse.

7) Derive the relationship between the nth order moment of $f(t)$ and its Laplace transform, $F(p)$, from the definition of the moment (eqn. II-57).

8) Under linear conditions, perform ten successive injections on a modern HPLC instrument and determine the first five moments of the bands recorded. Determine the relative standard deviation for each moment.

9) Demonstrate that, in the case of an n-component system, there are n characteristic velocities and n characteristic lines for the system of coupled hyperbolic PDEs of the ideal model. As an example, calculate the characteristic velocities when $n = 3$ (assume competitive Langmuir isotherms) and discuss the characteristic lines, in linear and nonlinear ideal chromatography.

LITERATURE CITED

[1] G. Guiochon, S. Golshan-Shirazi and A. M. Katti, *"Fundamentals of Preparative and Nonlinear Chromatography"*, Academic Press, Boston, MA, 1994.

[2] E. Kucera, *J. Chromatogr.*, **19** (1965) 237.

[3] I. Quiñones, C. M. Grill, L. Miller and G. Guiochon *J. Chromatogr. A*, **867** (2000) 1.

[4] J. C. Giddings, *"Dynamics of Chromatography"*, M. Dekker, New York, NY, 1965.

[5] S. Golshan–Shirazi and G. Guiochon, *Anal. Chem.*, **60** (1988) 2634.

[6] L. Lapidus and N. R. Amundson, *J. Phys. Chem.*, **56** (1952) 984.

[7] J. J. Van Deemter, F. J. Zuiderweg and A. Klinkenberg, *Chem. Eng. Sci.*, **5** (1956) 271.

[8] J. H. Seinfeld and L. Lapidus, *"Mathematical Methods in Chemical Engineering,"* Vol. 3, Prentice-Hall, Englewood Cliffs, NJ, 1974.

[9] R. Aris and N. R. Amundson, *"Mathematical Methods in Chemical Engineering,"* Prentice- Hall, Englewood Cliffs, NY, 1973.

[10] H.-K. Rhee, R. Aris and N. R. Amundson, *"First-Order Partial Differential Equations. II. Theory and Application of Hyperbolic Systems of Quasilinear Equations,"* Prentice Hall, Englewood Cliffs, NJ, 1989.

[11] E. Wicke, *Kolloid Z.*, **86** (1939) 295.

[12] J. N. Wilson, *J. Am. Chem. Soc.*, **62** (1940) 1583.

[13] D. DeVault, *J. Am. Chem. Soc.*, **65** (1943) 532.

[14] J. Weiss, *J. Chem. Soc.*, **1943**, 297.

[15] H. C. Thomas, *J. Am. Chem. Soc.*, **66** (1949) 1664.

[16] S. Goldstein, J. D. Murry, *Proc. Roy. Soc. (London)*, **A252** (1959) 360.

[17] G. Houghton, *J. Phys. Chem.*, **67** (1963) 84.

[18] E. Glueckauf, *J. Chem. Soc.*, **1947**, 1302.

[19] F. Helfferich and G. Klein, *"Multicomponent Chromatography. Theory of Interference,"* M. Dekker, New York, NY 1970.

[20] H.-K. Rhee, *NATO Adv. Study Inst. Ser.*, **Ser. E** 1981, 183.

[21] P. Lax, *"Contributions to Non-Linear Functional Analysis,"* E. H. Zarantello Ed., University of Wisconsin, Madison, WI, 1971, 603.

[22] G. B. Whitham, *"Linear and Nonlinear Waves,"* Wiley, New York, NY, 1974.

[23] B. Temple, *Trans. Am. Math. Soc.*, **280** (1983) 781.

[24] H.-K. Rhee, B. F. Bodin and N. R. Amundson, *Chem. Eng. Sci.*, **26** (1971) 1571.

[25] H.-K. Rhee and N. R. Amundson, *Chem. Eng. Sci.*, **27** (1972) 199.

[26] C.-C. Ting and P.-L. Chu, *Scientia Sinica*, **6**(9) (1962) 1269.

[27] Z. Ma, R. Whitley, and N.-H. L. Wang, *AIChE J.*, **42**, (1996), 1244.

[28] J. H. Knox and M. Pyper, *J. Chromatogr.*, **363** (1986) 1.

[29] A. Katti and G. Guiochon, *J. Chromatogr*, **449** (1988) 25.

[30] S. Golshan-Shirazi and G. Guiochon, *J. Chromatogr.*, **461**, (1989) 1.

[31] S. Golshan-Shirazi and G. Guiochon, *Anal. Chem.*, **61** (1989) 1276, 2464.

[32] G. B. Cox and L. R. Snyder, *Liq. Gas Chromatogr.*, **6** (1983) 894.

[33] G. V. Yeroshenkova, S. A. Volkov and K. I. Sakodynskii, *J. Chromatogr.*, **262** (1983) 19.

[34] J. L. Wade, A. F. Bergold and P. W. Carr, *Anal. Chem.*, **59** (1987) 1286.

[35] P. Sajonz, H. Guan–Sajonz, G. Zhong and G. Guiochon, *Biotechnol. Progr.*, **13** (1997) 170.

[36] S. Golshan–Shirazi and G. Guiochon, *J. Phys. Chem.*, **93** (1989) 4143.

CHAPTER IV

THE PROFILES OF
SINGLE-COMPONENT BANDS
IN LINEAR NONIDEAL CHROMATOGRAPHY

In this chapter, we discuss the influence of the various kinetic phenomena that influence the band profiles in linear chromatography. In this case, algebraic solutions of the mathematical models used can almost always be derived although they are sometimes too complex to be of any practical use. Linear chromatography takes place when the solute concentration is sufficiently low, the equilibrium behavior of the component in the phase system is practically linear, and the influence of the concentration on the shape of the band profile is negligible [1]. The actual meaning of the words *"sufficiently low"* in this context must be qualified because the term is too often used loosely. We mean by low that the maximum concentration of the injected pulse is such that the second term (or the higher terms in some rare cases) of an expansion of the isotherm around $C = 0$ is small and negligible compared to the first term (in practice, this second term should be less than a few percents of the first one). When this assumption is valid, the conditions of *linear chromatography* hold and the band profiles depend only on axial and eddy diffusion and on the mass-transfer resistances. They are independent of the concentration. Then, as we show later in the next section, the band profile is Gaussian or nearly Gaussian, provided that the kinetics of mass-transfer is sufficiently fast. At high concentrations, however, this is no longer true; the equilibrium isotherm is curved and the retention factor depends on the concentration. As a consequence, each concentration moves at a different velocity. This phenomenon has a considerable influence on the band profiles and it deserves a separate discussion which is given in Chapter V.

Both diffusion and the mass-transfer resistances are dissipative. They have a smoothing effect on both sides of the profiles. The sharper the profile, the higher the concentration gradient and the faster the diffusion (Fick's law). The sharper the profile, the more rapidly the concentration tends to vary when the band passes a location in the column, hence the higher the mass-transfer resistances for a given kinetics and the larger is the dispersive effect of diffusion and the mass- transfer resistances. In the simpler case of linear chromatography, a detailed study of the dispersive influence of diffusion and the mass-transfer resistances on the band profiles is possible. Some of its conclusions can be applied to the more complex case of nonlinear chromatography.

The first attempts at solving the linear, nonideal model were those of Wicke [2], Martin and Synge [3], and Craig [4]. The latter two methods lead to the plate models of chromatography. The lack of the proper mathematical tools at the time and the lack of any adequate calculation machine prevented the success of the first approach. Nevertheless, we should pay tribute to its foresight. It was the harbinger of most modern investigations. The plate models combine in one single concept a physical and a mathematical model of chromatography. They should be considered as crude attempts at papering over the actual difficulties of the theory of chromatography. Their popular success among analytical chromatographers is a tribute to their success at obfuscating the fundamental issues. Basically, the plate models divide the column into a string of successive stages or "theoretical plates", assume that equilibrium between the stationary and the mobile phase is reached at the exit of each of these stages, and calculate the resulting band profile. The difference between the two models is that the Martin and Synge model [3] is a continuous model while the Craig model is the very model accounting for the behavior of the Craig machine (which is not a chromatographic column) [4]. Although we do know the number of stages in an actual Craig machine, we do not know in advance the actual number of plates in a column. Even in the case of the Craig machine, because of various sources of nonideal behavior, the number of theoretical plates corresponding to the concentration profiles obtained is lower than the actual number of physical stages. The number of theoretical plates in a column cannot be predicted by the plate models but can only be determined afterward, by fitting the experimental results to the model (most often in a rather crude way). So, plate models have no predictive value. They will not be discussed in detail here. For further reference, the reader should consult previous publications where these models are described, calculated, discussed, and compared [1,5].

Suffice it to note here that the solution of the Martin and Synge plate model [3] is a Poisson distribution for the concentration profile along the column (*i.e.*, $C = f(x)$ at a certain time) and a Γ-density function for the elution profile (*i.e.*, $C = f(t)$ at $x = L$) [1]. The solution of the Craig model is a binomial distribution for the concentration profile along the column. All these distributions tend toward a Gaussian distribution with an increasing number of theoretical plates [6]. Guiochon *et al.* [1] demonstrated that the difference between the actual concentration distribution and a Gaussian profile is negligible for $N = 100$ and that it would be extremely difficult to acquire experimental data with a sufficient accuracy to demonstrate that the elution profiles are not actual Gaussian distributions for $N = 25$.

I - INFLUENCE OF AXIAL DIFFUSION[1]

The purpose of this section is to discuss the algebraic solution of the simplest model of linear, nonideal chromatography, the equilibrium–dispersive model. In this model, only axial dispersion is involved (axial dispersion includes both axial and eddy diffusion, see Chapter II, Section III) and it is characterized by an axial dispersion coefficient, D_a. There are no mass

---

[1]In this section, the axial diffusion coefficient is D_a because there is no mass transfer resistance and $D_R = D_a/(1 + Fa)$.

transfer resistances in this model, which assumes that phase equilibrium is instantaneous in the column. The linear isotherm equation is

$$q = aC \qquad \text{(IV-1)}$$

where a is a thermodynamic constant, the distribution coefficient or slope of the linear isotherm. The differential mass balance or chromatographic equation for the linear, nonideal model is then written

$$\frac{\partial C}{\partial t} + F\frac{\partial q}{\partial t} + u\frac{\partial C}{\partial x} = D_a\frac{\partial^2 C}{\partial x^2} \qquad \text{(IV-2a)}$$

or $\qquad (1+aF)\dfrac{\partial C}{\partial t} + u\dfrac{\partial C}{\partial x} = D_a\dfrac{\partial^2 C}{\partial x^2} \qquad \text{(IV-2b)}$

Equation IV-2b is obtained by combining eqns. IV-1 and IV-2a. aF is often replaced by k_0', a parameter that is called the retention factor in analytical chromatography.

As shown in Appendix A, eqn. IV-2b can be simplified into the diffusion equation by using the following transform:

$$C(x,t) = e^{\alpha x + \beta t} g(x,t)$$

The diffusion equation is written (see eqn. IV-A-3):

$$\frac{\partial g}{\partial t} = D_R\frac{\partial^2 g}{\partial x^2} \qquad \text{(IV-2c)}$$

with[2]

$$\alpha = \frac{u}{2D_a}$$

$$\beta = \frac{u^2}{4D_a\gamma}$$

$$\gamma = 1 + aF = 1 + k_0'$$

$$D_R = \frac{D_a}{\gamma} = \frac{D_a}{1 + k_0'}$$

In many cases, it is convenient to use the dimensionless form of the chromatographic mass balance equation [1]. It is written

$$\frac{\partial C_d}{\partial \tau} + \frac{\partial C_d}{\partial z} = \frac{1}{Pe}\frac{\partial^2 C_d}{\partial z^2} \qquad \text{(IV-2d)}$$

---

[2]See Symbols Used, at the end of this chapter.

$$\text{with} \qquad z = \frac{x}{L}$$

$$\tau = \frac{ut}{L(1 + k_0')} = \frac{t}{t_R}$$

$$C_d = \frac{C t_R}{A_p} = C \frac{\varepsilon S L(1 + k_0')}{n}$$

where L is the column length, $Pe = uL/D_a$ is the column Peclet number (proportional to the plate number, with $Pe = 2N$), $t_R = (1 + k_0')t_0$ is the retention time of the peak, t_0 is the hold-up time, $A_p = n/F_v$ is the area of the injected pulse (which is conservative, Chapter VII, Section IV), n is the number of mole of compound injected, F_v is the volume flow-rate of the mobile phase, S is the column cross-section area, and ε is the total porosity of the column (with $F = (1 - \varepsilon)/\varepsilon$ in eqns. IV-2a and IV-2b).

Note that $g(x, t)$ in eqn. IV-2c is a concentration. Accordingly there is a dimensionless version of eqn. IV-2c, written with the same reduced coordinates.

We now discuss the solution of the equations IV-2a to IV-2d for a series of different sets of initial and boundary conditions that correspond to different possible formulations of the two important practical problems in chromatography, those of frontal analysis (or rather of the breakthrough curve, since only one component is involved in this case) and of elution.

1 - Frontal Analysis. The Breakthrough Curve

If the stream of pure mobile phase pumped through a chromatographic column is suddenly replaced by a stream of a solution of the compound studied in the mobile phase, the column equilibrates progressively with this new solution and, after a while, the front of this solution breaks through the column. In mathematical terms, the boundary and initial conditions for the frontal analysis problem are usually written as follows:

$$C(0, t) = C_0 \qquad \text{for} \quad t > 0 \qquad\qquad \text{(IV-3a)}$$
$$C(x, 0) = 0 \qquad\qquad\qquad\qquad\qquad\qquad \text{(IV-3b)}$$
$$q(x, 0) = 0 \qquad\qquad\qquad\qquad\qquad\qquad \text{(IV-3c)}$$

The first equation states that the composition of the mobile phase pumped into the column changes abruptly at the time origin,[3] from $C = 0$ to $C = C_0$.

---

[3] In some cases, e.g., in SMB, the boundary condition is $C(0, t) = f(t)$. This case is much more complex and is not discussed here.

The other two equations state that the two phases were in equilibrium at the time origin and that the column was empty of solute. The first equation raises serious fundamental difficulties because diffusion would proceed at an infinite rate if the concentration profile at injection were to be discontinuous as assumed in eqn. IV-3a. Unfortunately, the mass balance equation of linear, nonideal chromatography cannot be integrated with the classical Danckwerts boundary condition (see Chapter III, section I-1d) which should be used in this case (see later).

To be able to understand and discuss the influence of diffusion on the band profile, we must solve eqn. IV-2d combined with eqns. IV-1 (the isotherm) and IV-3a-c (initial and boundary conditions). Equation IV-2d is a second-order linear partial differential equation. The classical method to solve it consists in transforming it into the conventional diffusion equation (eqn. IV-2c) whose solution is well known. The necessary transforms are given and discussed in Appendix A where the algebraic solution is derived. This solution is written

$$C(x,t) = \frac{C_0}{2} + \frac{C_0}{2}\mathrm{erf}\left(\sqrt{\frac{u_R^2 t}{4D_R}} - \sqrt{\frac{x^2}{4D_R t}}\right) +$$

$$\frac{C_0}{2}e^{\frac{u_R x}{D_R}}\mathrm{erfc}\left(\sqrt{\frac{u_R^2 t}{4D_R}} + \sqrt{\frac{x^2}{4D_R t}}\right) \qquad \text{(IV-4a)}$$

where $\mathrm{erf}(x)$ and $\mathrm{erfc}(x)$ are the error function and the complementary error function, respectively (see Symbols Used at the end of the chapter for u_R and D_R). These classical functions are defined by the following relationships

$$\mathrm{erf}(x) = \frac{2}{\sqrt{\pi}}\int_0^x \exp\left(-z^2\right)dz \qquad \text{(IV-5a)}$$

$$\mathrm{erfc}(x) = \frac{2}{\sqrt{\pi}}\int_x^{+\infty} \exp\left(-z^2\right)dz = 1 - \mathrm{erf}(x) \qquad \text{(IV-5b)}$$

Equation IV-4 gives the concentration profile, $C(x)$, along the column at any time t. If $x = L$ in eqn. IV-4, we obtain the concentration at the column exit, i.e., the elution profile of the breakthrough curve. This profile can be written under dimensionless form as

$$c = \frac{C}{C_0} = 0.5 + 0.5\mathrm{erf}\left[\frac{\sqrt{Pe}}{2}\left(\sqrt{\tau} - \frac{1}{\sqrt{\tau}}\right)\right] + 0.5e^{Pe}\mathrm{erfc}\left[\frac{\sqrt{Pe}}{2}\left(\sqrt{\tau} + \frac{1}{\sqrt{\tau}}\right)\right]$$

$$\text{(IV-4b)}$$

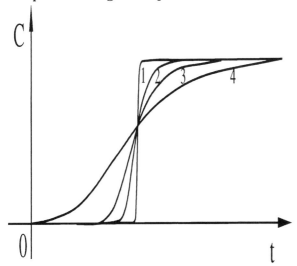

Figure IV-1   Illustration of Breakthrough Profiles. The profiles are given for the following values of the column Peclet number: 1, 1000; 2, 500; 3, 100; and 4, 20. Note that the column Peclet number is equal to twice the number of theoretical plates.

with $c = C/C_0$. The profile of the breakthrough curve given by eqn. IV-4b is illustrated in Figure IV-1 for different values of the column Peclet number.

When D_R tends toward 0 in eqn. IV-4, both the argument of the error function and that of the complementary error function become infinitely large. Actually, when $t > x/u_R$, the argument of the error function (second term in the RHS of eqn. IV-4) tends toward $+\infty$; when $t < x/u_R$, it tends toward $-\infty$. Since $\text{erf}(+\infty) = 1$ and $\text{erf}(-\infty) = -1$ while $\text{erfc}(+\infty) = 0$, we have for limits of C at infinitely fast diffusion:

$$C(x,t) = \begin{cases} C_0 & t > \frac{x}{u_R} \\ 0 & t < \frac{x}{u_R} \end{cases} \qquad \text{(IV-6)}$$

When the axial diffusion tends toward zero (ideal model), the breakthrough plateau tends to rise instantaneously from $C = 0$ to $C = C_0$ (a concentration equal to that of the stream injected into the column) at the breakthrough time, $t = x/u_R$. At column exit, the breakthrough time is the classical retention time, $t_R = (1 + k_0')t_0$. This is the solution of the equilibrium or ideal model (see Chapter V, section I).

By contrast, when $D \neq 0$, $C(x,t)$ is always smaller than C_0 but tends toward this limit when t increases indefinitely.

2 - Generalized Breakthrough Profile

The problem so far discussed is a particular case, in which the initial condition is $C(x,0) = 0$, whatever the value of x. The general case, encountered in various applications of chromatography or chemical engineering (*e.g.*, in simulated moving bed separations), or in the determination of equilibrium isotherms by the staircase method, is the one in which there is a concentration distribution in the column at the beginning of the experiment. Then, in the boundary condition above, eqns. IV-3b and IV-3c are replaced by the following two equations

$$C(x,0) = C_1(x) \tag{IV-7a}$$
$$q(x,0) = q_1(x) = aC_1(x) \tag{IV-7b}$$

The solution of eqn. IV-2 in this case is derived in Appendix B, following closely the similar derivation of eqn. IV-4.

In the most general case, we have as initial and boundary conditions the following relationships, respectively:

$$C(x,0) = \varphi(x) \qquad \text{for} \quad 0 < x < \infty \tag{IV-7c}$$
$$C(0,t) = \psi(t) \qquad \text{for} \quad t > 0 \tag{IV-7d}$$

As shown in Appendix B (see eqn. IV-B-17), the profile of the solution is given by the following equation

$$C(x,t) = \int_0^\infty \frac{\varphi(\xi)}{\sqrt{4\pi D_R t}} [e^{\frac{-(x-\xi)^2\gamma}{4D_R t}} - e^{\frac{-(x+\xi)^2\gamma}{4D_R t}}]d\xi +$$

$$\int_0^t \frac{x}{\sqrt{4\pi D_R t}} \frac{\psi(\tau)}{(t-\tau)^{\frac{3}{2}}} e^{\frac{-x^2\gamma}{4D_R(t-\tau)}} d\tau \tag{IV-8}$$

This equation is the general solution in linear chromatography of the equilibrium–dispersive model for the breakthrough profile. It is derived in Appendix B and is obtained as a combination of eqns. IV-B-8c and IV-B-16.

When $\varphi(x) = 0, \psi(t) = 0(t \leq 0), C_0(t > 0)$, we obtain the solution of the Riemann problem of chromatography (simple breakthrough curve, eqn. IV-4). If $\varphi(x) = 0, \psi(t) = B\delta(t)$, we have the solution of the Dirac problem (elution, see below).

3 - Elution

The system of equations of the elution problem again comprises eqns. IV-1 to IV-3 but the conditions to determine the solution are different. There is, obviously, a different solution for each set of conditions. Numerous different such solutions can be found in the literature, corresponding to as many different boundary conditions (see *e.g.*, Chapter III, Sections I-1c and I-1d) and to different column models. Some of these solutions are more important than others because the corresponding set of conditions is more general or more realistic. Without attempting to present an exhaustive review of these solutions, we will discuss now those that correspond to the most important combinations of conditions to determine the solution which can be used in the elution mode.

a - *The Conventional Dirac Injection and the Gaussian profile*

The most conventional boundary condition used in elution is the Dirac injection. It has a finite area, an infinite concentration, and a zero band-width. Although this condition seems to be quite unrealistic, this model is applied successfully in analytical chromatography because the width of an actual plug injection is often quite small compared to the width of the elution band. Thus, the influence of the finite width of this plug on the elution profile is negligible or nearly negligible in most analytical applications. There are two possibilities to model this injection. The first method consists of actually "injecting" nothing into the column but creating a Dirac concentration distribution along the column at $t = 0$, as expressed by the following conditions:

$$C(x, 0) = A\delta(x) \qquad \text{for} \quad -\infty < x < \infty$$
$$C(0, t) = 0$$

Since $\int_{-\infty}^{\infty} \delta(x)dx = 1$, the dimension of $\delta(x)$ is that of $1/dx$, hence $[\delta(x)] = [L]^{-1}$. So, the dimension of A is $[A] = [C][L]$, where $[C]$ and $[L]$ are the physical dimensions of a mobile-phase concentration and a length, respectively. This method actually consists in using the initial condition to simulate the injection. While this method is mathematically correct, using the boundary condition to model the injection is the only realistic approach, *i.e.*, the only one making physical sense. Note that the column is assumed to extend all along the x axis, from $-\infty$ to $+\infty$. So, this is an infinite problem, another unrealistic assumption.

The reason why we mention this quite unrealistic model of injection here is mainly because, as we discuss in the next chapter, Houghton adopted this model of conditions-to-solve in his study of the solution of the chromatographic equation under conditions of onset of column overloading (*i.e.*, with the isotherm $q = a_1C + a_2C^2$, with $a_2C << a_1$). In his paper, he assumed that $|x| < |L|$. However, for a true δ function, $|x| < L$ is the same as $|x| < \infty$. As shown in Appendix A (section 1), the equation of linear chromatography with axial diffusion can be reduced to the conventional diffusion equation. Accordingly, the solution of the chromatographic equation with the initial and boundary conditions just given is:

$$C(x,t) = \frac{A}{\sqrt{4\pi D_R t}} \exp\left(-\frac{(x - u_R t)^2}{4 D_R t}\right) \tag{IV-9}$$

This solution is a Gaussian concentration profile along the column at time t. It is illustrated in Figure IV-2. Obviously, the elution profile $C(x = L, t)$ is not a Gaussian profile since the pre-exponential coefficient is a function of time. It is not symmetrical since the part of the peak that is not yet eluted continues to broaden by axial diffusion. Since the dimension of Dt is $[Dt] = [L]^2$ and that of A is $[CL]$, the dimension of the RHS of the expression above is $[C]$, *i.e.*, that of a concentration.

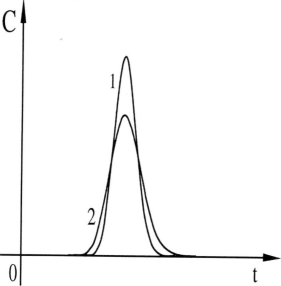

Figure IV-2   **Illustration of a Gaussian Profile along a Column. The equation of the profile is given by eqn. IV-9. Profiles are shown for two different values of D_R, 0.0005 (1) and 0.0010 (2) cm^2/s.**

The other possibility to model a Dirac injection is to inject an actual concentration pulse, $C = C_0\delta(t)$. This case, one that is quite realistic in contrast to the alternate case just treated above, is discussed below, after the rectangular pulse injections, as its limit for a zero pulse width.

b - Rectangular Pulse Injections

A very general representation of the injection process in chromatography consists of using a rectangular pulse of height C_0 and duration t_p (unless indicated otherwise, the initial condition corresponds to a column filled with pure mobile phase, equilibrated with the stationary phase). The boundary condition becomes

$$C(0,t) = 0 \qquad \text{for} \quad t < 0$$
$$C(0,t) = C_0 \qquad \text{for} \quad 0 < t \le t_p$$
$$C(0,t) = 0 \qquad \text{for} \quad t_p < t$$

Usually, t_p is finite but small compared to the hold-up time, $t_0 = L/u$. In extreme cases, the pulse is very narrow and can be represented by the δ function of Dirac. We must note, however, that in this case, the amount injected being finite, the maximum concentration of the pulse has to be infinite. Thus, for an initial period that may be quite short, the column is strongly overloaded.

If t_p is finite, the solution is obtained as the overlay of the solution of two successive breakthrough curves of opposite signs, made at times $t = 0$ and $t = t_p$ and of amplitude C_0 and $-C_0$, respectively. Obviously, the solution for the Dirac-injection boundary condition is the limit of this last solution when t_p tends toward zero and C_0 toward infinity while the pulse area remains constant. This solution is given by the sum of two appropriate expressions and is similar to the one in the RHS of eqn. IV-4.

Accordingly, there are two different but most important solutions for the chromatographic equation that correspond to these two different boundary conditions. We discuss them in the next two subsections.

c - Dirac Injection Condition

This boundary condition is the one that seems to best represent the injection process in analytical chromatography, at least as long as the experimental conditions are such that the contributions of the extra-column effects can be neglected. The boundary condition is written

$$C(x,0) = 0 \qquad \text{for} \quad 0 < x < \infty$$
$$C(0,t) = B\delta(t) \qquad \text{for} \quad t > 0$$

Note that in this case the column extends only from 0 to infinity. So, the problem is now a semi-infinite problem. Since the physical dimension of $\delta(t)$ is $[\delta(t)] = \frac{1}{[T]}$, the dimension of B is $[B] = [C][T]$, $[T]$ being the dimension of a time. Under this condition, the corresponding solution is:

$$C(x,t) = \frac{x}{t}\frac{B}{\sqrt{4\pi D_R t}} e^{-\frac{(x-u_R t)^2}{4D_R t}} \qquad \text{(IV-10)}$$

Equation IV-10 gives the concentration profile along the column $(C = f(x))$
at time t. This profile is illustrated in Figure IV-3. Note that the dimension
of the RHS of this equation is $[C]$, as it correctly should be. This concentration profile along the column is not a Gaussian curve; it is a Gaussian
curve multiplied by $\frac{x}{t}$. The difference between the solution of the two Dirac
problems, the one discussed above and this one, should not be a surprise.
These are two different mathematical problems and it should be expected
that they have different solutions. The differences between the two problems
are important. In the first problem (subsection a above), we had an infinite
problem, the column extending from $-\infty$ to $+\infty$ and the injection being
made in its center $(x = 0)$. Furthermore, it was an initial value problem, the
condition to determine the solution was an initial condition, $C(x, 0) = A\delta(x)$.
In the present case, we have a semi-infinite problem, the column extending
only from $x = 0$ to $x = +\infty$, and we have a boundary problem, and the
injection being made at the column origin, the condition to determine the
solution being a boundary condition, $C(0, t) = B\delta(t)$ (with B in eqn. IV-10
being the actual amount of sample injected). The end result is that the
Gaussian function is the solution of the first problem for the concentration
profile along the column while it is only a reasonable physical approximation
of the solutions of the other problem.

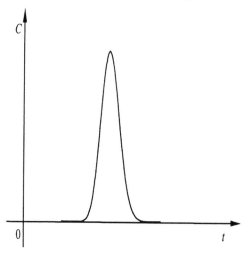

Figure IV-3   **Illustration of the Elution Profile arising from eqn. IV-10. Val-**
ues of the parameters: $B = 20$; $D = 0.0005$.

d - Rectangular Pulse Injection of Finite Width

This boundary condition best represents the injections made in preparative chromatography (although, in practice, the end of the injection profile

would be better approximated by an exponentional decay than by a vertical boundary [7]). The boundary condition for this type of injection is

$$C(x,0) = 0 \qquad \text{for} \quad 0 < x < \infty$$

$$C(0,t) = \begin{cases} C_0 & 0 < t < t_p \\ 0 & t > t_p \end{cases}$$

The general solution of the diffusion equation is given by eqn. IV-A-5 (see Appendix A), for the set of initial and boundary conditions considered here. In this case, however, the integral limits for the variable λ in the derivation of that equation are changed (see Appendix A and eqns. IV-A-8, IV-A-10a and IV-A-10b) and the general solution becomes (see eqn. IV-A-14 in Appendix A):

$$C(x,t) = \frac{C_0}{\sqrt{\pi}} \int_A^B e^{-\eta^2} d\eta - \frac{C_0}{\sqrt{\pi}} e^{\frac{u_R x}{D_R}} \int_C^D e^{-\eta'^2} d\eta' \qquad \text{(IV-11a)}$$

with the following integration limits

$$A = \sqrt{\frac{u_R^2(t - t_p)}{4D_R}} - \sqrt{\frac{x^2}{4D_R(t - t_p)}}$$

$$B = \sqrt{\frac{u_R^2 t}{4D_R}} - \sqrt{\frac{x^2}{4D_R t}}$$

$$C = \sqrt{\frac{u_R^2 t}{4D_R}} + \sqrt{\frac{x^2}{4D_R t}}$$

$$D = \sqrt{\frac{u_R^2(t - t_p)}{4D_R}} + \sqrt{\frac{x^2}{4D_R(t - t_p)}}$$

When $t_p \to 0$, the solution tends toward that of the Dirac problem and becomes eqn. IV-10 or

$$C(x,t) = \frac{x}{t} \frac{C_0 t_p}{\sqrt{4\pi D_R t}} \exp\left(-\frac{(x - u_R t)^2}{4D_R t}\right) \qquad \text{(IV-11b)}$$

where $C_0 t_p$ replaces the parameter B in the previous solution (eqn. IV-10). Note that the dimension of $C_0 t_p$ is the same as the dimension of B.

4 - Danckwerts condition

This boundary condition was discussed in Chapter III (eqn. III-12 and Figure III-2). It is the correct boundary condition for chromatography but, unfortunately, it is too complex for theoretical studies. The systems of equations used in chromatography cannot be integrated into closed forms with this boundary condition. In contrast however, the Danckwerts condition is frequently used in numerical calculations because (1) it is useful and realistic, particularly when the column efficiency is poor (*e.g.*, in the modeling of the operation of simulated moving bed separators); and (2) its implementation is easy in numerical calculations. For further information regarding the solution under Danckwerts conditions, the reader is referred to ref. [1], pp. 180 to 183.

5 - What is the Equation of the Elution Profile in Linear Chromatography?

In simple textbooks, it is often stated that the elution profile is Gaussian while the profile of the breakthrough curve is an error function. Strictly speaking, this is not correct, although in practice the difference between the actual profile supplied by the correct equation and a Gaussian profile is often small or negligible [1]. If we summarize the equations derived earlier in this section, we have obtained the following equations for the concentration profiles corresponding to a Dirac injection:

- For the initial value problem, *i.e.*, when the injection is a Dirac initial condition, $C(x,0) = C_0(x) = \delta(x)$, and if the column length in infinite, the profile is given by eqn. IV-9:

$$C(x,t) = \frac{A}{\sqrt{4\pi D_R t}} e^{-\frac{(x-u_R t)^2}{4D_R t}} \qquad \text{(IV-9)}$$

When t is constant, this gives a Gaussian concentration profile along the column. The elution profile is $C(x = L, t)$. Clearly, this is not a Gaussian profile since the variance of the profile increases linearly with increasing time. This makes eminent physical sense: as long as it is not eluted, the band broadens. The solution of the mass balance equation developed by Giddings [8] is the solution of an initial value problem as made clear by two earlier statements (beginning of sections 3a and 3c). It is identical to eqn. IV-9.

- A more realistic model of the injection treats it as a boundary problem ($C(0,t) = C(t) = \delta(t)$). Then, when the boundary condition corresponds to a Dirac injection made into a semi-infinite column, the profile is

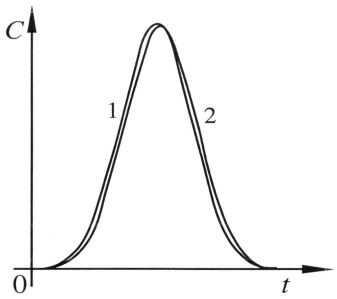

Figure IV-4   Comparison of the Elution Profiles given by Eqns. IV-9 and IV-10. Note that both are elution profiles (*i.e.*, $C(x = L) = f(t)$) so neither is a Gaussian profile. Profile 1 corresponds to the initial condition $C(x, 0) = \delta(x)$ and the boundary condition $C(0, t) = 0$ while profile 2 corresponds to the initial condition $C(x, 0) = 0$ and the boundary condition $C(0, t) = \delta(t)$. Values of the parameters: $A = 1$; $B = 20$; $D = 0.0005$.

given by eqn. IV-10 (or IV-11b):

$$C(x, t) = \frac{x}{t} \frac{B}{\sqrt{4\pi D_R t}} e^{-\frac{(x - u_R t)^2}{4 D_R t}} \qquad \text{(IV-10)}$$

This equation never gives a Gaussian profile, even as a profile along the column.

Figure IV-4 compares a true Gaussian profile (eqn. IV-9) and the elution profiles given by eqn. IV-10. The difference between these profiles is extremely small in the whole range of column efficiencies encountered in analytical and preparative chromatography.

II - INFLUENCE OF THE MASS-TRANSFER RESISTANCES[4]

In chromatography, the convection caused by the mobile phase stream percolating through the column displaces any equilibrium which could arise

---

[4]In this and subsequent sections, the axial dispersion coefficient is D_L and $D_R = D_L/(1 + Fa)$.

locally between the two phases by constantly moving the liquid phase with respect to the solid phase. If any local equilibrium could exist, even temporarily (*e.g.*, around a peak maximum), it does not last long. The incoming liquid phase, which pushes away the solution in equilibrium with the solid, has a different composition. A radial concentration gradient is formed between the fluid percolating through the column bed and the fluid stagnant in the particles and mass transfer takes place in the direction which tends to relax this gradient. Mass transfer kinetics is finite, however, and the mass transfer resistances prevent the establishment of an equilibrium.

The rate of the mass transfer kinetics depends on the nature of the packing material used and on the characteristics of the solute studied. This rate usually decreases with increasing molecular weight of the solute and also with increasing polarity of this solute, or rather with its ability to give selective, high-energy interactions with the stationary phase. The model discussed here is particularly suited to solute belonging to this last group, that is to relatively small molecules (*i.e.*, solutes with a large molecular diffusivity) which are able to interact strongly with the solid surface of an adsorbent. This is, for example, the case of strongly basic compounds eluted on RPLC phases, with an unbuffered liquid phase. Far from being exceptional, this situation is frequently encountered in routine analyses, particularly in the pharmaceutical industry. Two important cases in point are the elution of drugs and drug intermediates, because a large proportion of them are basic compounds, and chiral separations, because most chiral phases give strong, selective interactions with at least the more retained of the two enantiomers.

The system of equations of this model combines the mass balance equation of the ideal model (no axial diffusion) and a kinetic equation

$$u\frac{\partial C}{\partial x} + \frac{\partial C}{\partial t} + F\frac{\partial q}{\partial t} = 0 \qquad (IV\text{-}12)$$

$$\frac{\partial q}{\partial t} = -k_f\left[q - aC\right] \qquad (IV\text{-}13)$$

The simplest boundary condition corresponds to an arbitrary concentration profile $(C(0,t) = C_0(t))$ entering a column in which the stationary phase is in equilibrium with the pure mobile phase

$$C(x,0) = q(x,0) = 0 \qquad (IV\text{-}14a)$$

$$C(0,t) = \begin{cases} C_0(t) & 0 < t < t_p \\ 0 & t > t_p \end{cases} \qquad (IV\text{-}14b)$$

This formulation includes frontal analysis ($C_0(t)$ tends toward $C_0 \neq 0$ when t tends toward $+\infty$) and elution ($C_0(t)$ is a pulse of finite duration).

1 - Derivation of the Concentration Profile

Like other well-posed problems in which a linear, partial differential equation must be solved, this new model can be solved using the Laplace transform. The derivation of the transformed equation is trivial and the solution of this new equation in the Laplace domain is usually simple. As always when the Laplace transform is used, the main difficulty consists in returning into the time domain to obtain the solution of eqn. IV-12. This inversion is complex and requires the calculation of several integrals which are difficult if not impossible to solve. As we show below, the essential result of the Laplace analysis is to demonstrate the following statement: *when the injection is a narrow rectangular pulse (i.e., when it tends toward a Dirac function) and when the rate coefficient, k_f, tends toward infinity, $C(x,t)$ becomes equal to zero almost everywhere except in the neighborhood of the peak maximum.* Then, the solution is a narrow peak. By contrast, when k_f becomes small, the band becomes wide. It widens indefinitely when k_f tends toward 0. This suggests that the effect of the mass transfer resistances (resistances which increase with increasing value of $1/k_f$) is qualitatively similar to that of a diffusion term. This is also a dissipative factor. A demonstration of the equivalence of axial diffusion and mass transfer resistances in chromatography (*i.e.*, under such experimental conditions that the column efficiency is relatively high and axial dispersion is small) was given by Giddings [9] (see Chapter III, section I-2c).

In frontal analysis, the boundary condition is the injection of a stream of constant composition, C_0. Then, the same Laplace analysis shows that the concentration is nearly zero when $t \ll x(1+Fa)/u = x/u_R$, that it becomes close to C_0 when $t \gg x/u_R$ and that it is near $C_0/2$ at $t = x/u_R$. The range around this latter point is a transition zone during which the concentration varies rapidly. The larger k_f is, the narrower this transition zone.

The analytical solution of the model (eqns. IV-12 to IV-14) is derived by taking the Laplace transform of the chromatographic equations above. The transformed equations are

$$u\frac{d\overline{C}}{dx} + p\overline{C} + Fp\overline{q} = 0 \qquad \text{(IV-15)}$$

$$(p + k_f)\overline{q} - k_f a\overline{C} = 0 \qquad \text{(IV-16)}$$

where p is the parameter of the Laplace transform and where the new variables are

$$\overline{C} = \int_0^\infty C(x,t)e^{-pt}dt \qquad \text{(IV-17a)}$$

$$\overline{q} = \int_0^\infty q(x,t)e^{-pt}dt \qquad \text{(IV-17b)}$$

Eliminating $\overline{q}$ between eqns. IV-15 and IV-16 gives

$$u\frac{d\overline{C}}{dx} + p\left(1 + F\frac{k_f a}{p + k_f}\right)\overline{C} = 0 \qquad \text{(IV-18)}$$

Let

$$p\left(1 + F\frac{k_f a}{p + k_f}\right) = S_m(p) \qquad \text{(IV-19)}$$

Integrating in the Laplace domain the differential equation obtained by introducing eqn. IV-19 into eqn. IV-18 gives

$$\overline{C} = \overline{C_0}(p)\exp\left[-\frac{xS_m(p)}{u}\right] \qquad \text{(IV-20)}$$

where $\overline{C_0} = \overline{C}(0, p)$ is the boundary value in the Laplace domain. Now that we have solved the equation in the Laplace domain we need to derive the solution of the inverse Laplace transform in order to return into the time domain. We denote the inverse of the exponential function in the equation above as

$$L^{-1}\left(\exp\left[-\frac{xS_m(p)}{u}\right]\right) = H(x, t)$$

From the integration theorem of the Laplace transform (see Chapter III, section II-3), we have

$$C(x, t) = \int_0^t C_0(x, t - t')\, H(x, t - t')dt'$$

We now need to determine the function $H(x, t)$, which is a more complicated calculation. The solution can be found in the mathematical literature [10]. It is most complex. In the particular case of a Dirac pulse injection (i.e., when t_p tends toward 0), the solution is

$$C(x, t) = C_0 k_f t_p \sqrt{\frac{Fax}{u(t - \frac{x}{u})}}\; e^{-\frac{Fak_f x}{u} - K(t - \frac{x}{u})}\; I_1\left(2\sqrt{\frac{Fak_f^2 x(t - \frac{x}{u})}{u}}\right)$$

where $I_1(Z)$ is the first order Bessel function. When Z tends toward ∞, $I_1(Z)$ tends toward $e^Z(2\pi Z)^{-1/2}$. So, when the argument of I_1 is large, we have

$$C(x, t) = \frac{C_0 k_f t_p}{\sqrt{4\pi}}\sqrt{\sqrt{\frac{Fax}{uk_f^2(t - \frac{x}{u})^3}}}\; e^{-\left(\sqrt{k_f(t - \frac{x}{u})} - \sqrt{\frac{Fak_f x}{u}}\right)^2} \qquad \text{(IV- 21)}$$

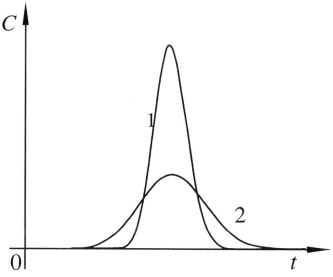

Figure IV-5   Illustration of the Profile Predicted by Eqn. IV-22. Values of the parameters: $A = 20$, $D = 0.0005$ **for the Gaussian peak;** $C_0 = 2.5$, $t_p = 0.04$, $Fa = 8$, $k_f = 50$ **in eqn. IV-22.**

This equation can be simplified further by observing that the exponential term is maximum and equal to 1 for $t = x(1 + Fa)/u$, *i.e.*, in the case of the elution profile, for $t = t_0(1 + Fa) = t_{R,0}$. Accordingly, the concentration is significantly different from 0 only in the region around $t = x(1 + Fa)/u$ (this is similar to the result of the analysis made by Van Deemter *et al.* [11]). As a first approximation, we may replace t by $x(1 + Fa)/u$ in eqn. IV-21 and this gives:

$$C(x,t) = \frac{C_0 t_p}{\sqrt{4\pi}} \sqrt{\frac{k_f}{Fat}} e^{-(\sqrt{k_f(t-\frac{x}{u})} - \sqrt{\frac{Fak_f x}{u}})^2} \qquad \text{(IV-22)}$$

This equation gives numerical results that are essentially the same as those of eqn. IV-21. It does not give a Gaussian profile. The profile is illustrated in Figure IV-5. Note that eqn. III-30c has already established the equivalence between the dispersive contribution of the mass transfer kinetics and that of axial dispersion.

2 - Derivation of the Moments

While it is difficult to obtain a general algebraic solution of the chromatographic equation, it is easy to derive from eqn. IV-20 the different moments of this solution (see definition of the moments and their calculation in Chapter III, section 3b). Note that the moments of a profile or a distribution are the only characteristics of this function which are relatively simple to derive from the solution of a problem in the Laplace domain. This is the reason why theoreticists usually like the approach and why more applied-minded scientists do not appreciate as much the results obtained through this approach.

The moments of the solution are derived through the use of the classical relationships [10] in eqns. III-64 to III-66. In this case, the first three moments, $\mathcal{M}_1, \mu_2,$ and μ_3, of the elution profile $(x = L)$ are:

$$\bar{t} = \mathcal{M}_1 = \frac{L}{u}(1 + Fa) \tag{IV-23}$$

$$\mu_2 = \frac{2L}{u}\frac{Fa}{k_f} \tag{IV-24}$$

$$\mu_3 = \frac{6L}{u}\frac{Fa}{k_f^2} \tag{IV-25}$$

These results are the same as those that will be obtained in the next section in the case when $D_a = 0$.

III - LINEAR CHROMATOGRAPHY WITH AXIAL DIFFUSION AND MASS-TRANSFER RESISTANCES

Now we combine the two effects studied in the previous two sections, axial diffusion (Section I) and mass transfer resistance (Section II). The system of equations of the model includes the mass balance of the nonideal model (*i.e.*, it includes a diffusive term different from 0 in the RHS) and a kinetic equation

$$u\frac{\partial C}{\partial x} + \frac{\partial C}{\partial t} + F\frac{\partial q}{\partial t} = D_a\frac{\partial^2 C}{\partial x^2} \tag{IV-26}$$

$$\frac{\partial q}{\partial t} = -k_f\left(q - aC\right) \tag{IV-27}$$

The initial and boundary conditions are respectively

$$C(x,0) = C_1(x) \tag{IV-28a}$$

$$q(x,0) = q_1(x) = f(C_1(x)) \tag{IV-28b}$$

$$C(0,t) = C_0(t) \tag{IV-28c}$$

where $f(C)$ is the isotherm equation ($f(C) = a_1 C$ in the linear case). This set of conditions is the most general possible. Its only practical application is in the modeling of simulated moving bed separations. Otherwise, it is rather unusual. It corresponds to a breakthrough curve, with a concentration of the stream pumped into the column which is not constant, and that enters a column along which there is already a concentration distribution of the solute studied. This problem is a two-element mathematical problem.

1 - Derivation of the Concentration Profile

To solve this problem, we first make a Laplace transform

$$\overline{C} = \overline{C}(x,p) = \int_0^\infty C(x,t)e^{-pt}dt$$

$$\overline{q} = \overline{q}(x,t) = \int_0^\infty q(x,t)e^{-pt}dt$$

$$\frac{\partial C(x,t)}{\partial t} = p\overline{C}(p) - C_1(x)$$

$$\frac{\partial q(x,t)}{\partial t} = p\overline{q}(p) - q_1(x)$$

and the two PDEs of the model become

$$p\overline{C} - C_1(x) + F[p\overline{q} - q_1(x)] + u\frac{d\overline{C}}{dx} = D_a\frac{d^2\overline{C}}{dx^2} \qquad \text{(IV-29)}$$

$$p\overline{q} - q_1(x) = -k_f(\overline{q} - a\overline{C}) \qquad \text{(IV-30)}$$

(with $C_1(x) = C_0(x,0)$ and $q_1(x) = q_0(x,0) = f_1(C_1(x))$). The initial condition of this equation is

$$\overline{C} = \overline{C_0}(p) \qquad \text{for} \qquad x = 0 \qquad \text{(IV-31)}$$

The Laplace function, $\overline{C}$, tends toward a finite limit when x increases indefinitely. Eliminating $\overline{q}$ between eqn. IV-29 and IV-30 gives

$$\frac{d^2\overline{C}}{dx^2} - \frac{u}{D_a}\frac{d\overline{C}}{dx} - \frac{\overline{C}}{D_a}\left[\frac{p(p+k_f)+pFk_fa}{p+k_f}\right] = -\frac{1}{D_a}\left[C_1(x) + \frac{Fk_fq_1(x)}{p+k_f}\right] \qquad \text{(IV-32)}$$

This equation is an ordinary differential equation which is easy to solve in the Laplace domain. The difficulty, as always with this method, is in calculating the inverse of the solution, to return from the Laplace domain into the real domain. Lapidus and Amundson [12] have discussed this inversion in detail.

We will discuss here the simplified case in which $C_1(x) = q_1(x) = 0$ and $C_0(t) = C_0 = $ constant, which corresponds to the classical breakthrough curve. The solution, derived by Lapidus and Amundson, is

$$C = C_0 \exp\left[\frac{ux}{2D_a}\right]\left[\varphi(t) + k_f \int_0^t \varphi(t)dt\right]$$
(IV-33)

with

$$\varphi(t) = e^{-k_1} \int_0^t I_0\left(2\sqrt{Fak_f^2\lambda(t-\lambda)}\right) \frac{x}{2\sqrt{\pi D_a\lambda^3}} \exp\left[-\frac{x^2}{4D_a\lambda} - \lambda\eta\right] d\lambda$$
(IV-34a)

and

$$\eta = \frac{u^2}{4D_a} + Fak_f - k_f$$
(IV-34b)

where λ is a dummy variable and $I_0(y)$ is the zero-order Bessel function of argument y. This equation is the solution of the problem (with the simplified initial and boundary conditions above). It represents the breakthrough profile in nonideal linear chromatography. Unfortunately, it is complex, certainly not intuitive, and it is difficult to calculate any numerical solution for a specific problem. Special programs for the calculation of Bessel functions are needed and they are relatively slow. As a matter of fact, a good computer program for the calculation of numerical solutions of the set of PDEs of this model (*i.e.*, of eqns. IV-23 and IV-24) may be faster than the calculation of these Bessel functions.

Although the solution given by eqns. IV-33 and IV-34 is not useful in practice, it is extremely important. It is at the origin of two of the most fruitful lines of investigations in chromatography in the fifty years that followed its publication. First, Van Deemter *et al.* [11] showed that this solution can be considerably simplified when the column efficiency is not very small. This assumption includes practically all the applications of analytical and preparative chromatography. From there arose the whole field of HETP investigations. Second, we observe that, in practical applications, the important results are the values of the parameters that define the peak position, its width and its asymmetry. This information is easily derived, at least in linear chromatography, from the values of the peak moments, which can be calculated either from the elution curve or from the solution of the Laplace transform. Thus began moment analysis of band profiles.

2 - Derivation and Use of the Moments

From the properties of the Laplace transform, it is easy to derive the moments of the band (see earlier, section II-2 and Chapter III, section II-3). From the equations of the chromatographic model (eqns. IV-26 and IV-27) and from their Laplace transform (eqn. IV-32, previous section), assuming as an initial condition that $C_1 = q_1 = 0$, we have

$$\bar{t} = \mathcal{M}_1 = \frac{L}{u_R} \tag{IV-35}$$

$$\mu_2 = \frac{L^2}{u_R{}^2} \left[\frac{2D_R}{u_R L} + 2\frac{Fa}{\gamma} \frac{u_R}{k_f L} \right] = \frac{2D_R L}{u_R{}^3} + \frac{2Fa}{1 + Fa} \frac{L}{k_f u_R} \tag{IV-36}$$

$$\mu_3 = \frac{L^3}{u_R{}^3} \left[\frac{12D_R{}^2}{u_R{}^2 L^2} + \frac{12D_R}{k_f L^2} \frac{Fa}{1 + Fa} + \frac{6Fa}{1 + Fa} (\frac{u_R}{k_f L})^2 \right] \tag{IV-37}$$

When D_R tends toward zero, the results of eqns. IV-35 to IV-37 tend toward those derived in the previous section (eqns. IV-22a to IV-22c).

These results demonstrate that the first-order moment is independent of the kinetic parameters, D_R and k_f, but depends only on the first isotherm coefficient, a_1. The peak variance depends linearly on both D_R and k_f, parameters that control separately the two contributions to band broadening, axial diffusion and the mass transfer resistances, respectively. Finally, it is obvious that, in the general case, the peak asymmetry, which is proportional to μ_3, is different from 0. In linear chromatography, all elution profiles are unsymmetrical unless both D_a and $1/k_f$ are equal to zero. The consequence is that, strictly speaking, the maximum concentration of a profile recorded under linear conditions is not eluted at $t = \mathcal{M}_1$. Because, in practice, the retention time is more usually defined as the elution time of the peak maximum, these kinetics parameters may contribute to some extent to define the apparent retention time, which is no longer necessarily nor entirely determined by the equilibrium isotherm. In practice, the chromatographer is presented with a difficult choice. Using the retention time of the peak maximum causes a model error of unknown magnitude. On the other hand, the determination of the first moment of the peak is less precise than that of the peak maximum, as illustrated in Figure III-3.

More generally, it might seem at first that the calculation of all the moments of a chromatographic profile is easy once it has been recorded. On the other hand, it is much easier to derive the moments of a chromatographic peak from its equation in the Laplace domain than to calculate the inverse of this distribution in the time domain. Hence, it might be tempting to compare not the concentration distributions, that theory cannot afford, but merely the moments of the calculated and experimental distributions. This hope lured many chromatographers into trying to characterize experimental peak profiles through their moments in order to compare theoretical solutions and experimental data [13-15]. The great difficulties encountered in all the attempts made at carrying out precise determinations of the higher moments of experimental peaks [16,17] show the limits of the usefulness of this approach. Moments higher than the second one are poorly precise because the signal noise causes serious uncertainties regarding the times when the peak integration should begin and end. Since the relative contributions of the sides or wings of the peak to a moment increases rapidly with increasing order, the precision on the determination of the third and fourth moments is poor (see Figure III-3). The difficulties also encountered when trying to account for actual band profiles by using classical results of fundamental statistics to relate these profiles and their moments (*e.g.*, the Gram–Charlier series) [18] is a further reason why this approach has now been abandoned.

3 - The Van Deemter Concept of HETP

If we assume that the boundary condition is a rectangular pulse of height C_0 and width t_p and that the initial condition corresponds to a column containing only the mobile and stationary phase in equilibrium but without sample, hence, if:

$$C(0, t) = \{ \begin{matrix} C_0 & t < t_p \\ 0 & t > t_p \end{matrix}$$

$$C(x, 0) = 0$$

the solution of Lapidus and Amundson for nonideal, linear chromatography (eqns. IV-26 and IV-27) becomes:

$$C(x, t) = C_0 \left[\frac{t_p}{\sqrt{4\pi D_a t^3}} e^{-\frac{(x - ut)^2}{4D_a t} - \frac{k_f t}{\varepsilon}} + \int_0^t \frac{x t_p}{\sqrt{4\pi D_a t'^3} e^{\frac{-(x - ut')^2}{4D_a t'}}} F(t') dt' \right]$$

where

$$F(t') = \sqrt{\frac{k_f^2 a t'}{\varepsilon \mu (t - t')}} e^{\frac{-k_f a(t - t')}{\mu} - \frac{k_f t'}{\varepsilon}} I_1 [2 \sqrt{\frac{k_f^2 a t'(t - t')}{\varepsilon \mu}}]$$

where ε is the column total porosity and $\mu = 1 - \varepsilon$. Van Deemter *et al.* [11] demonstrated that, provided the rate coefficient of mass transfer is not

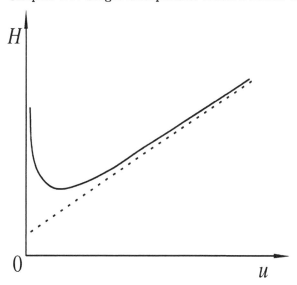

Figure IV-6   A Typical Van Deemter Curve. Plot of the HETP of a column versus the mobile phase flow velocity.

very small, the elution profile at the end of the column can be considerably simplified and becomes

$$C(x,t) = \frac{C_0 t_p}{\sqrt{4\pi\left(\frac{D_R x}{u_R^3} + Fa^2\mu\frac{x}{k_f u}\right)}} e^{-\frac{(x-u_R t)^2}{4\left(\frac{D_R x}{u_R} + \frac{Fa^2\mu}{1+Fa}\frac{x u_R}{k_f}\right)}} \tag{IV-38}$$

When the mass transfer resistances become small and k_f increases indefinitely, $1/k_f$ tends toward 0 and eqn. IV-38 becomes

$$C(x,t) = \frac{C_0 t_p}{\sqrt{4\pi D_R x/u_R^3}} e^{-\frac{(x-u_R t)^2}{4D_R x/u_R}} \tag{IV-38a}$$

Comparison between the profile given by eqn. IV-38 and the one obtained in the case in which there is only axial diffusion (eqn. IV-9) shows that there is an additive relationship between the variance of the elution profile and the contributions of the different processes which broaden the band:

$$H = \frac{2D_a}{u} + 2\left(\frac{k_0'}{1+k_0'}\right)^2 \frac{u}{k_f k'} \tag{IV-39}$$

where H is the column height equivalent to a theoretical plate or HETP ($H = L/N$) and ε the column total porosity. This was the first HETP equation. The literature contains a huge number of papers dealing with issues related to the column HETP. This discussion, however, is outside the scope of this book. The dependence of the HETP on the flow velocity is illustrated in Figure IV-6.

From series of HETP measurements, it is most difficult to derive useful values of the physicochemical parameters related to the mass transfer kinetics. There are many important sources of errors, some of which are difficult to eliminate, control, or assess. Most of the data found in the literature is empirical at best.

IV - INFLUENCE OF THE PACKING IRREGULARITIES ON CHROMATOGRAPHIC PERFORMANCES

The irregularity of the packing distribution in the column affects the migration rate of the solute bands as well as their axial dispersion along the column and the resistances to mass transfer. The packing distribution, *i.e.*, the local density of the packing material, is heterogeneous at all levels of observation, from the particle level to the level of the entire column [9]. Consequently, the local densities of the mobile and the stationary phases are not constant but fluctuate along and across the column. They are stochastic functions of the position in the column. In this section, we consider the column to be radially homogeneous, hence to be defined by a single space coordinate. We will further assume that the densities of the two phases averaged over sufficiently long distances are constant and do not exhibit a significant trend. Actually, it has been shown experimentally that the density of the solid phase varies systematically along the column [19,20]. For analytical columns, packed by slurry consolidation, it tends to decrease progressively from the column inlet to its outlet [21]. For preparative columns, packed by mechanical compression, it decreases from the end that is compressed by the piston to the opposite end. Furthermore, actual columns are not radially homogeneous but the packing density varies systematically in the radial direction. This fact is not taken into account in this discussion. The origins of these effects are in the friction between the particles themselves and between the packed bed and the column wall [22,23].

Let the stochastic functions $\varepsilon_\omega(x)$ and $\mu_\omega(x)$ be the local porosity (or volume fraction available to the mobile phase) and the local volume fraction occupied by the stationary phase, respectively [24]. The subscript ω refers to the stochastic process involved. In chromatography, it is usually assumed that the sum of the volume fractions available to the mobile phase and occupied by the stationary phase is unity, hence that $\varepsilon_\omega(x)+\mu_\omega(x) = 1$. This is not necessary, however, and the definition of two independent stochastic variables given by the authors of the only study of this problem [24] can be reconciled with common sense by relating $\mu_\omega(x)$ to the local value of the density of surface area of the adsorbent.

The random process was shown to be stationary [25]. We can write

$$\varepsilon_\omega(x) = <\varepsilon_\omega(x)> +\varepsilon_{1,\omega}(x) = \varepsilon_0 + \varepsilon_{1,\omega}(x)$$

(IV-40a)

$$\mu_\omega(x) = <\mu_\omega(x)> +\mu_{1,\omega}(x) = \mu_0 + \mu_{1,\omega}(x)$$

(IV-40b)

where $<\varepsilon_\omega(x)>$ and $<\mu_\omega(x)>$ are the average values of $\varepsilon_\omega(x)$ and $\mu_\omega(x)$, respectively. Because these functions are stochastic, there is a certain probability distribution for their value along the column. In general, chromatographic columns are long and this distribution may be assumed to be Gaussian. If η denotes the deviation from the average value and σ^2 the variance of the distribution, the probability density, W, is given by

$$W(\eta) = \frac{1}{\sqrt{\pi\sigma^2}}e^{-\frac{\eta^2}{\sigma^2}}$$

(IV-41)

This distribution states that the probability of a deviation decreases rapidly with increasing amplitude of the deviation, which is in agreement with physical experience. It also states that the width of the distribution decreases with decreasing variance. For $\sigma = 0$, we have $\eta = \varepsilon_{1,\omega}(x) = \mu_{1,\omega}(x) = 0$.

A nonhomogeneous packing density has several consequences. First, the local permeability fluctuates along the column since the local permeability is related to the local porosity. Hence the local pressure gradient will also fluctuate. Note that the mass flow rate is constant along the column because the liquid phase is not compressible and the mass balance of the mobile phase requires that its mass flux remains constant along the column. However, the fluctuations in ε cause fluctuations in the local mobile phase flow velocity because the latter is related to the (constant) flow rate by

$$F_v = u_\omega \varepsilon_\omega S$$

(IV-42)

where S is the column cross-section area ($S = \pi d_c^2/4$, with d_c the column inner diameter). Obviously, when $\sigma = 0$ in eqn. IV-41, u is constant, as conventionally assumed in most studies.

A second consequence of the fluctuations of the local packing density will be to affect the concentration distribution between the two phases, hence the migration velocity of the band. The concentration wave is modulated by the local fluctuations of the packing density. If we ignore the axial dispersion and assume no mass-transfer resistances, the fundamental mass balance of chromatography can be written

$$\varepsilon_\omega \frac{\partial C_\omega}{\partial t} + u_\omega \frac{\partial q_\omega}{\partial t} + \mu_\omega \varepsilon_\omega \frac{\partial C_\omega}{\partial x} = 0$$

(IV-43)

where C_ω and q_ω stand for the local mobile and stationary-phase concentrations of the solute when the packing density is nonhomogeneous. This mass balance equation shows the influence of the packing heterogeneity on the concentration wave, on the concentration profile along the column, and on the elution profile of the chromatographic band.

The solution of this problem was discussed in detail by Yeroshenkova *et al.* [24] using the model of linear, ideal chromatography. These authors also derived an expression accounting for the effect of the fluctuations of the packing density on the peak profile in nonlinear, nonideal chromatography. Their essential result is that statistical fluctuations of the packing density cause a broadening of the band which is similar to that arising from axial dispersion. More exactly, the effect of these fluctuations is equivalent to the addition of a second dispersion term to the mass balance equation, $\frac{\tilde{D}_L \lambda}{2u\varepsilon_0} \frac{\partial^2}{\partial t^2}$, where $\tilde{D}_L \lambda$ is proportional to σ^2, the variance of the stochastic distribution of the porosity (eqn. IV-41) and λ is the correlation length (see later). Because of the addition of this term to the equation of the ideal model, a term in $1/\sqrt{x}$ will arise in the concentration wave function, which leads to the asymmetry of the band profile along the column. If the second-order term is related to $\partial^2/\partial x^2$, the additional factor in the concentration wave function will be proportional to $1/\sqrt{t}$, which leads to an asymmetry along the time axis. Note that these asymmetries can only be very small as long as the concentration is low.

The method used to derive these results consists of introducing the stochastic process in the mass balance equation, of solving the new equation, and of calculating the statistical average of the results. In order to determine the contribution of the packing irregularities to band broadening in chromatography, we assume in this section a linear, ideal model. In other words, in the absence of the contribution of the packing irregularities, the only effect of the column would be to introduce a delay in the propagation of the injected profile.

The initial and boundary conditions of the problem are the same as for earlier problems

$$C_\omega(x, 0) = 0 \qquad\qquad 0 \le x < \infty \qquad\qquad \text{(IV-44a)}$$
$$C_\omega(0, t) = \varphi(t) \qquad\qquad\qquad\qquad \text{(IV-44b)}$$

where $\varphi(t)$ is the injection function and M the amount of substance injected into the column, with

$$\int_{-\infty}^{+\infty} \varphi(t)dt = M < \infty \qquad\qquad \text{(IV-44c)}$$

Finally, for the sake of simplicity in this discussion, we assume a linear isotherm with

$$q_\omega(x,t) = aC_\omega(x,t) \qquad \text{(IV-44d)}$$

The mass balance equation becomes

$$(\varepsilon_\omega(x) + a\mu_\omega(x)) \frac{\partial C_\omega}{\partial t} + u_p \frac{\partial C_\omega}{\partial x} = 0 \qquad \text{(IV-45)}$$

where $u_p = u_\omega \varepsilon_\omega$. This equation is a stochastic differential equation. To integrate it, we need to transform its coefficients into non-stochastic coefficients. For this purpose, we introduce the stochastic elapsed time, $t_\omega(x,t)$, and the stochastic migration distance, $x_\omega(x,t)$, as follows

$$t_\omega = t - \frac{1}{u_p} \int_0^x (\varepsilon_\omega(\xi) + a\mu_\omega(\xi)) \, d\xi \qquad \text{(IV-46)}$$

with a similar equation for $x_\omega(x,t)$. Accordingly, the concentration profile, which is a function of the distance along the column and the time, is given by

$$\tilde{C}_\omega = \tilde{C}_\omega (x_\omega(x,t), t_\omega(x,t)) \qquad \text{(IV-47)}$$

Differentiation of eqn. IV-47 with respect to time and space gives:

$$\frac{\partial \tilde{C}_\omega}{\partial t} = \frac{\partial \tilde{C}_\omega}{\partial t_\omega} \frac{\partial t_\omega}{\partial t} + \frac{\partial \tilde{C}_\omega}{\partial x_\omega} \frac{\partial x_\omega}{\partial t}$$

$$\frac{\partial \tilde{C}_\omega}{\partial x} = \frac{\partial \tilde{C}_\omega}{\partial t_\omega} \frac{\partial t_\omega}{\partial x} + \frac{\partial \tilde{C}_\omega}{\partial x_\omega} \frac{\partial x_\omega}{\partial x}$$

With these new relationships, we have

$$\frac{\partial}{\partial t} = \frac{\partial t_\omega}{\partial t} \frac{\partial}{\partial t_\omega} + \frac{\partial x_\omega}{\partial t} \frac{\partial}{\partial x_\omega} = \frac{\partial}{\partial t_\omega} \qquad \text{(IV-48a)}$$

$$\frac{\partial}{\partial x} = \frac{\partial t_\omega}{\partial x} \frac{\partial}{\partial t_\omega} + \frac{\partial x_\omega}{\partial x} \frac{\partial}{\partial x_\omega} = \frac{-1}{u_p} (\varepsilon_\omega(x) + a\mu_\omega(x)) \frac{\partial}{\partial t_\omega} + \frac{\partial}{\partial x_\omega} \qquad \text{(IV-48b)}$$

where a is the slope of the linear equilibrium isotherm. From these two equations, we derive

$$\frac{\partial C_\omega}{\partial t} = \frac{\partial \tilde{C}_\omega}{\partial t_\omega}$$

and

$$\frac{\partial C_\omega}{\partial x} = \frac{-1}{\partial u_p} (\varepsilon_\omega(x) + a\mu_\omega(x)) \frac{\partial \tilde{C}_\omega}{\partial t_\omega} + \frac{\partial \tilde{C}_\omega}{\partial x_\omega}$$

Combination of these relationships with eqn. IV-45 gives

$$(\varepsilon_\omega(x) + a\mu_\omega(x))\frac{\partial \tilde{C}_\omega}{\partial t_\omega} + u_p\left[\frac{-1}{u_p}(\varepsilon_\omega(x) + a\mu_\omega(x))\frac{\partial \tilde{C}_\omega}{\partial t_\omega} + \frac{\partial \tilde{C}_\omega}{\partial x_\omega}\right] = 0$$

Hence, we have

$$\frac{\partial \tilde{C}_\omega}{\partial x_\omega} = 0 \qquad\qquad\qquad\text{(IV-49a)}$$

or

$$\tilde{C}_\omega(x_\omega, t_\omega) = \tilde{C}_\omega(0, t_\omega) \qquad\qquad\qquad\text{(IV-49b)}$$

In other words, $\tilde{C}(x_\omega, t_\omega)$ is only a function of t_ω. From eqns. IV-44b and IV-46, we have

$$\tilde{C}_\omega(0, t_\omega) = \tilde{C}_\omega(0, t) = \varphi(t) = \varphi(t_\omega) \qquad\qquad\text{(IV-50)}$$

and so

$$\tilde{C}_\omega(x_\omega, t_\omega) = \varphi(t_\omega) \qquad\qquad\qquad\text{(IV-51)}$$

Combining eqns. IV-46 and IV-47 gives

$$C_\omega(x, t) = \varphi\left(t - \frac{1}{u_p}\int_0^x [\varepsilon_\omega(\xi) + a\mu_\omega(\xi)]d\xi\right) \qquad\text{(IV-52)}$$

For an homogeneous packing, we would have $\varepsilon_\omega(x) = \varepsilon$, $\mu_\omega(x) = \mu$, $u_\omega = u$, and $u_p = u\varepsilon$. Then, eqn. IV-52 would give

$$C(x, t) = \varphi(t) - \frac{1}{u\varepsilon}\int_0^x (\varepsilon + a\mu)d\xi = \varphi(t - \frac{x}{u_R}) \qquad\text{(IV-53)}$$

which is the classical solution of the ideal, linear model of chromatography.

Now, considering that the more realistic stochastic model, assuming statistical fluctuations of the packing density along the column, leads to $\varepsilon_\omega(x) = \varepsilon_0 + \varepsilon_{1,\omega}(x)$ and $\mu_\omega(x) = \mu_0 + \mu_{1,\omega}(x)$, we have

$$C_\omega(x, t) = \varphi\left(t - \frac{\varepsilon_0 + a\mu_0}{u_p}, x - \frac{\nu_\omega(x)}{u_p}\right) \qquad\text{(IV-54)}$$

where $\nu_\omega(x)$ is also a random process with a zero mean,

$$\nu_\omega(x) = \int_0^x [\varepsilon_{1,\omega}(\xi) + a\mu_{1,\omega}(\xi)]d\xi \qquad\qquad\text{(IV-55)}$$

Equation IV-54 expresses the fact that the stochastic fluctuations of $C_\omega(x, t)$ depend essentially on the fluctuations of $\nu_\omega(x)$.

The elution profile, $C(L, t)$, depends on both the average and the variance of $\nu_w(x)$, which we must now calculate. Obviously, when the column length, L, is very large, the concentration profile is the average of $C_w(L, t)$:

$$C(L, t) = \langle C_w(L, t) \rangle \tag{IV-56}$$

Since the fluctuation distribution is Gaussian, as explained in the beginning of this section, we can substitute $\nu_w(x)$ for η in $W(\eta)$ (eqn. IV-41), giving

$$W(\eta) = \frac{1}{\sqrt{\pi \sigma^2}} e^{-\frac{\eta^2}{\sigma^2}} \tag{IV-57a}$$

$$\sigma^2 = \langle \nu_w(x), \nu_w(x) \rangle \tag{IV-57b}$$

The statistical average of the concentration profile along the column is

$$C(x, t) = < C_w(x, t) > = \int_{-\infty}^{\infty} \varphi[t - \frac{\varepsilon_0 + a\mu_0}{u_p} x - \frac{\eta}{u_p} W(\eta)] d\eta \tag{IV-58}$$

As it results from the discussion of the properties of the solution of the mass balance equation derived in this case [25], the variance of $\nu_w(x)$ is given by

$$\sigma^2 = 2\tilde{D}_L \lambda x \tag{IV-59}$$

where $\tilde{D}_L$ is the following average

$$\tilde{D}_L = \langle [\varepsilon_{1,w}(x) + a\mu_{1,w}(x)]^2 \rangle \tag{IV-60}$$

and where λ is the correlation length. Thus

$$W(\nu_w(x)) = \frac{1}{\sqrt{2\pi \tilde{D}_L \lambda x}} \exp\left(-\frac{\nu_w^2(x)}{2\tilde{D}_L \lambda x}\right) \tag{IV-61}$$

which, combined with eqn. IV-58 gives

$$C(x, t) = \frac{1}{\sqrt{2\pi \tilde{D}_L \lambda x}} \int_0^{\infty} \varphi\left(t - \frac{\varepsilon_0 + a\mu_0}{u_p} x - \frac{\eta}{u_p}\right) e^{-\frac{\eta^2}{2\tilde{D}_L \lambda x}} d\eta \tag{IV-62}$$

When $\tilde{D}_L = 0$, the variance is zero (eqn. IV-59) and $W(\eta)$ becomes the Dirac δ function, $\delta(\eta)$. Thus, the result is different from 0, except in the case when $\eta = 0$. Since $\varepsilon_w = \varepsilon_0$ and $\mu_w = \mu_0$ when $\tilde{D}_L = 0$, we have

$$C(L, t) = \varphi\left(t - \frac{\varepsilon_0 + a\mu_0}{u_p} L\right) = \varphi\left(t - \frac{1 + Fa}{u_p} L\right) \tag{IV-63}$$

This means that if the variance of the stochastic process is zero, the concentration distribution obtained is the same as the classical result derived for an homogeneous column.

When $\tilde{D}_L \neq 0$, then $\lambda \neq 0$ and $W(\eta)$ is a Gaussian function. It acts as a dispersion factor and the product $\tilde{D}_L \lambda$ controls the extent of this apparent dispersion. If the injection is a narrow pulse:

$$\varphi(t) = M\delta(t) \tag{IV-64}$$

and if x is large enough, then the concentration profile is given by

$$C(x,t) = \frac{M}{\sqrt{2\pi \tilde{D}_L \lambda x}} \int_{-\infty}^{\infty} \delta\left(t - \frac{\varepsilon_0 + a\mu_0}{u_p}x - \frac{\eta}{u_p}\right) e^{-\frac{\eta^2}{2\pi \tilde{D}_L \lambda x}} d\eta$$

$$= \frac{Mu_p}{\sqrt{2\pi \tilde{D}_L \lambda x}} e^{\frac{[u_p t - (\varepsilon_0 + a\mu_0)x]^2}{2\tilde{D}_L \lambda x}} \tag{IV-65}$$

It is easy to show that this equation is a solution of the following partial differential equation

$$u_p \frac{\partial C}{\partial x} + (\varepsilon_0 + a\mu_0)\frac{\partial C}{\partial t} = \frac{\tilde{D}_L \lambda}{2u_p}\frac{\partial^2 C}{\partial t^2} \tag{IV-66a}$$

or

$$\frac{u_p}{\varepsilon_0}\frac{\partial C}{\partial x} + (1 + Fa)\frac{\partial C}{\partial t} = \frac{\tilde{D}_L \lambda}{2u_p\varepsilon_0}\frac{\partial^2 C}{\partial t^2} \tag{IV-66b}$$

If we assume that, as a first approximation, $C(x,t) = C(t - x/u_{R0})$, with $u_{R,0} = u/(\varepsilon_0 + \mu_0 a)$, eqn. IV-66b can be rewritten as

$$\frac{u_p}{\varepsilon_0}\frac{\partial C}{\partial x} + (1 + Fa)\frac{\partial C}{\partial t} = \frac{\tilde{D}_L u_p \lambda}{2(\varepsilon_0 + a\mu_0)^2 \varepsilon_0}\frac{\partial^2 C}{\partial x^2} \tag{IV-66c}$$

or, taking $u_p/\varepsilon_0 \simeq u$, we have

$$u\frac{\partial C}{\partial x} + (1 + Fa)\frac{\partial C}{\partial t} = \frac{\tilde{D}_L u \lambda}{2(\varepsilon_0 + \mu_0 a)^2}\frac{\partial^2 C}{\partial x^2} \tag{IV-67}$$

This result shows that the influence of the stochastic irregularities of a packed bed is equivalent to the addition of a new dispersive term to the mass balance equation. The additional dispersion coefficient is equal to $\tilde{D}_L u \lambda/(2(\varepsilon_0 + a\mu_0)^2)$. This conclusion is approximate, insofar as the factor in eqn. IV-65 is $1/\sqrt{2\pi \tilde{D}_L \lambda x}$ while the second differential term in the RHS of eqn. IV-166 is a $\partial^2/\partial t^2$ term, not a $\partial^2/\partial x^2$ term. Thus, the effect on the elution profile is different from the one calculated in the general dispersion theory. It will result in an asymmetry of the peak along the x axis.

APPENDIX A
Derivation of the Simple Breakthrough
Profile (Solution of Equation IV-2)

As indicated in the beginning of Section I (Influence of Axial Diffusion), the first step in the derivation of the solution of the equilibrium–dispersive model under linear conditions is to transform the mass balance equation, eqn. IV-1, into the diffusion equation. Then the diffusion equation can be solved for the initial and boundary conditions corresponding to those useful in chromatography.

1. Transformation of Eqn. IV-2 into the Diffusion Equation

The difference between eqn. IV-2 and the diffusion equation is that the later does not contain the partial differential $\partial C/\partial x$ while the former does. So, we need to find a transformation eliminating this derivative. Let

$$C(x,t) = g(x,t) \, \exp{(\alpha x + \beta t)} \tag{IV-A-1}$$

where α and β are parameters to be determined. If we introduce eqn. IV-A-1 into eqn. IV-1 (the mass balance equation), we find that, if we chose

$$\alpha = \frac{u}{2D_a} \tag{IV-A-2a}$$

$$\beta = -\frac{u_R{}^2}{4D_R} \tag{IV-A-2b}$$

we transform the mass balance equation into the following equation

$$\frac{\partial g}{\partial t} = D_R \frac{\partial^2 g}{\partial x^2} \tag{IV-A-3}$$

which is the conventional form of the diffusion equation. The initial and boundary conditions of the chromatography equation become respectively

$$g(x,0) = 0 \tag{IV-A-4a}$$

$$g(0,t) = C_0(t) \exp{\left(\frac{u_R{}^2 t}{4D_R}\right)} \tag{IV-A-4b}$$

The combination of eqns. IV-A-3 and IV-A-4a,b constitutes a well-posed problem of diffusion theory. Its solution is obtained using the principle of Duhamel, which we present now.

2. The Duhamel Principle

The Duhamel principle establishes a relationship between the solution of constant boundary values of a problem and the solution of the same problem with variable boundary values. This principle says that

If $f_0(x,t)$ is a solution of the following problem:

$$\begin{cases} LC(x,t) &= 0 \\ C(x,0) &= 0 \\ C(0,t) &= 1 \end{cases}$$

where L is an operator (for example $L = \frac{\partial}{\partial t} - D_a \frac{\partial^2}{\partial x^2}$), *then, when $C(0, t) = \varphi(t)$, we have*

$$C(x, t) = -\int_0^t \varphi(\tau) \frac{\partial f_0(x, \tau)}{\partial \tau} d\tau$$

The demonstration of this important theorem is as follows.
(1) When

$$C(0, t) = \begin{cases} 0 & \text{for} & t \leq \tau \\ 1 & \text{for} & t > \tau \end{cases}$$

we have

$$C(x, t) = \begin{cases} 0 & \text{for} & t \leq \tau \\ f_0(x, t, \tau) & \text{for} & t > \tau \end{cases}$$

(2) When

$$C(0, t) = \begin{cases} 0 & \text{for} & t \leq \tau, t \geq \tau + d\tau \\ \varphi(t) & \text{for} & \tau < t < \tau + d\tau \end{cases}$$

we have

$$C(x, t) = \varphi(t)[f_0(x, t, \tau) - f_0(x, t, \tau + d\tau)] = -\varphi(\tau)\frac{\partial f_0}{\partial \tau}$$

when $t \in (\tau, \tau + d\tau)$.
(3) In the interval $(0, t)$, we have

$$C(x, t) = -\int_0^t \varphi(\tau)\frac{\partial f_0(x, t, \tau)}{\partial t} d\tau$$

This is the expression of Duhamel principle which is used for the following derivation.

3. Solution of the Diffusion Equation

We consider frontal analysis, as performed with the boundary condition of the Riemann problem (eqn. IV-3a), and apply the Duhamel principle to the function $g(x, t)$, which is a solution of eqn. (IV-2) and has the proper boundary conditions, and we have

$$g(x, t) = \frac{x}{2}\sqrt{\frac{1}{\pi D_R}}\int_0^t C_0(\lambda)\exp\left(\frac{u_R^2\lambda}{4D_R} - \frac{x^2}{4D_R(t - \lambda)}\right)\frac{d\lambda}{(t - \lambda)^{3/2}} \qquad \text{(IV-A-5)}$$

where λ is a dummy variable. The combination of eqns. IV-A-1 and IV-A-2a,b gives the solution of the problem as

$$C(x, t) = g(x, t)\exp\left(\frac{ux}{2D_a} - \frac{u_R^2t}{4D_R}\right) \qquad \text{(IV-A-6)}$$

To obtain the solution, we need to calculate the integral in eqn. IV-A5. For this, we observe that, in frontal analysis, the concentration of the stream entering the column is constant (see boundary condition, eqn. IV-3a). So, by combining eqns. IV-A-2 and IV-A-5 we obtain

$$C(x, t) = \frac{x}{2}\sqrt{\frac{1}{\pi D_R}}C_0\int_0^t \exp\left(\frac{ux}{2D_a} - \frac{u_R^2t}{4D_R} + \frac{u_R^2\lambda}{4D_R} - \frac{x^2}{4D_R(t - \lambda)}\right)\frac{d\lambda}{(t - \lambda)^{3/2}}$$

$$\text{(IV-A-7)}$$

To calculate the integral in this equation, let $v = t - \lambda$. Then

$$C(x, t) = \frac{x}{2}\frac{C_0}{\sqrt{\pi D_R}}\int_0^t \exp\left(\frac{ux}{2D_a} - \frac{u_R^2v}{4D_R} - \frac{x^2}{4D_Rv)}\right)\frac{dv}{v^{3/2}} \qquad \text{(IV-A-8)}$$

We can rearrange eqn. IV-A-8 into the sum of two integrals

$$C(x,t) = \frac{C_0}{\sqrt{\pi D_R}} \left\{ \int_0^t \exp\left(\frac{u_R x}{2D_R} - \frac{u_R^2 v}{4D_R} - \frac{x^2}{4D_R v} \right) \left(\frac{x}{4\sqrt{D_R} v^{3/2}} - \frac{\sqrt{\frac{u_R^2}{4D_R}}}{2v^{1/2}} \right) dv \right.$$

$$\left. + \int_0^t \exp\left(\frac{u_R x}{2D_R} - \frac{u_R^2 v}{4D_R} - \frac{x^2}{4D_R v} \right) \left(\frac{x}{4\sqrt{D_R} v^{3/2}} + \frac{u_R}{4\sqrt{D_R} v^{1/2}} \right) dv \right\} \quad \text{(IV-A-9)}$$

Now, we make the following two different changes of variable, one in each of the two parts of the equation above, respectively, as follows

$$\eta = \sqrt{\frac{u_R^2 v}{4D_R}} - \sqrt{\frac{x^2}{4D_R v}} \quad \text{(IV-A-10a)}$$

$$\eta' = \sqrt{\frac{u_R^2 v}{4D_R}} + \sqrt{\frac{x^2}{4D_R v}} \quad \text{(IV-A-10b)}$$

accordingly, we have

$$d\eta = \left(\frac{x}{4\sqrt{D_R} v^{3/2}} + \frac{u_R}{4\sqrt{D_R} v} \right) dv \quad \text{(IV-A-11a)}$$

$$e^{-\eta^2} = \exp\left(\frac{u_R x}{2D_R} - \frac{u_R^2 v}{4D_R} - \frac{x^2}{4D_R v} \right) \quad \text{(IV-A-12a)}$$

$$d\eta' = \left(-\frac{x}{4\sqrt{D_R} v^{3/2}} + \frac{u_R}{4\sqrt{D_R} v} \right) dv \quad \text{(IV-A-11b)}$$

$$e^{\frac{vx}{D_a} - \eta'^2} = \exp\left(\frac{u_R x}{2D_R} - \frac{u_R^2 v}{4D_R} - \frac{x^2}{4D_R v} \right) \quad \text{(IV-A-12b)}$$

The lower and upper integration boundaries of η are $-\infty$ and $\sqrt{\frac{u_R^2 t}{4D_R}} - \sqrt{\frac{x^2}{4D_R t}}$, respectively; those of η' are $\sqrt{\frac{u_R^2 t}{4D_R}} + \sqrt{\frac{x^2}{4D_R t}}$ and $+\infty$, respectively.

Introducing eqns. IV-A-10 to IV-A-12 into eqn. IV-A-9, we obtain

$$C(x,t) = \frac{C_0}{\sqrt{\pi}} \int_{-\infty}^0 e^{-\eta^2} d\eta + \frac{C_0}{\sqrt{\pi}} \int_0^{\sqrt{\frac{u_R^2 t}{4D_R}} - \sqrt{\frac{x^2}{4D_R t}}} e^{-\eta^2} d\eta +$$

$$\frac{C_0}{\sqrt{\pi}} e^{\frac{u_R x}{D_R}} \int_{\sqrt{\frac{u_R^2 t}{4D_R}} + \sqrt{\frac{x^2}{4D_R t}}}^{\infty} e^{-\eta'^2} d\eta' \quad \text{(IV-A-13)}$$

where the boundary interval of η is divided into two parts, from $-\infty$ to 0 and from 0 to $\sqrt{u_R^2 t/(4D_R)} - \sqrt{x^2/(4D_R t)}$. Introducing the definition of the error function $(\text{erf}(z) = 2/\sqrt{\pi} \int_0^t \exp(-z^2) dz)$ and the complementary error function $(\text{erfc}(z) = 1 - \text{erf}(z))$, leads to the following equation:

$$C(x,t) = \frac{C_0}{2} + \frac{C_0}{2} \text{erf}\left(\sqrt{\frac{u_R^2 t}{4D_R}} - \sqrt{\frac{x^2}{4D_R t}} \right) +$$

$$\frac{C_0}{2} e^{\frac{u_R x}{D_R}} \text{erfc}\left(\sqrt{\frac{u_R^2 t}{4D_R}} + \sqrt{\frac{x^2}{4D_R t}} \right) \quad \text{(IV-4)}$$

This equation is reported as eqn. IV-4 in the text of Section I (p. 74).

4. Transition from a Rectangular Pulse to a Dirac Boundary Condition

The general type of solution of the diffusion equation is given by eqn. (IV-A-5). When a Riemann injection condition (frontal analysis) is replaced by a rectangular pulse injection, the integration boundaries of λ change from $(0, t)$ to $(0, t_p)$. Accordingly, the integration limits for the new variable $\nu(= t - \lambda)$ will be changed into $(t, t - t_p)$. Since η and η' are function of ν (see eqns. IV-A-10a and IV-A-10b), the general solution becomes

$$
C(x,t) = \frac{C_0}{\sqrt{\pi}} \int_{\sqrt{\frac{u_R^2(t-t_p)}{4D_R}} - \sqrt{\frac{x^2}{4D_R(t-t_p)}}}^{\sqrt{\frac{u_R^2 t}{4D_R}} - \sqrt{\frac{x^2}{4D_R t}}} e^{-\eta^2} d\eta -
$$

$$
\frac{C_0}{\sqrt{\pi}} e^{\frac{u x}{D_a}} \int_{\sqrt{\frac{u_R^2 t}{4D_R}} + \sqrt{\frac{x^2}{4D_R t}}}^{\sqrt{\frac{u_R^2(t-t_p)}{4D_R}} + \sqrt{\frac{x^2}{4D_R(t-t_p)}}} e^{-\eta'^2} d\eta' \qquad \text{(IV-A-14)}
$$

(see eqn. IV-11 in Section I). This equation can be written more clearly as follows:

$$
C(x,t) = \frac{C_0}{\sqrt{\pi}} \int_A^B e^{-\eta^2} d\eta \ - \ \frac{C_0}{\sqrt{\pi}} e^{\frac{u_R x}{D_R}} \int_C^D e^{-\eta'^2} d\eta'
$$

with the integration limits given by

$$
A = \sqrt{\frac{u_R^2(t - t_p)}{4D_R}} - \sqrt{\frac{x^2}{4D_R(t - t_p)}}
$$

$$
B = \sqrt{\frac{u_R^2 t}{4D_R}} - \sqrt{\frac{x^2}{4D_R t}}
$$

$$
C = \sqrt{\frac{u_R^2 t}{4D_R}} + \sqrt{\frac{x^2}{4D_R t}}
$$

$$
D = \sqrt{\frac{u_R^2(t - t_p)}{4D_R}} + \sqrt{\frac{x^2}{4D_R(t - t_p)}}
$$

When the width of the rectangular injection decreases toward that of a Dirac pulse injection, i.e., when t_p tends toward 0, we have

$$
\sqrt{\frac{u_R^2(t - t_p)}{4D_R}} \approx \sqrt{\frac{u_R^2 t}{4D_R}}(1 - \frac{t_p}{2t})
$$

$$
\sqrt{\frac{x^2}{4D_R(t - t_p)}} \approx \sqrt{\frac{x^2}{4D_R t}}(1 + \frac{t_p}{2t})
$$

and

$$
\sqrt{\frac{u_R^2(t - t_p)}{4D_R}} - \sqrt{\frac{x^2}{4D_R(t - t_p)}} \approx (\sqrt{\frac{u_R^2 t}{4D_R}} - \sqrt{\frac{x^2}{4D_R}}) - \frac{t_p}{2}(\sqrt{\frac{u_R^2 t}{4D_R}} + \sqrt{\frac{x^2}{4D_R}})
$$

$$
\sqrt{\frac{u_R^2(t - t_p)}{4D_R}} + \sqrt{\frac{x^2}{4D_R(t - t_p)}} \approx (\sqrt{\frac{u_R^2 t}{4D_R}} + \sqrt{\frac{x^2}{4D_R}}) - \frac{t_p}{2}(\sqrt{\frac{u_R^2 t}{4D_R}} - \sqrt{\frac{x^2}{4D_R}})
$$

so when $t_p \to 0$, the boundary condition is given by $C(0,t) = C_0 t_p \delta(t)$ and the solution becomes

$$
C(x,t) \approx \frac{C_0}{\sqrt{\pi}} e^{-\left(\sqrt{\frac{u_R{}^2 t}{4D_R}} - \sqrt{\frac{x^2}{4D_R}}\right)^2} \frac{t_p}{2t}\left(\sqrt{\frac{u_R{}^2 t}{4D_R}} + \sqrt{\frac{x^2}{4D_R t}}\right)
$$

$$
+ \frac{C_0}{\sqrt{\pi}} e^{-\frac{u_R x}{D_R}\left(\sqrt{\frac{u_R{}^2 t}{4D_R}} + \sqrt{\frac{x^2}{4D_R}}\right)^2} \frac{t_p}{2t}\left(\sqrt{\frac{u_R{}^2 t}{4D_R}} - \sqrt{\frac{x^2}{4D_R t}}\right) \qquad \text{(IV-A-15a)}
$$

The elution profile is nearly 0 everywhere, except in a central region where $x \approx u_R t$. Then, we have $\sqrt{\frac{u_R{}^2 t}{4D_R}} \approx \sqrt{\frac{x^2}{4D_R t}}$. Compared to the first term, the second term of the equation becomes negligible and we have

$$
C(x,t) \approx \frac{C_0}{\sqrt{\pi}} e^{-\left(\sqrt{\frac{u_R{}^2 t}{4D_R}} - \sqrt{\frac{x^2}{4D_R}}\right)^2} \frac{t_p}{t}
$$

$$
= \frac{x}{t} \frac{C_0 t_p}{\sqrt{4\pi D_R t}} e^{-\frac{(x - u_R t)^2}{4D_R t}} \qquad \text{(IV-A-15b)}
$$

5. Solution for a Dirac Boundary Condition

With this boundary condition, eqns. IV-A-1 to IV-A-4 become

$$
C(x,t) = e^{\alpha x + \beta t} g(x,t)
$$

$$
\frac{\partial q}{\partial t} = D_R \frac{\partial^2 q}{\partial x^2}
$$

$$
g(x,0) = 0
$$

$$
g(0,t) = B\delta(t)e^{-\beta t}
$$

$$
\alpha = \frac{u}{2D_a}
$$

$$
\beta = -\frac{u_R{}^2}{4D_R}
$$

From Appendix B, eqn. IV-B-16, we have

$$
g(x,t) = \frac{x}{\sqrt{4\pi D_R t}} \int_0^t \frac{g(0,\tau)}{(t-\tau)^{\frac{3}{2}}} e^{-\frac{x^2}{4D_R(t-\tau)}} d\tau
$$

Putting $g(0,t) = B\delta(t)e^{-\beta t}$ into the previous equation, we have

$$
g(x,t) = \frac{xB}{t\sqrt{\frac{4\pi D_R}{t}}} e^{\frac{-x^2}{4D_R t}}
$$

From $C(x,t) = e^{\alpha x + \beta t} g(x,t)$, we have

$$
C(x,t) = \frac{x}{t} \frac{B}{\sqrt{4\pi D_R t}} e^{-\frac{(x - u_R t)^2}{4D_R t}}
$$

This equation is identical to eqn. IV-A-15 above. It was given earlier as eqn. IV-10 in Section I.

6- Derivation of the Solution for the Dirac Pulse Injection using the Laplace transform

Actually the solution of the mass balance equation IV-2 with the Dirac boundary condition can also be obtained using the Laplace transform. Applying this transform gives

$$p\overline{C} + u_R \frac{d\overline{C}}{dx} = D_R \frac{d^2\overline{C}}{dx^2} \tag{IV-A-16}$$

where $\overline{C}(x,p) = \int_0^\infty C(x,t)e^{-pt}dt$.

Obviously, the solution of this new equation is

$$\overline{C} = \alpha e^{k_1 x} + \beta e^{k_2 x}$$

where $k_{1,2} = \frac{1}{2D_R}(u_R \pm \sqrt{u_R^2 + 4D_R p})$ and α, β are constants.

Since $C(\infty, t)$ and $\overline{C}(\infty, p)$ are finite, $\alpha = 0$. Since $C(0,t) = B\delta(t)$, $\overline{C}(0,p) = B$, and $\beta = B$, so

$$\overline{C} = Be^{\frac{ux}{2D_a}} e^{\frac{-x}{2D_R}\sqrt{u_R^2 + 4D_R p}}$$

From $L^{-1}[\overline{C}(p - p_0)] = c(t)e^{-p_0 t}$ and taking the inverse of the Laplace transform, we have

$$C(x,t) = \frac{x}{t} \frac{B}{\sqrt{4\pi D_R t}} e^{\frac{-(x - u_R t)^2}{4 D_R t}}$$

This is the same result as the one obtained with the previous method.

APPENDIX B
Derivation of the Generalized Breakthrough
Profile Equation (General Solution of Equation IV-2)

The problem is similar to the one discussed in the previous Appendix, but now the initial conditions are given by eqns. IV-6a and IV-6b instead of eqns. IV-3a and IV-3b. The boundary condition remains unchanged. The initial steps of the derivation of the breakthrough profile are the same and the starting point is now the diffusion equation (eqn. IV-A-3). The system to be solved is given by the following set of equations

$$\frac{\partial g}{\partial t} = D_R \frac{\partial^2 g}{\partial x^2} \tag{IV-B-1a}$$

$$g(0, t) = C_0(t) \exp\left(\frac{u_R^2 t}{4D_R}\right) \tag{IV-B-1b}$$

$$g(x, 0) = C_1(x) \exp\left(\frac{-u_R x}{2D_R}\right) \tag{IV-B-1c}$$

This problem can be solved by decomposing it into two simpler problems. Let $g = f_1 + f_2$ in which f_1 and f_2 are the solutions of the following two problems

$$\frac{\partial f_1}{\partial t} = D_R \frac{\partial^2 f_1}{\partial x^2} \tag{IV-B-3a}$$

$$f_1(x, 0) = C_1(x) \exp\left(-\frac{ux}{2D_a}\right) \tag{IV-B-3b}$$

$$f_1(0, t) = 0 \tag{IV-B-3c}$$

and

$$\frac{\partial f_2}{\partial t} = D_R \frac{\partial^2 f_2}{\partial x^2} \tag{IV-B-4a}$$

$$f_2(x, 0) = 0 \tag{IV-B-4b}$$

$$f_2(0, t) = C_0 \exp\left(\frac{u_R^2 t}{4D_R}\right) \tag{IV-B-4c}$$

The solution of the second problem is the same as that of the problem discussed in Appendix A. We need to discuss here only the solution of the first of the two problems.

1. Calculation of $f_1(x, t)$

The main difficulty encountered in the solution of this problem is that the diffusion problem is defined in an infinite interval of the variable x, $-\infty < x < +\infty$, while, with the initial and boundary conditions selected, the chromatography problem is defined only in a semi-infinite interval, $0 < x < \infty$. Other authors have selected different conditions, including a column of infinite length, extending from $-\infty$ to $+\infty$, with an injection made at $x = 0$ [1].

The conventional diffusion problem is defined by the following PDE and boundary and initial conditions

$$\frac{\partial f_1}{\partial t} = D_a \frac{\partial^2 f_1}{\partial x^2} \qquad \text{with} \qquad -\infty < x < +\infty, \qquad t > 0 \tag{IV-B-5a}$$

$$f_1(x, 0) = \varphi(x) \tag{IV-B-5b}$$

$$f_1(0, t) = 0 \tag{IV-B-5c}$$

Its solution is

$$f_1(x,t) = \int_{-\infty}^{+\infty} \frac{\varphi(\xi)}{\sqrt{4\pi D_R t}} e^{\frac{-(x-\xi)^2}{4\pi D_R}} d\xi \qquad \text{(IV-B-6)}$$

where ξ is a dummy variable. Now, we must transform our chromatographic problem, defined in a semi-infinite interval (eqns. IV-B-3a, 3b, and 3c) into an equivalent diffusion problem, defined in an infinite interval. An extension is needed. It is provided by the following definition

$$\varphi(-x) = -\varphi(x) \qquad \text{for} \quad x < 0 \qquad \text{(IV-B-7)}$$

This extension of $f(x)$ is called an *odd* extension. The solution of our problem becomes

$$f_1(x,t) = \int_0^{+\infty} \varphi(\xi) \frac{1}{\sqrt{4\pi D_R t}} e^{\frac{-(x-\xi)^2}{4 D_R t}} d\xi + \int_{-\infty}^0 \varphi(\xi) \frac{1}{\sqrt{4\pi D_R t}} e^{\frac{-(x-\xi)^2}{4 D_R t}} d\xi \quad \text{(IV-B-8a)}$$

In the second integral, we have $\xi < 0$. So, let $\eta = -\xi$. Then, $\varphi(\xi) = \varphi(-\eta) = -\varphi(\eta)$ and $d\eta = -d\xi$; the boundaries of the second integral change from $(-\infty, 0)$ to $(+\infty, 0)$. Hence eqn. IV-B-8a becomes

$$f_1(x,t) = \int_0^{+\infty} \frac{\varphi(\xi)}{\sqrt{4\pi D_R t}} e^{\frac{-(x-\xi)^2}{4 D_R t}} d\xi - \int_0^{+\infty} \frac{\varphi(\eta)}{\sqrt{4\pi D_R t}} e^{\frac{-(x+\eta)^2}{4 D_R t}} d\eta \qquad \text{(IV-B-8b)}$$

This equation may be rewritten as

$$f_1(x,t) = \int_0^{+\infty} \varphi(\xi) \frac{1}{\sqrt{4\pi D_R t}} \left(e^{\frac{-(x-\xi)^2}{4 D_R t}} - e^{\frac{-(x+\xi)^2}{4 D_R t}} \right) d\xi \qquad \text{(IV-B-8c)}$$

Obviously, when $x = 0$, then $f_1(x,t) = 0$.

Since $\text{erf}(z) = \frac{1}{\sqrt{\pi}} \int_{-\infty}^t e^{-z^2} dz$ and letting $\varphi(x) = \varphi_0$ (see eqn. IV-B-5b) and

$$\alpha = (\xi - x) \frac{1}{\sqrt{D_R t}} \qquad\qquad \beta = (\xi + x) \frac{1}{\sqrt{4 D_R t}} \qquad \text{(IV-B-9)}$$

we have

$$f_1(x,t) = \frac{\varphi_0}{\sqrt{\pi}} \int_{-\frac{x}{\sqrt{4 D_a t}}}^{+\infty} e^{-\alpha^2} d\alpha - \frac{\varphi_0}{\sqrt{\pi}} \int_{\frac{x}{\sqrt{4 D_a t}}}^{+\infty} e^{-\beta^2} d\beta$$

$$= \frac{\varphi_0}{\sqrt{\pi}} \int_{-\frac{x}{\sqrt{4 D_R t}}}^{+\frac{x}{\sqrt{4 D_R t}}} e^{-\eta^2} d\eta = 2\frac{\varphi_0}{\sqrt{\pi}} \int_0^{\frac{x}{\sqrt{4 D_R t}}} e^{-\eta^2} d\eta$$

$$= \varphi_0 \, \text{erf}\left(\frac{x}{\sqrt{4 D_R t}} \right) \qquad \text{(IV-B-10)}$$

2. Calculation of $f_2(x,t)$

We have to find the function $f_2(x,t)$, for solution of the following problem

$$\frac{\partial f_2}{\partial t} = D_R \frac{\partial^2 f_2}{\partial x^2} \qquad \text{(IV-B-3a)}$$

$$f_2(x,0) = 0 \qquad \text{for} \quad 0 < x < \infty \qquad \text{(IV-B-3b)}$$

$$f_2(0,t) = \varphi(t) \qquad \text{for} \quad 0 < t \qquad \text{(IV-B-3c)}$$

This is the same problem as the one solved in Appendix A (see eqn. IV-A-3 and IV-A-4). We discuss first the case in which $\varphi(t) = \varphi_0$, a constant (this is the conventional boundary condition of frontal analysis) and $f_2(x,t) = f_{2,0}(x,t)$. Let us define $\bar{f}_{2,0}$ as $\bar{f}_{2,0} = \varphi_0 - f_{2,0}$. Then, we have

$$\frac{\partial \bar{f}_{2,0}}{\partial t} = D_R \frac{\partial^2 \bar{f}_{2,0}}{\partial x^2} \tag{IV-B-11a}$$

$$\bar{f}_{2,0}(x,0) = \varphi_0 \qquad \text{for} \quad 0 < x < \infty \tag{IV-B-11b}$$

$$\bar{f}_{2,0}(0,t) = 0 \qquad \text{for} \quad 0 < t \tag{IV-B-11c}$$

The solution is given by

$$\bar{f}_{2,0} = \varphi_0 \mathrm{erf}\left(\frac{x}{\sqrt{4D_a t}}\right) \tag{IV-B-12}$$

and

$$f_{2,0} = \varphi_0 \left[1 - \mathrm{erf}\left(\frac{x}{\sqrt{4D_R t}}\right)\right] = \varphi_0 \mathrm{erfc}\left(\frac{x}{\sqrt{4D_R t}}\right) \tag{IV-B-13}$$

Assume now that φ is a function of time. From Duhamel principle and since $f_2(0,t) = \varphi(t)$, we have

$$f_2(x,t) = -\int_0^t \varphi(\tau) \frac{\partial f_{2,0}(t-\tau)}{\varphi_0 \partial \tau} d\tau \tag{IV-B-14}$$

where

$$\frac{\partial f_{2,0}(t-\tau)}{\varphi_0 \partial \tau} = \frac{\partial}{\partial \tau}\left[\mathrm{erfc}\left(\frac{x}{\sqrt{4D_R(t-\tau)}}\right)\right] \tag{IV-B-15a}$$

$$= \frac{\partial}{\partial \tau}\left[1 - \frac{2}{\sqrt{\pi}}\frac{-1}{2}\frac{x(-1)}{\sqrt{4D_R(t-\tau)^3}}e^{-\frac{x^2}{4D_R(t-\tau)}}\right] \tag{IV-B-15b}$$

$$= \frac{-x}{\sqrt{4D_R}(t-\tau)^{3/2}}e^{-\frac{x^2}{4D_R(t-\tau)}} \tag{IV-B-15c}$$

So, finally, we obtain

$$f_2 = \frac{x}{\sqrt{4\pi D_R}}\int_0^t \frac{\varphi(\tau)}{(t-\tau)^{3/2}}e^{-\frac{x^2}{4D_R(t-\tau)}} d\tau \tag{IV-B-16}$$

When $\varphi(\tau) = C_0(\tau)\exp\frac{u_R^2 \tau}{D_R}$, eqn. IV-B-16 becomes the same as eqn. IV-A-5.

3. Solution of the Generalized Breakthrough Problem

As explained in the introduction of this Appendix, the solution is the function $g(x,t) = f_1(x,t) + f_2(x,t)$. These two functions are given by eqns. IV-B-8c and IV-B-16, respectively. Accordingly, the solution of the problem is:

$$C(x,t) = \int_0^\infty \frac{\varphi(\xi)}{\sqrt{4\pi D_R t}}[e^{\frac{-(x-\xi)^2 \gamma}{4D_R t}} - e^{\frac{-(x+\xi)^2 \gamma}{4D_R t}}]d\xi +$$

$$\int_0^t \frac{x}{\sqrt{4\pi D_R t}}\frac{\psi(t)}{(t-\tau)^{\frac{3}{2}}}e^{\frac{-x^2 \gamma}{4D_R(t-\tau)}} d\tau \tag{IV-B-17}$$

This is equation IV-8 in section I-2.

SYMBOLS USED

The theory of chromatography involves different parts of physical chemistry which use their own conventional symbols. It has evolved a set of familiar notations. In several instances, complex equations have to be written. In an attempt to simplify the writing of these equations, we use here some symbols that group several conventional expressions. A few definitions are useful.

D_a stands for the coefficient of axial dispersion, resulting from the combined effects of molecular diffusivity and of convection in a packed bed. The former involves the contributions of the tortuosity of the bed and of its constriction, the latter that of eddy diffusion.

D_L stands for the apparent coefficient of axial dispersion, combining the effects of axial dispersion (see above) and of the mass transfer resistances. It is used only when the latter are small and can be considered as a minor contribution to the band profiles (equilibrium dispersive model).

D_m stands for the molecular diffusivity in the bulk solution or coefficient of diffusion.

k'_0 stands for the retention factor, with $k'_0 = Fa$ where $F = \varepsilon/(1 - \varepsilon)$ is the phase ratio and a is the slope of the linear isotherm or, under nonlinear conditions (see later Chapters), the initial slope of the isotherm. For the sake of simplifying complex equations, we set:

$$D_R = \frac{D_a}{1+Fa} = \frac{D_a}{1+k'_0}$$
$$u_R = \frac{u}{1+Fa} = \frac{u}{1+k'_0}$$
$$\gamma = 1 + Fa = 1 + k'_0$$

EXERCISES

1) Derive the PDE equation of the equilibrium–dispersive model of chromatography for a single component. Write the set of equations for two components.

2) Derive the solution of equations IV-2b and IV-3 using the Laplace transform.

3) Derive from equations IV-2b and IV-3 the first, second, and third-order moments of a chromatographic band, using the Laplace transform. Explain why the third moment, μ_3, is different from 0 for a Gaussian profile.

4) Derive the solution of the equilibrium–dispersive model of chromatography for a breakthrough profile (boundary conditions of frontal analysis;

first one component, then two components) and draw profiles corresponding to different values of the experimental conditions, using a computer.

5) The solution of the equation of the equilibrium–dispersive model for the elution of a wide rectangular pulse was derived in section I-3-d (eqn. IV-11). This solution could also be obtained as the sum of the solutions for the two successive Riemann problems corresponding to the following boundary condition:

$$C(0,t) = 0 \qquad \text{for} \quad t < 0$$
$$C(0,t) = C_0 \qquad \text{for} \quad 0 < t < t_p$$
$$C(0,t) = 0 \qquad \text{for} \quad t_p < 0$$

Show that the solution obtained is the same as eqn. IV-11.

6) Derive from eqns. IV-12 and IV-13 a first order approximation of the expression giving q when $k_f \to \infty$. Show that the effect of k_f is the same when $D = \frac{u^2 Fa}{k_f}$.

7) Compare eqns. IV-9 and IV-38 and show that the former is the limit of the latter when k_f tends toward infinity and the mass transfer resistances tends toward 0. Use the relationship between the time, the position of the mass center of the band, and its velocity.

LITERATURE CITED

[1] G. Guiochon, S. Golshan-Shirazi and A. M. Katti, *"Fundamentals of Preparative and Nonlinear Chromatography "*, Academic Press, Boston, MA, 1994.

[2] E. Wicke, *Kolloid Z.*, **86** (1939) 295.

[3] A. J. P. Martin and R. L. M. Synge, *Biochem. J.*, **35** (1941) 1358.

[4] L.C. Craig, *J. Biol. Chem.*, **155** (1944) 519.

[5] A. Klinkenberg and F. Sjenitzer, *Chem. Eng. Sci.*, **5** (1956) 258.

[6] M. Kendall and A. Stewart, *"The Advanced Theory of Statistics,"* 4th ed., Chas. Griffen, London, UK, 1977.

[7] I. Quiñones, C. M. Grill, L. Miller and G. Guiochon, *J. Chromatogr. A*, **867** (2000) 1-21

[8] J. C. Giddings, *"Unified Separation Science"*, J. Wiley, New York, NY, 1991.

[9] J. C. Giddings, *"Dynamics of Chromatography"*, M. Dekker, New York, NY, 1965.

[10] R. Aris and N. R. Amundson, *"Mathematical Methods in Chemical Engineering"*, Prentice Hall, Englewood Cliffs, NJ, 1973.

[11] J. J. van Deemter, F. J. Zuiderweg and A. Klinkenberg, *Chem. Eng. Sci.*,**271** (1956) 5.

[12] L. Lapidus and N. R. Amundson, *J. Phys. Chem.*, **56** (1952) 984.

[13] E. Kucera, *J. Chromatogr.*, **19** (1965) 237.

[14] M. Kubin, *Collect. Czech Chem. Commun.*, **30** (1965) 2900.

[15] E. Grushka, *J. Phys. Chem.*, **76** (1972) 2586.

[16] S. N. Chester and S. P. Cram, *Anal. Chem.*, **43** (1971) 1922.

[17] M. Goedert and G. Guiochon, *Anal. Chem.*, **45** (1972) 1188.

[18] C. Vidal-Madjar and G. Guiochon, *J. Chromatogr.*, **142** (1977) 61.

[19] B. G. Yew, E. C. Drum and G. Guiochon, *Proc. 4th Int. Conf. Constitutive Laws for Eng. Materials: Experiment, Theory, Computation and Applications*, Troy, NY, 1999, pp. 513.

[20] B. G. Yew, E. C. Drumm and G. Guiochon, *AIChE J.*, In Press.

[21] M. Sarker, A. M. Katti and G. Guiochon, *J. Chromatogr. A*, **719** (1996) 275.

[22] G. Guiochon, T. Farkas, H. Guan - Sajonz, J.-H. Koh, M. Sarker, B. J. Stanley and T. Yun, *J. Chromatogr. A*, **762** (1997) 83.

[23] G. Guiochon, E. Drumm and D. E. Cherrak, *J. Chromatogr. A*, **835** (1999) 41.

[24] G. V. Yeroshenkova, S. A. Volkov and K. I. Sakodynskii, *J. Chromatogr.*, **262** (1983) 19.

[25] L. Arnold, *"Stochastic Differential Equations"*, New York, NY, 1972.

CHAPTER V
SINGLE-COMPONENT
IDEAL NONLINEAR CHROMATOGRAPHY

Some of the most fruitful concepts in the theory of nonlinear chromatography arose from investigations into the ideal model, its solutions and its properties. This model deals directly with the most important consequences of the nonlinear behavior of the equilibrium isotherm. In this model, we assume that the column has an infinite efficiency and that there are no axial dispersion and no mass transfer resistances. Accordingly, no dissipative mechanisms are involved in this model nor will any be discussed in this Chapter, in contrast with the previous one. We assume that equilibrium is constantly and instantaneously achieved. We also assume in this chapter that the feed contains a single component, so there is only one mass balance equation. The more complex case of multicomponent separations will be considered in Chapter VIII. The essential feature of the ideal nonlinear model is the nonlinear behavior of the isotherm. This introduces some entirely new properties which are ignored in linear chromatography and which explain the profound difference between linear and nonlinear chromatography. These properties are discussed in detail in this Chapter.

In chromatography, (1) concentration changes made at the column inlet migrate along the column as waves; and (2) a concentration migrates as a wavelet, the amplitude of this wavelet being the concentration of the band at the corresponding position and time. The most important features of nonlinear chromatography are that (1) the migration velocity associated with each concentration or wavelet depends on this concentration; and (2), as a consequence of this dependence, self-sharpening takes place on one side (and one side only) of an elution band. Usually, this self-sharpening leads to the formation of a concentration shock[1] and causes the retention times of pulses to depend strongly on the sample size. Admittedly, the ideal model is a simplification of the actual chromatographic phenomenon. Still, it remains quite realistic in the numerous cases in which the highly efficient columns of modern HPLC are used under overloaded conditions, especially in the preparative separations of low molecular weight compounds, which have a high molecular diffusivity and, usually, exhibit relatively low mass transfer resistances. It has often been shown that in these cases the dissipative mechanisms have a nearly negligible influence on band profiles at high concentrations and that the experimental bands profiles obtained for large sample sizes are often relatively close to those predicted by the ideal model [1].

In Chapter VI, we will extend the discussion of the ideal model to the important case of reaction chromatography, a case in which a reaction takes place during the separation of the reaction products. The reaction causes a dampening of the concentration profile, although a concentration discontinuity remains an important feature of the solution of the ideal model in this case also.

I. ANALYSIS OF THE MATHEMATICAL MODEL OF NONLINEAR CHROMATOGRAPHY

The definitive solution of this problem, in pure mathematical terms, was given by James [2]. Major contributions to the solution of this problem were made first by Wilson [3] and DeVault [4], later by Amundson et al. [5-8] and Jacob et al. [9-11], and finally by Golshan–Shirazi et al. [12]. A comparison between the solutions of James [2] and of Golshan–Shirazi et al. [12] illustrates the gulf between mathematics and physical chemistry. A detailed review of all these publications and of numerous other important contributions is available [1].

---

[1]A shock is a stable discontinuity. A concentration discontinuity can be either a shock or an unstable discontinuity. When a discontinuity becomes unstable, it collapses into a diffuse boundary.

The mass balance equation of one compound in ideal, nonlinear chromatography is a hyperbolic partial differential equation:

$$F\frac{\partial q}{\partial t} + \frac{\partial C}{\partial t} + u\frac{\partial C}{\partial x} = 0 \qquad \text{(V-1a)}$$

where q and C are the concentrations in the solid and liquid phase at equilibrium, $F = (1 - \varepsilon)/\varepsilon$ is the phase ratio, with ε the column total porosity, and u is the mobile phase velocity. The ideal model assumes constant and instantaneous equilibrium between the two phases everywhere in the column, with no axial dispersion and no mass transfer resistances. Accordingly, in eqn. V-1a, the concentrations q and C are related by the isotherm equation, $q = f(C)$, where $f(C)$ is the isotherm equation (see Chapter II, section I). So, we can write that $\frac{\partial q}{\partial t} = \frac{df(C)}{dC}\frac{\partial C}{\partial t}$ and rewrite eqn. V-1a accordingly

$$\frac{1 + F\frac{dq}{dC}}{u}\frac{\partial C}{\partial t} + \frac{\partial C}{\partial x} = 0 \qquad \text{(V-1b)}$$

Equations V-1a and V-1b are wave equations. The corresponding wave is a *kinetic wave*. Thus, the propagation of a chromatographic band can be regarded as the propagation of a concentration wave, $C(x, t)$. The general wave equation is

$$\frac{\partial^2 C}{\partial t^2} - a^2\frac{\partial^2 C}{\partial x^2} = 0 \qquad \text{(V-2a)}$$

where a is the velocity of propagation of the wave. In operator form, this equation is written:

$$\left(\frac{\partial^2}{\partial t^2} - a^2\frac{\partial^2}{\partial x^2}\right)C = 0 \qquad \text{(V-2b)}$$

We may separate the operator in this equation into two operators and write

$$\left(\frac{\partial}{\partial t} + a\frac{\partial}{\partial x}\right)\left(\frac{\partial}{\partial t} - a\frac{\partial}{\partial x}\right)C = 0 \qquad \text{(V-2c)}$$

so the equation

$$\left(\frac{\partial}{\partial t} + a\frac{\partial}{\partial x}\right)C = 0 \qquad \text{(V-3)}$$

is a branch of the wave equation (eqn. V-2a). Equation V-3 is equivalent to the mass balance equation in eqn. V-1b, with the propagation velocity of the concentration wave being a in eqn. V-3 and $u/(1+F\frac{df}{dC})$ in eqn. V-1b. Under linear conditions, the isotherm is $f = aC$ and we have $\frac{df}{dC} = a = $ constant, so the propagation velocity is constant. In nonlinear chromatography, however, since dq/dC is a function of C, the propagation velocity associated with concentration C is also a function of C.

Equation V-1b can also be written as

$$\frac{\partial Q(C)}{\partial t} + \frac{\partial C}{\partial x} = 0 \qquad \text{(V-4)}$$

where $Q(C) = [C + Ff(C)]/u$. Equation V-4 is called the *hyperbolic conservation law* in mathematics and it is of great importance. The nonlinear term in this equation is the time differential. Equation V-1 is the fundamental equation of nonlinear chromatography.

1. Characteristics analysis, shocks, and shock velocity

Let us consider a certain point of a concentration wave (*i.e.*, a concentration slice), $C(x, t)$, moving along a column. The point, x, t, at which the concentration is equal to C moves along the column and, since the concentration remains constant in this point, we have for this concentration

$$\frac{dC}{dt} = 0 \qquad \text{(V-5)}$$

This total differential is equal to

$$\frac{dC}{dt} = \frac{\partial C}{\partial t} + \frac{\partial C}{\partial x}\frac{dx}{dt} \qquad \text{(V-6)}$$

Comparing eqns. V-1b and V-6, we see that, when the profile moves, we must have for each value of C

$$u_z = \frac{dx}{dt} = \frac{u}{1 + F\frac{df}{dC}} \qquad \text{(V-7)}$$

hence, u_z is the velocity associated with the concentration C. So, eqn. V-1b is equivalent to the combination of eqns. V-5 and V-7 and, for a given concentration, the RHS of eqn. V-7 is constant and

$$x - \frac{u\,t}{1 + F\frac{df}{dC}} = \text{constant} \qquad \text{(V-8)}$$

Equation V-8 is called *the characteristics equation*. The velocity u_z given by eqn. V-7 is the slope of the corresponding characteristic line. In physics, this corresponds to the velocity of a wavelet whose concentration is C. Since $\frac{df}{dC}$ is a function of C in nonlinear chromatography, so also is u_z. This means that wavelets of different concentrations propagate at different velocities. *This phenomenon is the fundamental reason why band profiles in nonlinear chromatography depend so much on the concentration range sampled by the band profile.* The concentration dependence of the velocity associated with a concentration is particularly important when the isotherm curvature is strong. This causes either the dispersion or the sharpening of the concentration profile, depending on the side of the profile considered and on the sign of the isotherm curvature. For a convex upward isotherm, *e.g.*, for a Langmuirian[2] isotherm, the curvature is convex upward and we have $\frac{d^2 f}{dC^2} < 0$, i.e. $\frac{df}{dC}$ decreases with increasing C. Therefore, u_z increases with increasing C. The velocity associated with a concentration increases with increasing concentration. The phenomenon of self-sharpening of the front is illustrated in Figure V-1. It takes place on the front boundary of a concentration wave $C(x, t)$ that propagates along the column. A converse effect, the self-sharpening of the rear part of the profile, takes place in the case of an anti-Langmuirian isotherm, for which the curvature is convex downward, $\frac{d^2 f}{dC^2} > 0$, and the velocity associated with a concentration decreases with increasing concentration.

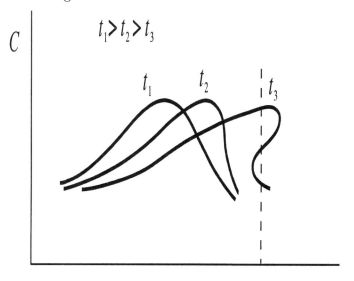

Figure V-1   Propagation of a Nonlinear Wave. The profiles at times t_1 and t_2 illustrate the self-sharpening of the front during its migration. The third profile, at time t_3, is a physical impossibility: High concentrations cannot pass low ones.

---

[2]We call Langmuirian an isotherm that is convex upward in the region around the origin, like the Langmuir isotherm. In practice, most isotherms are Langmuirian.

   The first two profiles in Figure V-1 illustrate the self-sharpening evolution
of the front of a band. The third curve corresponds to the hypothetical case
in which high concentrations, moving faster, attempt to pass low concentrations, which move more slowly. This assumption is a physical impossibility,
however, since there cannot be three different values of the concentration
of a compound at the same time and in the same position. Physically, it is
impossible for the band profile to be similar to the third curve in Figure V-1.
So, since the fast high concentrations cannot pass the slow low concentrations, they all pile up and a discontinuity builds up [3]. This discontinuity,
being stable, is a shock. The need to explain this type of discontinuous
solution of the mass balance equation will lead us to investigate the *weak
solutions* of hyperbolic, quasi-linear PDEs (Section II).
   A physical explanation of the very steep profiles observed in high concentration bands is sometimes given. Although it appears to be valid in
practice, it is important to underline that this solution relies mostly on the
lack of realism of the ideal model. Therefore, it can explain the band profiles
actually observed, but not the discontinuous nature of the band profiles that
are the solutions of the system of equations of the ideal model. The efficiency
of an actual column is never infinite. There is always a certain amount of
axial dispersion and the mass transfer resistances are never entirely negligible. We know from Fick's law [13,14] that the mass flux caused by diffusion
depends on the concentration gradient. Therefore, there will be two opposite effects on a self-sharpening front, those of thermodynamics which tend
to sharpen the front of the profile and to create a high concentration gradient and those of dispersion and the mass transfer resistances which tend
to relax this gradient. The effects of both dispersion and the mass transfer
resistances increase with increasing concentration gradient. When the self-
sharpening effect reaches a certain stage, the diffusional mass flux becomes
very large and, eventually, the dispersive effects balance the self-sharpening
caused by the nonlinear behavior of the isotherm. A very sharp front profile
forms, with a very thin layer in which the concentration varies rapidly (see
Chapter VII, Section V - Shock Layer Analysis). The slope of this front is
(as a first approximation) proportional to the inverse of the column HETP
under linear conditions. If the axial dispersion were really negligible (*i.e.*,
with $D = 0$), the thickness of this layer would approach zero. Then, a true
concentration "discontinuity" would arise. This rationale is at the basis of
the shock layer theory (Chapter VII, Section V).
   Note also that since the slopes of the characteristic lines corresponding to
different concentration wavelets are different, these lines intersect, as shown
in Figure V-2. The intersection point of two characteristic lines, corresponding to two different concentrations which originate from different times, corresponds to a multi-values solution, such as the one represented by the third

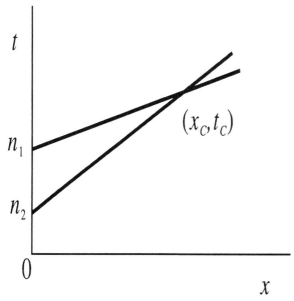

t

n_1

n_2

(x_C, t_C)

0

x

Figure V-2   Intersection of Characteristic Lines. A shock takes place when two characteristics intersect. A concentration located at a given position, at a certain time, cannot propagate at two different velocities.

profile in Figure V-1. The intersection of the characteristic lines also means that there are different propagation velocities in the same time and at the same space point. Like a multi-values solution, this is a physical impossibility. In the practical case, this can only correspond to the discontinuity (and the discontinuity follows a migration law that is different from that of the wave, see next section). Obviously, these explanations regarding the front of the peak and the intersection of the characteristic lines are based on the assumption of a convex upward isotherm (*e.g.*, a Langmuir isotherm). If the isotherm is convex downward, *i.e.*, toward the mobile phase concentrations, the wavelets in the front of the chromatographic band in Figure V-1 will disperse, the low concentrations moving faster than the high ones. The characteristic lines originating from the band front will diverge instead of converge. At the same time, the rear of the band will exhibit a discontinuity, as illustrated later in Figure V-6. For the sake of convenience, we assume in most of the following discussions that the isotherm is convex.

The analysis given above shows that there will almost always be a concentration discontinuity in the profiles obtained in ideal, nonlinear chromatography. The only cases in which the solution of the model will not exhibit a shock are those in which a shock does not have time to build up, *e.g.*, because of the shape of the boundary condition (see subsection I-2) and of the column length. The propagation velocity of the discontinuity, however, is different from the velocity of the characteristics. Where there is a discontinuity, the partial differential equation cannot be satisfied because the partial differentials are not defined and the mass balance equation cannot be used to discuss the properties of the discontinuity. This discussion should be made using an integral of the profile around the discontinuity [15]. The mass conservation relation corresponding to eqn. V-4 in the case of a discontinuous profile is given by

$$\frac{d}{dt}\int_{x_1}^{x_2} Q\,dx = C_2 - C_1 \tag{V-9}$$

where C_1 and C_2 are the values of the concentration C in the positions x_1 and x_2, respectively, at the same time and Q is defined in eqn. V-4. If X_s is the location of the concentration discontinuity at this time, eqn. V-9 can be rewritten as

$$\frac{d}{dt}\int_{x_1}^{X_s(t)} Q\,dx + \frac{d}{dt}\int_{X_s(t)}^{x_2} Q\,dx = C_2 - C_1 \tag{V-10a}$$

The fundamental importance of eqn. V-10 stems from the fact that the derivation involves not only the integrated function $Q(x,t)$ but also the integration boundary, $X_s(t)$. From the definition of derivatives, the first term in eqn. V-10 can be written as

$$\frac{d}{dt}\int_{x_1}^{X_s(t)} Q\,dx = \lim_{\Delta t \to 0} \frac{1}{\Delta t}\left[\int_{x_1}^{X_s(t)}(Q(x,t+\Delta t) - Q(x,t))\,dx + \right.$$
$$\left.\left(\int_{x_1}^{X_s(t+\Delta t)} Q\,dx - \int_{x_1}^{X_s(t)} Q\,dx\right)\right]$$
$$= \int_{x_1}^{X_s(t)} \frac{\partial Q}{\partial t}\,dx + Q(x_1)\frac{dX_s}{dt} \tag{V-10b}$$

Finally, eqn. V-10a becomes

$$\int_{x_1}^{X_s(t)} \frac{\partial Q}{\partial t}\,dx + Q_1\frac{dX_s}{dt} + \int_{X_s(t)}^{x_2} \frac{\partial Q}{\partial t}\,dx - Q_2\frac{dX_s}{dt} = C_2 - C_1 \tag{V-11}$$

where Q_1 and Q_2 are the values of Q at x_1 and x_2, respectively. This equation remains valid when x_1 and x_2 both tend toward X_s and $x_2 - x_1$ toward zero. Under these conditions, the two integrals in eqn. V-11 tend toward 0 and, at the limit, we have

$$u_s = \frac{dX_s}{dt} = \frac{C_2 - C_1}{Q_2 - Q_1} = \frac{u}{1 + F\frac{[f]}{[C]}} \tag{V-12}$$

where $[f] = \Delta q$ and $[C] = \Delta C$ denote the differences between the values of f and C, respectively, on the two sides of the shock and u_s is the velocity of the shock. Equation V-12 gives the relationship between the velocity of the shock and its amplitude in the two phases in equilibrium (obviously, the assumption of equilibrium at the shock is physically impossible in an actual column but it is a logical consequence of the stricture of the ideal model). Equation V-12 results from the principle of mass conservation. For this reason, in chromatography, it plays a role similar to that of the Rankine–Hugoniot condition in hydrodynamics.

If we assume that $q = f(C)$ is convex upward (*i.e.*, that we have a Langmuirian isotherm), the following inequation is valid at the front of the band

$$u_{z,left} > u_s > u_{z,right} \qquad \text{(V-13)}$$

Equation V-13 is easy to demonstrate. From eqn. V-4, from the definition of Q, and from eqn. V-7, *i.e.*, from the definition of u_z, it follows simply that

$$u_z = \frac{dC}{dQ}$$

From eqn. V-13, it follows also that

$$\left(\frac{dQ}{dC}\right)_{left} < \frac{[Q]}{[C]} < \left(\frac{dQ}{dC}\right)_{right}$$

The combination of eqns. V-7 and V-10 with these inequations leads to the relationship V-13 above. Equation V-13 is called the developing condition or Lax entropy condition. A discontinuous solution that satisfies the Lax condition is stable and this discontinuity is called a *"shock"*. The corresponding velocity is called the shock velocity. Discontinuities that do not satisfy this condition may possibly exist, *e.g.*, as a consequence of certain boundary conditions (for example in the case of a rectangular pulse injection), but they are not stable and collapse immediately ("into a flight of characteristics" [1]).

The shock velocity (eqn. V-12) can also be obtained through a different derivation. Since the compound considered is retained, the shock velocity is smaller than the mobile phase velocity, u. So, the mobile phase passes continuously through the shock. It carries with it a certain amount of compound. The amount of this compound which enters into the shock during time Δt is $S\varepsilon(u - u_s)C_{left}\Delta t$, with S column cross-section area. The amount of the compound that exits from the shock, in front of it, is $S\varepsilon(u - u_s)C_{right}\Delta t$. The difference or net mass of compound accumulated in the shock during that time is

$$\Delta M = S\varepsilon(u - u_s)(C_{left} - C_{right})\Delta t \qquad \text{(V-14)}$$

The net mass of component that is transported by the shock during the same time because the shock moves with respect to the stationary phase is

$$\Delta M = S(1 - \varepsilon) u_s \Delta t (q - q_s) \tag{V-15}$$

Because of the principle of mass conservation, the expressions giving ΔM in eqns. V-14 and V-15 must be equal, hence

$$\varepsilon (u - u_s)(C_{left} - C_{right}) = (1 - \varepsilon) u_s (q - q_s)$$

or

$$u_s = \frac{u}{1 + F \frac{[q]}{[C]}} \tag{V-16}$$

This is the same equation as eqn. V-12, since $F = (1 - \varepsilon)/\varepsilon$.

So long as there is a shock in the band profile, it will propagate at the velocity u_s. This velocity is a function of the ratio of the amplitudes of the concentration jumps in the two phases. In elution chromatography, the band broadens and dilutes during its migration. Its height decreases and the shock velocity decreases (see later, subsection 4, *Development of the Band Profile*). In most cases, the peak maximum is at the top of a shock (this will be demonstrated later). Therefore, the calculation of the retention time of the band or elution time of its maximum will be more complex in nonlinear than in linear chromatography. By contrast, in frontal chromatography, the discontinuity is formed at the injection of the concentration step into the column, it is stable if $\Delta C > 0$ for a Langmuirian isotherm, and its height is constant. So is its velocity. Therefore, the retention time of the breakthrough front can be calculated easily, with $t_R = L/u_s$. In staircase frontal analysis, the concentrations of the successive steps are $C_1, C_2, \cdots, C_n$, the corresponding concentrations in the stationary phase are $q_1, q_2, \cdots, q_n$ and the retention times are $t_{R,1}, t_{R,2}, \cdots, t_{R,n}$, respectively. These retention times are given by the following relationships, easily derived from eqn. V- 16. Assuming $C_0 = q_0 = 0$, we have

$$t_{R,1} = t_0 \left(1 + F \frac{q_1}{C_1} \right)$$

$$t_{R,2} = t_0 \left(1 + F \frac{q_2 - q_1}{C_2 - C_1} \right)$$

$$\vdots$$

$$t_{R,n} = t_0 \left(1 + F \frac{q_n - q_{n-1}}{C_n - C_{n-1}} \right)$$

where $t_0 = L/u$ is the hold-up time. If C_i and $t_{R,i}$ $(i = 1, 2, \cdots, n)$ are measured in an experiment, it is easy to derive from the experimental data the corresponding value of q, $i.e.$, q_i. This result is the basis of the most popular method of determination of equilibrium isotherms (it is accurate, reliable, and reasonably easy to implement; unfortunately it is expensive and wasteful of chemicals). Technical difficulties arise, obviously, if the equilibrium isotherm happens to have an inflection point [1].

2 - The Different Velocities Encountered in Chromatography

It is useful to elaborate at this stage on a statement made in the previous section, that there are three different velocities that should be considered in nonlinear chromatography. These velocities are:

(1) The actual velocity of the mobile phase, u. There are, unfortunately, three different possible definitions of this velocity: (a) the superficial velocity, F_v/S, where F_v is the volume flow rate of the mobile phase and S the geometrical cross-section of the column; (b) the actual average velocity of the stream percolating through the bed, $F_v/(S\varepsilon_e)$, where ε_e is the external porosity of the bed; and (c) the chromatographic velocity, $L/t_0 = F_v/(S\varepsilon_T)$, where ε_T is the total bed porosity. In most cases, chromatographers use the last definition. Chemical engineers tend to prefer one of the first two definitions.

(2) The characteristics velocity, $u_z(C)$, given by eqn. V-7 above, $i.e.$, the migration velocity associated with concentration C on the continuous part of the concentration wave.

(3) The shock velocity, u_s, given by eqn. V-12. It is the migration velocity of the discontinuity of the concentration wave. It is also the material velocity, the velocity of the material that is contained in the infinitely narrow slice of the profile at concentration C.

It is important to realize the physical difference between the shock velocity and the velocity associated with a concentration. If we consider the point at concentration C on the diffuse part of the profile of a zone moving along a column, we can use eqn. V-7 to calculate the trajectory of this concentration, hence the position of this concentration on the zone profile in the column at any future time, and its elution time (see later, eqn. V-28a). However, the migration of a given element of mass of the solute follows a different law. The mass of compound found in the slice of thickness δx, between concentrations C and $C + dC$ in the profile at time t is not in the slice of the same thickness nor of the same concentration of the profile obtained at a later time: the concentration $C + dC$ moves faster than the concentration C (eqn. V-7), the slice broadens during its migration, hence it dilutes and the material is now found in a broader slice corresponding to a lower concentration. The actual slice of solute moves at the velocity u_s (eqn. V- 12), a velocity that is lower than u_z for a Langmuirian isotherm [1,9]. This broadening and dilution of the slice during the migration of the zone explains the changes in the whole profile (see subsection 4).

Finally, the difference between the velocities u_z, the velocity associated with a concentration, or wave velocity, or velocity of the information, and u_s, the shock velocity, or velocity of mass transport, is illustrated by the Helfferich Paradox [16]. Assume that a column is equilibrated with a stream of a solution of one compound in the mobile phase, at constant concentration. At the column exit, we have two detectors. One detector is nonselective, *e.g.*, it is a UV or a refractive index detector. The other detector responds only to one isotope of an atom, *e.g.*, it is a mass spectrometer locked on M = 13 or a radioactivity detector. Now suppose that the compound and the mobile phase contain only ^{12}C atoms. If, on the plateau concentration, we inject a pulse of a solution of the same compound but made only of ^{13}C or ^{14}C, the signals detected by the two detectors will take place at different times. The selective detector will give a signal corresponding to the elution of the actual amount of the labeled compound that was injected. This signal is recorded when the labeled molecules actually elute out of the column. It travels at the shock velocity [1,16,17]. The nonselective detector records a signal that corresponds to the perturbation of the equilibrium of the compound in the column, at the plateau concentration. It travels at the velocity u_z [1,16]. These two signals are obviously recorded at different times (we assume that the physicochemical properties of the compound are not affected by isotopic substitution). The consequences of the Helfferich Paradox have recently been illustrated by a most informative series of experiments [18].

3 - Time Needed to Form the Shock

One of the characteristic properties of nonlinear systems is that, under certain conditions, during the migration of a band, a concentration discontinuity can arise and grow on a profile that was initially continuous (*e.g.* that resulted from a continuous injection profile) while, under other experimental conditions, a discontinuity can collapse into a continuous profile. For example, in the case of a rectangular injection of finite length, the boundary condition has two discontinuities, one in the front, the other in the rear. With a nonlinear isotherm, only one of these discontinuities is stable [1,2,7,8,9,11]. It becomes a shock and propagates along the column. Usually, the front discontinuity is stable (Langmuirian isotherm).[3] The other discontinuity becomes a continuous profile. It is important to understand that, in this case, the continuous rear profile corresponds to the second discontinuity, the one that constitutes the rear of the rectangular injection pulse. The front of this pulse is usually taken as initial time $t = 0$. Thus, the rear discontinuity (and the rear, diffuse profile) is injected at time $t = t_p$, t_p being the width of the rectangular injection pulse. Because the top of the rear diffuse profile moves faster (at the velocity u_z) than the top of the front shock (at the velocity u_s), it will take a certain time for the injection plateau, between the two discontinuities, to be eroded away and disappear. This question will be discussed in the next subsection. We are concerned here with the spontaneous formation of a shock during the evolution of a continuous profile.

If the injection profile is not a rectangular pulse but a continuous profile, a concentration shock will eventually form but this will not take place instantaneously. The continuous front part of profile is going to become steeper and steeper, to sharpen, while is propagates. For this reason, this effect is called the self-sharpening of the front. An important practical question then arises, how long does it take for the shock to form? Obviously, the answer to this question depends in part on the shape of the boundary condition. We discuss it here in the case of a trapezoidal boundary condition (see Figure V-3). The same rationale would apply to any other continuous boundary condition, although the algebraic calculations might be impossible to carry out to a closed-form solution. Let us assume the injection of a concentration plateau, as in frontal chromatography, except that this plateau does not begin with a concentration jump but, instead, with a linear concentration gradient, as shown in Figure V-3. This new boundary condition can be written

$$C(0,t) = \begin{cases} C_0 t/t_h & t \le t_h \\ C_0 & t > t_h \end{cases} \qquad \text{(V-17)}$$

with t_h defined as the time when the concentration becomes constant and equal to C_0. Note that, in practice, the front injection boundary looks more often like an error function profile while the rear injection boundary tend to adopt an exponential profile [19].

Assuming a Langmuir isotherm, $f(C) = aC/(1 + bC)$, we have

$$\frac{df}{dC} = \frac{a}{(1 + bC)^2} \qquad \text{(V-18)}$$

---

[3]Note that the opposite is true in the case of an anti-Langmuirian isotherm. The front discontinuity is unstable and becomes a continuous boundary while the rear discontinuity propagates as a stable shock.

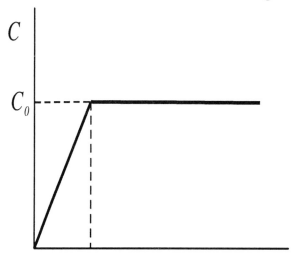

C

C_0

t

Figure V-3   Nonrectangular Injection Function with a Linear Gradient Ramp

In this particular case, the mass balance equation of the ideal model (eqn. V-1b) becomes

$$\frac{1}{u}\left[1 + \frac{Fa}{(1+bC)^2}\right]\frac{\partial C}{\partial t} + \frac{\partial C}{\partial x} = 0 \qquad (V\text{-}19)$$

A characteristic line is a straight line in the (x, t) plane, issued from a point of the boundary condition and having a slope given by eqn. III-43. The general equation of the characteristic line, which starts from a point at $x = 0$ (*i.e.*, on the injection profile or boundary condition), at time $t = \eta t_h$, $(0 < \eta < 1)$, is

$$t = \eta t_h + \left(\frac{1}{u} + \frac{Fa}{u(1+b\eta C_0)^2}\right)x \qquad (V\text{-}20)$$

The shock begins to form as soon as two different characteristic lines, having this general equation, intersect for the first time. The condition for this intersection is that the two characteristic lines have a common point of coordinates x, t. In other words, there must be two values, η_1 and η_2, of η such that we have

$$\eta_1 t_h + \left(\frac{1}{u} + \frac{Fa}{u(1+b\eta_1 C_0)^2}\right)x = \eta_2 t_h + \left(\frac{1}{u} + \frac{Fa}{u(1+b\eta_2 C_0)^2}\right)x \qquad (V\text{-}21)$$

Solving eqn. V-21 for x and porting the solution into eqn. V-20 gives the coordinates of the intersection point of these two characteristics. These are

$$x_c = \frac{ut_h(1 + \eta_1 bC_0)^2 (1 + \eta_2 bC_0)^2}{FabC_0(2 + (\eta_1 + \eta_2)bC_0)} \tag{V-22}$$

$$t_c = \eta_1 t_h + \left(\frac{1}{u} + \frac{Fa}{u(1 + b\eta_1 C_0)^2}\right) \left\{\frac{ut_h(1 + \eta_1 bC_0)^2 (1 + \eta_2 bC_0)^2}{FabC_0(2 + (\eta_1 + \eta_2)bC_0)}\right\} \tag{V-23}$$

When η_1 and η_2 tend toward a common value, η, these equations give the coordinates of the intersection point of two characteristic lines which are infinitely close to each other. These limit equations are

$$x_c = \frac{ut_h(1 + \eta bC_0)^3}{2FabC_0} \tag{V-24}$$

$$t_c = \eta t_h + \left[\frac{1}{u} + \frac{Fa}{u(1 + bC_0\eta)^2}\right]\frac{ut_h(1 + \eta bC_0)^3}{2FabC_0} \tag{V-25}$$

These two equations define a curve, $x_c(\eta), t_c(\eta)$, which is the envelope of the characteristic lines (see Figure V-4). From eqns. V-24 and V-25, it is easy to prove that the ratio $\frac{dx_c}{d\eta}/\frac{dt_c}{d\eta}$ is the direction of the characteristic lines, $\left(\frac{1}{u} + \frac{Fa}{u(1+b\eta C_0)^2}\right)^{-1}$: it coincides with eqn. V-20. Both x_c and t_c increase with increasing η. So, the envelope begins at the point corresponding to $\eta = 0$, hence of coordinates

$$x_{0,s} = \frac{ut_h}{2FabC_0} \tag{V-26}$$

$$t_{0,s} = \frac{(1 + Fa)t_h}{2FabC_0} \tag{V-27a}$$

The shock begins to form at time $t = t_{0,s}$ and location $x = x_{0,s}$ given by eqns. V-27 and V-26, respectively. These points correspond to $\eta = 0$, hence to $C = 0$. The shock always begins to form on the base line, at the base of the peak. At that moment, the profile is a curve that has a vertical tangent and the initial amplitude of the shock is 0. Then, the shock amplitude increases until the shock height is equal to the peak height. The time $t_{0,s}$ is called the relaxation time of the shock formation. This time is proportional to t_h (i.e., it is inversely proportional to the slope of the concentration gradient at the front of the boundary condition). It decreases with increasing value of the term bC_0 that characterizes the degree of deviation of the isotherm from linear behavior under the specific experimental conditions. If $t_h = 0$, we have a rectangular pulse injection and the front shock is stable, so it is formed instantaneously. If $t_h \neq 0$, the relaxation time is finite as long as

t

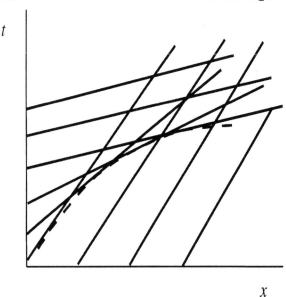

x

Figure V-4   The Characteristics and their Envelope in the x, t Plane

the equilibrium behavior is nonlinear. However, if bC_0 tends toward 0, the relaxation time increases indefinitely. Eventually, for a small but finite value of bC_0, the relaxation time for the formation of the shock becomes longer than the retention time of the breakthrough curve and the shock does not have time to form before the moment when the profile leaves the column.

This calculation can be applied to the determination of the time after which the strong solvent will break through the column in gradient elution chromatography. In this method, the composition of the mobile phase varies with time. Usually, a strong solvent is added to the weak solvent at a concentration that increases linearly with passing time. The retention factors of the sample components decrease rapidly with increasing strong solvent concentration and this method allows the elution in a reasonable time of strongly retained components. In many cases, the strong solvent is retained on the stationary phase (the retention factor of methanol in pure water is of the order of 1, that of acetonitrile is around 10). At the concentrations typically used in gradient elution, the isotherm of the strong solvent is nonlinear. Therefore, the gradient profile of the strong solvent, which is linear at the column inlet, becomes delayed and its initial slope increases progressively until a shock appears on its front. As time passes, the shock height increases and the shock velocity increases. The breakthrough time of the strong solvent is obtained from eqn. V- 27a which can be rewritten as:

$$t_{0,s} = \frac{1 + k'_s}{2\, k'_s\, b_s\, \beta} \qquad \text{(V-27b)}$$

where β is the gradient slope (equal to C_0/t_h in eqn. V-27a), k'_s is the retention factor of the strong solvent and b_s the second coefficient of its Langmuir isotherm.

In principle, the same calculation method as discussed above can be used to derive the time when the shock appears on a continuous profile with more realistic but more complex continuous boundary conditions, e.g., a Gaussian or an error function profile. Similarly, other isotherm models could be used. Unfortunately, the intermediate mathematical expressions become too complex to be tractable and the relaxation time for the formation of the shock cannot be given under closed form.

4 - Development of the Band Profile (Rectangular Pulse Injection)

One of the most important features of nonlinear chromatography is that a migration velocity (u_z, eqn. V-7) is associated with each concentration and that this velocity is a function of the concentration. A direct consequence of this property is that the shape of the profile changes continuously during its migration along the column. One side of the profile is self-sharpening. Even if the injection is a continuous boundary condition and the profile of the peak is also continuous at the beginning of its migration, sooner or later a shock forms as the band migrates along the column (see previous section). The other side of the profile is a diffuse boundary that becomes less and less steep as time passes.

If the isotherm is convex upward, i.e., toward the concentrations in the stationary phase like the Langmuir isotherm, its curvature is negative. The velocity u_z associated with the concentration C increases with increasing concentration because the first-order differential of the isotherm, $\partial f/\partial C$, decreases monotonically with increasing concentration, the second-order differential being always negative for a Langmuirian isotherm ($\partial^2 f/\partial C^2 = -2ab/(1+bC)^3$ for the Langmuir isotherm). Then, the front of the band is self-sharpening, a front shock is formed eventually, and the rear of the band is a diffuse boundary because the velocity associated with a concentration decreases with decreasing concentration and tends toward $u/(1+k'_0)$. In the case of an anti-Langmuirian isotherm, which is convex toward the axis of the mobile phase concentrations, the curvature is positive, $\partial^2 f/\partial C^2$ is positive, the velocity associated with a concentration decreases with increasing concentration and the converse is true: the peak front is a diffuse boundary while its rear is self-sharpening and, eventually, becomes a concentration shock.

When the injection profile is the conventional wide rectangular pulse, there are two discontinuities. However, the peak of a pure compound can

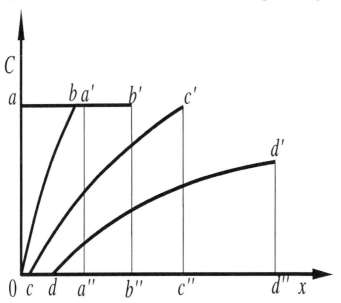

Figure V-5   The Processes of Formation and Collapse of Shocks in the Case of a Rectangular Pulse injection. Successive profiles are shown: Oaa'a" (rectangular injection), Obb'b" (the rear discontinuity has collapsed and the plateau is shrinking); cc'c" (the injection plateau is reduced is reduced to just one point), and dd'd".

have only one stable discontinuity [1,2,7,8,9,11]. Only one of the two discontinuities of the injection profile can become a stable shock. The other one becomes a diffuse profile. The process of this second discontinuity turning into a diffuse front is called the collapse of this discontinuity. The process of such a collapse is illustrated in Figure V-5 which shows the progressive evolution of the band after injection. In this case, the equilibrium isotherm is convex upward. The injection profile is shown in Oaa'a". The front shock of the injection (a'a") is stable. However, point a is unstable because the rear discontinuity is unstable, so Oa is actually a limit. The concentration increases gradually along Oa, albeit steeply. Since the higher concentrations are associated with higher velocities, the form of the band changes progressively, spreading in the process. The velocity of each concentration wavelet is given by eqn. V-7 and, accordingly, its retention time is given by

$$t(C) = t_p + \frac{L}{u_z(C)} \qquad \text{(V-28a)}$$

where t_p is the width of the rectangular pulse injection and is, consequently, the injection time of the rear shock of the rectangular pulse, the discontinuity that collapses into the diffuse rear boundary of the band (line Ob). The shape of the diffuse rear of the peak depends on u_z and, consequently, on the isotherm equation. From eqn. V-7, the diffuse boundary of the peak is given by

$$t(C) = t_p + \frac{L}{u}\left(1 + F\frac{df}{dC}\right) \tag{V-28a}$$

This equation was first derived by DeVault [4] and is the basis for the ECP method (elution by characteristic points) of derivation of isotherms [20]. Note that, in all its practical applications, this method suffers from a model error because it applies the ideal model, which assumes that there is absolutely no axial dispersion, to an actual band profile, which is always somewhat dispersed by axial diffusion, eddy diffusion, and the mass transfer resistances.

For a convex upward isotherm, the longest retention time is associated with the concentration $C = 0$ that is associated with the lowest velocity. It is given by

$$t(0) = t_p + \frac{L}{u}(1 + Fa) = t_p + t_0(1 + Fa) = t_{R,0} \tag{V-29}$$

where $t_0 = L/u$ is the retention time of an unretained tracer. The time $t_{R,0}$ is the retention time observed in linear chromatography. Beyond this time, the elution is entirely finished. This property results from the hyperbolic character of the mass balance equation of the ideal model.

As we show later (Chapter VII, section V), a true discontinuity never forms in actual chromatography. Axial dispersion is never entirely negligible and the elution of a true shock never actually takes place. Instead of a shock, a special equilibrium layer, the shock layer, forms. A true concentration discontinuity cannot ever take place but the concentration varies very rapidly during the passage of the shock layer, which moves at the same velocity as the shock and has a thickness depending on both the ratio $\Delta f/\Delta C$ of the shock amplitudes in the two phases and the column HETP, i.e., the coefficient of axial dispersion. Because the properties of the shock layer are so important in practice and so closely related to those of the shock, the study of the latter is most important.

When a rectangular band (height $C = C_0$, width t_p) is injected and the isotherm is given by the Langmuir equation $q = f(C) = aC/(1 + bC)$, the front discontinuity of the injection becomes a stable shock that migrates at the velocity $u_s(C_0) = u/(1 + Fa/(1 + bC_0))$. It reaches the point of abscissa z at the time:

$$t_s = \frac{z}{u_s} = \frac{z}{u}\left(1 + \frac{\Delta f}{\Delta C}\right) = \frac{z}{u}\left(1 + \frac{f}{C}\right) = \frac{z}{u}\left(1 + \frac{Fa}{1 + bC_0}\right) \tag{V-30a}$$

(see eqn. V-12 and note that here $\Delta f/\Delta C = f/C = q_0/C_0$ since the shock extends from the base line to the injection plateau at $C = C_0$) Behind this shock, the band profile has a plateau region at $C = C_0$, followed by the rear diffuse boundary. The point of this continuous profile at the top of this boundary moves at the velocity $u_z(C_0) = u/(1 + Fa/(1 + bC_0)^2)$. It reaches the point of abscissa z at the time:

$$t_z = t_p + \frac{z}{u_z} = t_p + \frac{z}{u}(1 + \frac{df}{dC}) = t_p + \frac{z}{u}(1 + \frac{Fa}{(1 + bC_0)^2}) \qquad \text{(V-30b)}$$

(see eqn. V-7) Obviously, for a convex upward isotherm, $us(C_0) < u_z(C_0)$ (the slope of the isotherm chord is steeper than that of its tangent). Therefore, during the band migration, the plateau shrinks progressively. It eventually vanishes. This happens for a distance $x = L_c$ such that the shock at $C = C_0$ and the concentration C_0 on a diffuse boundary arrive at the same time, so the times t_s and t_z are equal and

$$\frac{L_c}{u_s} = t_p + \frac{L_c}{u_z}$$

Hence:

$$L_c = t_p u \frac{(1 + bC_0)^2}{FabC_0}$$

If the column length is shorter than L_c, the shock moves throughout the column at a constant velocity $u_s(C_0)$ and the eluted band still has a plateau at $C = C_0$. If the column length is longer than L_c, the eluted band is a curvilinear triangular peak with a height smaller than C_0. During the migration beyond $x = L_c$, the peak height decreases continuously and so does the velocity of its front shock. The retention time becomes more difficult to calculate. We address this issue now.

5 - Retention Time of the Shock on an Elution Band

We will assume here again that the isotherm is convex upward (*i.e.*, is a Langmuirian isotherm) and that the front shock is stable. We assume also that the width of the rectangular pulse injected is small, so that the plateau is eroded quickly and disappears. Accordingly, we neglect L_c compared to the column length. The retention time of the zone in nonlinear chromatography is the elution time of the shock. To obtain this time, *i.e.*, the time when the shock reaches position $x = L$ along the column, we need to know the velocity of the shock (see eqns. V-12 and V-30a). When the plateau has disappeared, the shock velocity varies with the band height. Since, for a given concentration, $u_z(C) > u_s(C)$, the top of the diffuse profile propagates faster than the discontinuity. Accordingly, the discontinuity is constantly eroded. The shock height decreases with increasing time and migration distance and the shock velocity keeps decreasing.

The shock connects the characteristic lines

$$t = \frac{1 + F\frac{\partial f}{\partial C}}{u} x + t_p \qquad (V\text{-}31)$$

at C and at $C = 0$ while each wavelet of the diffuse boundary moves along its characteristic lines, at the velocity associated with its concentration.

Both $\partial f/\partial C$ in eqn. V-31 and f/C in eqn. V-30a depend on the mobile phase concentration through the isotherm equation. Eliminating C between these two equations gives the relationship between x and t, i.e., the position of the shock at any time, hence the retention time of the shock, $t_R(L)$, or time at which the shock is in $x = L$. However, when the isotherm is nonlinear, this elimination may be difficult or even impossible. As a general rule, the solution is easy to derive if df/dC is a function of f/C (which is the case of the Langmuir isotherm, see below, subsection 2a) [21]. Then, eqns. V-30a and V-31 can, respectively, be rewritten as

$$\frac{d(t - t_p - \frac{x}{u})}{d(\frac{x}{u})} = F\frac{f}{C} \qquad (V\text{-}32)$$

$$\frac{t - t_p - \frac{x}{u}}{\frac{x}{u}} = F\frac{df}{dC} \qquad (V\text{-}33)$$

If df/dC is a function of f/C,

$$\frac{df}{dC} = \phi(\frac{f}{C}) \qquad (V\text{-}34)$$

we can derive from eqns. V-33 and V-34 that

$$\frac{f}{C} = \phi^{-1}(\frac{df}{dC}) = \phi^{-1}\left(\frac{t - t_p - \frac{x}{u}}{F\frac{x}{u}}\right) \qquad (V\text{-}35)$$

where $x = \phi^{-1}(y)$ is the inverse function of $y = \phi(x)$. Combining eqn. V-32 and V-35 gives

$$\frac{d(t - t_p - \frac{x}{u})}{d(\frac{x}{u})} = F\phi^{-1}\left(\frac{t - t_p - \frac{x}{u}}{F\frac{x}{u}}\right) \qquad (V\text{-}36)$$

Equation V-36 is an homogeneous differential equation, the general solution of which is

$$\ln\left(F\frac{x}{u}\right) = \int \frac{d\left[(t - t_p - \frac{x}{u})/(F\frac{x}{u})\right]}{\phi^{-1}\left[(t - t_p - \frac{x}{u})/(F\frac{x}{u})\right] - \left[(t - t_p - \frac{x}{u})/(F\frac{x}{u})\right]} + \text{constant}$$

$$(V\text{-}37)$$

It is now difficult to pursue, even in this particular case, and a general solution of eqn. V-37 can be written only in the case in which

$$\phi(\frac{f}{C}) = K\left(\frac{f}{C}\right)^n \tag{V-38}$$

Then

$$\phi^{-1}(\frac{f}{C}) = \left(\frac{f}{KC}\right)^{1/n} \tag{V-39}$$

Combining eqns. V-37 and V-39 gives

$$\ln(F\frac{x}{u}) = \int \frac{d\left[(t - t_p - \frac{x}{u})/(F\frac{x}{u})\right]}{\left[(t - t_p - \frac{x}{u})/(KF\frac{x}{u})\right]^{1/n} - (t - t_p - \frac{x}{u})/(F\frac{x}{u})} + \text{constant} \tag{V-40}$$

For the sake of simplicity, let $y = \left[(t - t_p - \frac{x}{u})/(F\frac{x}{u})\right]^{1/n}$ and we have

$$\ln(F\frac{x}{u}) = \int \frac{dy^n}{yK^{-1/n} - y^n} + \text{constant} = \int \frac{ny^{(n-1)}dy}{y(K^{-1/n} - y^{(n-1)})} + \text{constant}$$

$$= \int \frac{ny^{(n-2)}dy}{K^{-1/n} - y^{(n-1)}} + \text{constant}$$

$$= \frac{-n}{n-1} \int \frac{d(K^{-1/n} - y^{(n-1)})}{K^{-1/n} - y^{(n-1)}} + \text{constant}$$

$$= \frac{-n}{n-1} \ln\left[\frac{1}{K^{1/n}} - y^{n-1}\right] + \text{constant}$$

or

$$F\frac{x}{u} = \text{constant} \left[\frac{1}{K^{1/n}} - \left(\frac{t - t_p - \frac{x}{u}}{F\frac{x}{u}}\right)^{\frac{n-1}{n}}\right]^{\frac{-n}{n-1}} \tag{V-41}$$

Equation V-41 gives the trajectory of the shock for the particular class of isotherms considered so far. The Langmuir isotherm is the most prominent of this class.

a - Langmuir Isotherm

For the Langmuir isotherm we have

$$f = \frac{aC}{1 + bC} \qquad\qquad \frac{f}{C} = \frac{a}{1 + bC} \qquad\qquad \frac{df}{dC} = \frac{a}{(1 + bC)^2}$$

Thus

$$\frac{df}{dC} = \frac{1}{a}\left(\frac{f}{C}\right)^2$$

This is the simplest form of eqn. V-38. For $n = 2$ in eqn. V-41, we have

$$F\frac{x}{u} = \text{constant} \left[\sqrt{a} - \sqrt{\frac{t - t_p - \frac{x}{u}}{F\frac{x}{u}}}\right]^{-2}$$

or $\quad\sqrt{aF\frac{x}{u}} - \sqrt{t - t_p - \frac{x}{u}} = \text{constant}$ $\quad\quad$ (V-42)

Let $x = L$ in eqn. V-42 and replace the constant by λ

$$t_R = t_p + t_0 \left[1 + Fa\left(1 - \sqrt{\frac{\lambda}{Fat_0}}\right)^2\right] \quad\quad \text{(V-43)}$$

We must now determine the value of λ.

The plateau disappears because the point at the top of the diffuse profile $(C = C_0)$ moves faster than the shock. However, the shock has entered earlier than the rear into the column. It entered at time $t = 0$ while the rear diffuse profile of the band results from the collapse of the rear discontinuity of the rectangular injection profile which entered the column at time $t = t_p$. The initial distance between the shock and the top of the diffuse profile is t_p. As long as the concentration plateau has not been entirely destroyed, its height just behind the shock is C_0 and the shock migrates at the constant velocity $u_s = u/[1 + F(f/C_0)] = u/[1 + (Fa)/(1 + bC_0)]$. In this case, the migration distance of the shock is

$$L_c = x_s = u_s t \quad\quad \text{(V-44a)}$$

The migration distance of the wavelet at $C = C_0$, on the rear of the profile is

$$x_z = u_z(t - t_p) \quad\quad \text{(V-44b)}$$

with

$$u_z = \frac{u}{1 + F\frac{df}{dC}\big|_{C=C_0}} = \frac{u}{1 + F\frac{a}{(1+bC_0)^2}} \quad\quad \text{(V-45)}$$

At the moment when the plateau disappears, we have

$$x_z = x_s$$

$$\frac{u}{1 + F\frac{a}{1+bC_0}}t = \frac{u}{1 + F\frac{a}{(1+bC_0)^2}}(t - t_p) \quad\quad \text{(V-46)}$$

Since the velocities were constant during this time, we can derive the time and space coordinates of this event

$$t_d(C_0) = \frac{t_p\left(1 + \frac{Fa}{1+bC_0}\right)(1+bC_0)^2}{FabC_0} \tag{V-47}$$

$$L_c = x_d(C_0) = \frac{ut_p(1+bC_0)^2}{FabC_0} \tag{V-48}$$

Beyond this time and place, the peak height decreases constantly; its velocity does likewise. Introducing the values of t_d and x_d in eqn. V-42 gives the value of the integration constant in eqns. V-42 and V-43

$$\lambda = \sqrt{bC_0 t_p}$$

Porting this value into eqn. V-43 gives the retention time of the shock of the band obtained for a rectangular injection of a solute with a concentration $C = C_0$ and a width t_p, hence for a sample size $F_v t_p C_0$:

$$t_R = t_p + t_0\left[1 + Fa\left(1 - \sqrt{\frac{bC_0 t_p}{Fat_0}}\right)^2\right] \tag{V-49}$$

It is easy to show that in the case in which b is negative (an anti-Langmuir isotherm), the shock is now at the rear of the band and we have

$$t_R = t_0\left[1 + Fa\left(1 + \sqrt{\frac{-bC_0 t_p}{Fat_0}}\right)^2\right] \tag{V-50}$$

Thus, if $b > 0$, the isotherm is convex upward, there is a front shock, the retention time at high concentrations is smaller than under linear conditions, $t_{R,0} = t_p + t_0(1 + Fa)$, and it decreases with increasing pulse concentration. In this case, the front discontinuity of the rectangular pulse injection (the first shock to be injected) is stable. The rear discontinuity of this injection pulse is unstable and collapses into the rear diffuse boundary. This boundary extends from the shock to the point at $C = 0$ and $t = t_{R,0}$, this last point of the boundary remaining unchanged, independent of the sample size.

Conversely, if $b < 0$, the isotherm is convex downward, the rear discontinuity of the rectangular pulse injection is stable and gives the rear shock of the profile while the front discontinuity of the injection pulse collapses into the diffuse front boundary. The retention time under nonlinear conditions is longer than under linear conditions and it increases with increasing pulse concentration. Note that the band height is higher with $b > 0$ than with $b < 0$ (see below, eqns. V-53a for the Langmuir isotherm and V-53b for an anti-Langmuir isotherm). Finally, if $b = 0$, both discontinuities are stable (or, more correctly, metastable). They propagate unchanged within the framework of the ideal model.

These results are in agreement with those of the theoretical analysis made for an isotherm deviating weakly from linear behavior and from the perturbation analysis (see Chapter VII, Section III and Figure VII-1). Profiles derived from eqns. V-49 and V-50 are illustrated in Figure V-6. Comparisons between experimental profiles and profiles calculated with the equilibrium model are discussed in Chapter IX, section XI.

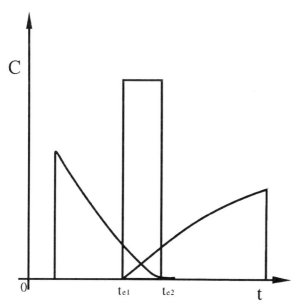

Figure V-6   Typical Chromatographic Profiles obtained through Numerical Calculations.   Left curve, Langmuir isotherm, $q = \frac{25C}{1+0.25C}$; **Central curve, linear isotherm,** $q = 25C$; **Right curve, anti- Langmuir isotherm,** $q = \frac{25C}{1-0.25C}$.

The peak height can be derived from the retention time, eqn. V-31, and given as an algebraic equation since the isotherm equation is known. We obtain, for the profile of the diffuse boundary of the band, the following relationship that can be solved for the time at which a concentration C passes at the distance x along the column (eqn. V-51a), for the retention time t_R of the concentration C_R (time at $x = L$, eqn. V-51b), for the location at which this concentration is at time t, or for the concentration C_R that elutes at time t_R (eqn. V-52):

$$t(C) = \frac{1 + \frac{Fa}{(1+bC)^2}}{u} x + t_p \qquad \text{(V-51a)}$$

$$t_R(C) = \left[1 + \frac{Fa}{(1+bC)^2}\right] t_0 + t_p \qquad \text{(V-51b)}$$

$$C = \frac{1}{b}\left[\sqrt{\frac{Fat_0}{t_R - t_p - t_0}} - 1\right] \qquad \text{(V-52)}$$

This equation is also valid for the maximum of the band, if considered as the point at the top of the diffuse profile. If we replace C by C_R, the maximum concentration of the band or height of its front shock and combine eqns. V-49 and V-52, we obtain

$$C_R = \frac{1}{b}\left[\frac{1}{1 - \sqrt{\frac{bC_0 t_p}{Fat_0}}} - 1\right] \qquad \text{(V- 53a)}$$

In the case when $b < 0$, a different equation is obtained:

$$C_R = \frac{1}{b}\left[\frac{1}{1 + \sqrt{\frac{-bC_0 t_p}{Fat_0}}} - 1\right] \qquad \text{(V-53b)}$$

Comparison between eqns. V-53a and V-53b explains why, for a given sample size, the band is higher with a Langmuir than with an anti-Langmuir isotherm (see Figure V-6).

When the sample size is small and $bC_0 t_p \ll Fat_0$, a Taylor expansion gives a simplified equation

$$C_R = \sqrt{\frac{C_0 t_p}{Fabt_0}} \qquad \text{(V-54)}$$

The derivation of the retention time and the maximum concentration of the band obtained with the simplest isotherm, the Langmuir model, is already long and tedious. A similar calculation is possible only for few other isotherms [1,21]. Fortunately, there is a simpler and far more general solution (see later, section I-4c).

b - The Loading Factor

Many of the previous equations are simpler when expressed as a function of the loading factor. This factor is the ratio of the sample size to the monolayer capacity of the isotherm estimated for the whole column. This capacity is simply related to the parameters of the isotherm. In the case of the Langmuir isotherm, the monolayer capacity is the product of q_s and the volume of the stationary phase, $(1-\varepsilon)SL$. Hence, the loading factor is given by

$$L_f = \frac{n}{(1-\varepsilon)SLq_s} = \frac{nb}{\varepsilon SLk_0'} \qquad (V\text{-}55)$$

where n is the sample size, ε the total porosity of the column, S its cross-section area and L its length. Although eqn. V-55 is valid only for the Langmuir isotherm, the concept is readily extended to any isotherm with a saturation capacity, using the first part of eqn. V-55. The loading factor is the only meaningful parameter to characterize the degree of overloading of a chromatographic column.

In the case of the Langmuir isotherm, eqns. V-51 and V-53 become respectively:

$$t_R = t_p + t_0 \left[1 + Fa(1 - \sqrt{L_f})^2\right] \qquad (V\text{-}51b)$$

$$C_R = \frac{\sqrt{L_f}}{b(1 - \sqrt{L_f})} \qquad (V\text{-}53)$$

Equation V-53 shows that the band height does not increase in proportion to the sample size under nonlinear conditions (in contrast to what it does under analytical, *i.e.*, linear conditions). This dependence is far more complex.

c - General Solution

Golshan-Shirazi and Guiochon [1,12] derived a general solution of the problem of the shock position. This elegant solution avoids the complex calculation of the shock velocity during the elution of the band. The method is based on the conservation of the sample size, hence of the peak area during the migration of the zone along the column. It merely assumes that the flow rate remains constant, which is valid in liquid chromatography, since there is no sorption effect [1]. The area of the rectangular injection pulse of concentration height C_0 and duration t_p is given by

$$A_p = C_0 t_p = \frac{n}{F_v} = \frac{n}{\varepsilon Su} \qquad (V\text{-}56)$$

where n is the sample size (number of moles, equal to the product of the flow rate, the feed concentration and the duration of the injection), ε the total porosity of the column, S the column cross-section area, and u the mobile phase velocity.

From the velocity, u_z, associated with a certain concentration C (eqns. V-7 and V- 45), we derive the retention time of this concentration

$$t(C) = t_p + \frac{L}{u}\left(1 + F\frac{dq}{dC}\right)$$

This equation is the same as eqn. V-28b. It gives the profile of the diffuse part of the elution profile. The maximum concentration of the band is such that the peak area is equal to A_p (eqn. V-56). Integration of the diffuse front of the elution profile (eqn. V-28b) gives

$$|q - C\frac{dq}{dC}|_{C=C_M} = \frac{n}{F_v t_0 F} \qquad (V\text{-}57)$$

The solution of this algebraic equation gives the concentration of the shock. Introduction of this concentration in eqn. V-28b gives the retention time of the concentration C_M on the diffuse boundary of the band, hence the retention time of the shock. Obviously, this method gives the same equations as those derived above in the case of the Langmuir isotherm. Solutions can also be obtained easily for the biLangmuir and the Freundlich isotherm [1,12]. When eqn. V-57 cannot be integrated algebraically, a numerical solution is easily obtained.

II. WEAK SOLUTIONS

In the previous section we discussed the position and the orientation of the characteristic lines originating from the concentration discontinuity which eventually arises on one side of the profile of any chromatographic band eluted under nonlinear conditions. However, as explained earlier (see end of subsection 1, *Characteristics Analysis, Shock and Shock Velocity*), when a shock is formed, the mass balance around this discontinuity does not follow the differential mass balance equation of chromatography [11]. There is a serious mathematical problem here and, in order to avoid inconsistencies or errors, we must give a mathematical definition of the discontinuity. The following analysis will show how it is possible to derive from the chromatography equation an integral equation that accepts the discontinuity as a solution when a shock takes place but otherwise has the same continuous solution as the differential mass balance equation when the solution is continuous [7].

1 - Definition of the Weak Solution of a PDE

The mathematical model of ideal, nonlinear chromatography can be written as

$$\frac{\partial Q}{\partial t} + \frac{\partial C}{\partial x} = 0 \qquad (V\text{-}58)$$

$$C(0, t) = C_0(t) \qquad 0 < t < \infty$$

$$C(x, 0) = 0 \qquad 0 < x < \infty$$

where $Q = [C + Ff(C)]/u$ with $f(C)$ adsorption or partition isotherm. Equation V-58 is identical to eqn. V-4 derived at the beginning of this chapter.

Let us define a class of functions $W(x, t)$ which are continuous, can be differentiated once and tend toward 0 at the boundary of their interval of definition, *i.e.* such that

$$W(\infty, t) = 0$$

$$W(x, \pm\infty) = 0 \qquad (V\text{-}59)$$

In order to obtain the weak solution, $W(x, t)$, as a function belonging to this class, we must construct an integral equation of which it will be a solution because, the weak solution being discontinuous in some isolated points, it cannot be the solution of a differential equation. This integral equation will also contain both $C(x, t)$ and $Q(x, t)$. Because $\frac{\partial W}{\partial x}$ and $\frac{\partial W}{\partial t}$ are defined everywhere, we consider the following equation:

$$\int_0^\infty dx \int_{-\infty}^\infty \left[\frac{\partial W}{\partial x} C + \frac{\partial W}{\partial t} Q \right] dt + \int_{-\infty}^\infty W(0, t) C_0(t) dt = 0 \qquad (V\text{-}60a)$$

This equation is an integral equation which has two important properties that will be demonstrated in the next two subsections. First, when its solution, $C(x, t)$, is continuous and can be differentiated, it is also a solution of eqn. V-58. Second, when its solution is discontinuous in a finite number of points but is continuous along the segments between these points and can be differentiated there, although it does not satisfy eqn. V-58, we still have

$$u_s = \frac{\Delta C}{\Delta Q} = \frac{[C]}{[Q]}$$

where u_s is the migration velocity of such a discontinuity. Note that this last equation is the same as eqn. V-12. So, although a discontinuous solution of eqn. V-60 is not a true solution of eqn. V-58, it still has a certain meaning for this equation. Such a solution of eqn. V-60 is called a *weak solution* or a generalized solution of eqn. V-58. We now demonstrate the two properties of eqn. V-60 just stated. We will use these properties in the following section (section III).

2 - Proof of the First Property of Equation V-60a

In the range (x, t) of the variables within which the function $C(x, t)$ is continuous and can be differentiated, we can write

$$\frac{\partial W}{\partial x}C + \frac{\partial W}{\partial t}Q = \frac{\partial(WC)}{\partial x} - W\frac{\partial C}{\partial x} + \frac{\partial(WQ)}{\partial t} - W\frac{\partial Q}{\partial t}$$

Accordingly, eqn. V-60a can be written as

$$\int_0^\infty dx \int_{-\infty}^\infty \left[\frac{\partial(WQ)}{\partial t} - W\frac{\partial Q}{\partial t} + \frac{\partial(WC)}{\partial x} - W\frac{\partial C}{\partial x}\right] dt +$$

$$\int_{-\infty}^\infty W(0, t)C_0(t)dt = 0 \quad \text{(V-60b)}$$

Integrating eqn. V-60b by part, taking eqn. V-59 into account, gives:

$$\int_{-\infty}^\infty (WC)|_{x=0}^{x=\infty} dt - \int_0^\infty dx \int_{-\infty}^\infty W\frac{\partial C}{\partial x} dt + \int_{-\infty}^\infty (WQ)|_{t=-\infty}^{t=\infty} dx -$$

$$\int_0^\infty dx \int_{-\infty}^\infty W\frac{\partial Q}{\partial t} dt + \int_{-\infty}^\infty W(0, t)C_0(t)dt = 0 \quad \text{(V-61)}$$

Considering the boundary conditions of eqn. V-58 and eqn. V-59, we have

$$\int_{-\infty}^\infty (WC)|_{x=0}^{x=\infty} dt = 0 - \int_{-\infty}^\infty W(0, t)C_0(t)dt \quad \text{(V-62)}$$

From eqn. V-59 we derive

$$\int_{-\infty}^\infty [WQ]|_{t=-\infty}^{t=\infty} dx = 0 \quad \text{(V-63)}$$

Combining eqns. V-61 to V-63, we have

$$\int_0^\infty \int_{-\infty}^\infty \left[\frac{\partial Q}{\partial t} + \frac{\partial C}{\partial x}\right] W dt dx = 0 \quad \text{(V-64)}$$

Because the function $W(x, t)$ is arbitrary, this relationship is valid for any W, so we must have

$$\frac{\partial Q}{\partial t} + \frac{\partial C}{\partial x} = 0$$

$$C(0, t) = C_0(t) \qquad 0 < t < \infty$$

$$C(x, 0) = 0 \qquad 0 < x < \infty$$

Therefore, when a solution of eqn. V-60 is continuous and can be differentiated, it is also a solution of eqn. V-58.

3 - Proof of the Second Property of Equation V-60

Now we assume that $C(x,t)$ is continuous almost everywhere but is discontinuous in some points of the space x, t, *i.e.*, at certain times and positions, along a curve G in that plane (see Figure V-7). Everywhere else, it is continuous and can be differentiated. Let us call (a) $D+$ the domain above curve G and D_- the domain below this curve, (b) G_+ a path along curve G in domain D_+ and, similarly, G_- a path along curve G in domain D_-. The respective boundaries of the domains D_+ and D_- are (G_+, R_+, I_+) and (G_-, R_-, I_-), R and I being arbitrary straight segments parallel to the axis of coordinates.

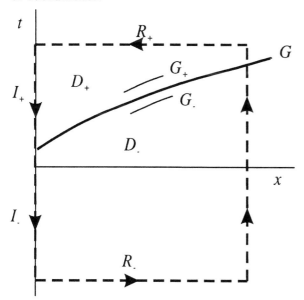

Figure V-7   Trajectory of the Shock of a Band in the x, t Plane.

To proceed further, we need to apply Green relationship. It applies to the coordinates R_x and R_y of a vector $\vec{V}$ in the plane x, y. It states that:

$$\iint_D \left[\frac{\partial R_x}{\partial y} - \frac{\partial R_y}{\partial x} \right] dx dy = \oint [R_x dx + R_y dy]$$

We apply the Green relationship to a vector defined in the plane x, t and having the coordinates $R_x = -WQ$ and $R_t = WC$. Then, we have

$$\iint_D \left[\frac{\partial(WQ)}{\partial t} + \frac{\partial(WC)}{\partial x} \right] dx dt = \oint W [C dt - Q dx] \tag{V-65}$$

The circular integral is calculated along the boundaries of D in the plane x, t (see Figure V-7). The LHS of eqn. V-65 can be rewritten

$$\iint\limits_{D_+ + D_-} \left[\frac{\partial(WQ)}{\partial t} + \frac{\partial(WC)}{\partial x} \right] dx dt = \iint\limits_{D_+ + D_-} W \left[\frac{\partial Q}{\partial t} + \frac{\partial C}{\partial x} \right] dx dt +$$

$$\iint\limits_{D_+ + D_-} \left[\frac{\partial W}{\partial t} Q + \frac{\partial W}{\partial x} C \right] dx dt \quad \text{(V-66)}$$

where D_+ and D_- are the two areas defined in Figure V-7, as indicated above.

Since the function $C(x, t)$ is continuous and can be differentiated in both domains D_+ and D_-, we have in each of these domains

$$\frac{\partial Q}{\partial t} + \frac{\partial C}{\partial x} = 0$$

So, the first term in the RHS of eqn. V-66 is equal to 0. From the integral equation V-60, the second term becomes

$$\iint\limits_{D_+ + D_-} \left[\frac{\partial W}{\partial t} Q + \frac{\partial W}{\partial x} C \right] dx dt = - \int_{-\infty}^{\infty} W(0, t) C_0(t) dt \quad \text{(V-67)}$$

The term in the RHS of eqn. V-65 can be rewritten

$$\oint W \left[C dt - Q dx \right] = \int_{I_+ + I_-} W(C dt - Q dx) + \int_{R_+ + R_-} W(C dt - Q dx) +$$

$$\int_{G_+ + G_-} W(C dt - Q dx) \quad \text{(V-68)}$$

The path R_+, R_- extends from $-\infty$ to ∞. Because of eqn. V-59 ($W(x, \pm\infty)$ $= 0$), the second term in the RHS of this equation is equal to 0. The first term in the RHS of eqn. V-68 is an integral along the path I_+, I_-. Since $x = 0$ along this path, $C(x, t) = C(0, t) = C_0(t)$. Since the path is chosen from top to bottom in Figure V- 7, the first term in the RHS of eqn. V-68 can be expressed as

$$\int_{I_+ + I_-} W(C dt - Q dx) = - \int_{-\infty}^{\infty} W(0, t) C_0(t) dt \quad \text{(IV-69)}$$

Combination of eqns. V-68, V-65, V-66, V-67, and V-69 gives

$$\int_{G_+} W(C_+dt - Q_+dx) + \int_{G_-} W(C_-dt - Q_-dx) = 0$$

where C_+, Q_+, C_- and Q_- are the values of C and Q on either sides of G, respectively. Since the directions of the integration paths along G_+ and G_- are opposite, we must have

$$\int_G W\left[(C_+ - C_-)dt - (Q_+ - Q_-)dx\right] = 0 \tag{V-70}$$

Since the function W is arbitrary, this relationship is valid for any such function and we have just shown that we have

$$\frac{dx}{dt} = \frac{C_+ - C_-}{Q_+ - Q_-} = \frac{[C]}{[Q]} \tag{V-71}$$

As a consequence, we have demonstrated that eqn. V-60 allows concentration discontinuities, that the migration equation of these discontinuities is eqn. V-70, and that their velocity is given by eqn. V-71, $i.e.$, is equal to u_s.

In summary, a shock or concentration discontinuity does not satisfy the mass balance of chromatography and cannot be a solution of eqn. V-1. However, it satisfies the integral equation V-60a. A weak solution, that is a solution of eqn. V-60a, contains a shock and a continuous, diffuse boundary. Both satisfy eqn. V-60a. The diffuse boundary that is the solution of eqn. V-60a is also a solution of the mass balance equation, eqn. V-1. It is in this sense that a weak solution is a generalized solution of eqn. V-1.

The introduction of the concept of *weak solution* of the chromatographic equation allows the enlargement of the class of functions among which we may search for a possible solution of this equation. Solutions considered do not need to be differentiable everywhere. They may contain discontinuities. In such a case, the solution of the chromatographic equation will no longer be a *classical solution*, it will be a *generalized solution*. We discuss these generalized solutions in the next section.

III. LAX SOLUTION OF THE CHROMATOGRAPHIC EQUATION

In the first section of this chapter (subsections I-4 and I-5), we derived the complete elution profile of a compound in the case of a Langmuir isotherm when the equilibrium model applies. We also showed how this approach can be extended, first to the class of isotherms for which $df/dC = \phi(f/C)$, then to a more general case of isotherms without an inflection point (see eqn. V-57). These solutions are usually considered satisfactory by physical chemists and chemical engineers, but not so by mathematicians. It was impossible to give a rigorous, general solution to the problem. The most important investigation of the mathematical properties of the hyperbolic conservation law (of which the most important example is the fundamental equation of aerodynamics, with the chromatographic equation of the ideal, nonlinear model a close second) was made by Lax [22-24]. We draw now on the results of his investigations of the asymptotic solution, through the convex function theory, and apply them to nonlinear chromatography. This approach will lead to a more general solution of the chromatographic problem than those discussed earlier.

We begin again with eqn. V-58, the general form of the mass balance equation in the ideal model

$$\frac{\partial Q}{\partial t} + \frac{\partial C}{\partial x} = 0$$
$$C(0, t) = C_0(t) \qquad 0 < t < \infty$$
$$C(x, 0) = 0 \qquad 0 < x < \infty$$

where $Q = [C + Ff(C)]/u$ with $f(C)$ adsorption or partition isotherm.

Let us assume that $Q(C)$, hence $f(C)$, is a convex function in the general, mathematical sense (Figure V-8). Accordingly, the solution derived in this section is more general than many conventional investigations of the ideal model that are, more or less implicitly, restricted to the Langmuir isotherm model. The definition of a convex function is that it is a function to which the following relationship applies

$$Q(C) = Q(b) + (C - b)Q'(b) - r$$

where b and $r > 0$ are parameters and $Q' = dQ/dC$. We can also write this equation

$$Q(C) = \min_b[Q(b) + (C - b)Q'(b)] \qquad (V-72)$$

The meaning of this relationship is illustrated in Figure V-8[4]. The equation of the tangent to the function $Q(C)$ in point $C = b, Q(b)$ is

$$\tilde{Q}(C) = Q(b) + (C - b)Q'(b) \qquad (V-73)$$

---

[4]In practical, physical terms, this means that the curve $Q(C)$ is always below its tangent.

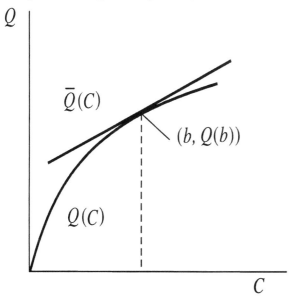

Figure V-8   Definition of a Convex Function.

When $C = b$, $Q(C) = \tilde{Q}(C) = Q(b)$. Equation V-72 shows that

$$\tilde{Q}(C) \geq Q(C)$$

the equality being verified at the point of contact between the curve and its tangent.

Now, we introduce the solution $C(x,t)$ of the chromatographic equation from the point of view of the properties of convex functions. The method used by Lax is the linearization of eqn. V-58 by replacing $Q(C)$ by $\tilde{Q}(C)$, followed by the search of the solution corresponding to $\tilde{Q}(C)$, a comparison between $Q(C)$ and $\tilde{Q}(C)$, and the derivation of the actual solution $C(x,t)$. The derivation is presented in Appendix A. The result obtained is that the solution of the chromatography equation (eqn. V-58) is given by eqn. V-A-18:

$$C(x,t) = C_0(\bar{t}_b)$$

where $C_0(t) = C(0,t)$ and $\bar{t}_b$ is the solution of the equation

$$P\left(\frac{t - \bar{t}_b}{x}\right) = C_0(\bar{t}_b) \qquad \text{(V-A-17)}$$

where $P(s)$ is the inverse function of $Q'(C)$, i.e., $Q'(P(s)) = s$. So, in order to obtain the Lax solution, it is necessary to calculate the inverse function of $Q'(C)$ (see Appendix A).

The Lax solution is extremely important for all forms of the hyperbolic conservation law, not only for the one that states mass conservation in chromatography. For example, it was demonstrated first in the case of the aerodynamic equation and the propagation of pressure signals in gases. It is a complete solution that extends to all convex or concave upward isotherms, at least with the formulation presented here. The extension to an isotherm equation with an inflection point is far more difficult and outside the scope of this book.

IV - THE ASYMPTOTIC SOLUTION

The asymptotic solution of eqn. V-4 is the limit toward which the profile of a band tends when its migration distance tends toward infinity. Obviously, under the assumption of the equilibrium theory of chromatography, this solution cannot be a Gaussian profile because there are no sources of axial diffusion or dispersion in the ideal model. However smooth the injection profile can be, a shock will eventually form, as explained earlier (Section I-2) and the profile tends toward a limit that is independent of the boundary condition. Using results by Lax [23], Golshan–Shirazi and Guiochon [12] derived this asymptotic solution. Note that Lax discussed the solution of the aerodynamic equation, similar to the mass balance equation of chromatography but different. The time and space variables are exchanged in these two equations. The explicit equation for the asymptotic solution derived by Lax [23] must be modified accordingly [12].

First that the mass balance equation of the ideal model (eqn. V-4) is written:

$$\frac{\partial Q(C)}{\partial t} + \frac{\partial C}{\partial x} = 0$$

with

$$Q(C) = \frac{C + Ff(C)}{u} \tag{V-74a}$$

$$Q'(C) = \frac{1 + Ff'(C)}{u} \tag{V-74b}$$

$$Q''(C) = \frac{Ff''(C)}{u} \tag{V-74c}$$

where $f(C)$ is the equation of the isotherm model. In writing eqns. V-74b and V-74c, we assume implicitly that $f(C)$ is a continuous function of the concentration that can be differentiated twice. Equation V-4 is a hyperbolic

PDE. Accordingly, if the boundary condition is a bounded, measurable function as is the elution boundary condition in chromatography, its solution is finite within a limited interval of distance along the column (concentration profile) or time (elution profile). Everywhere else, it is zero.

The asymptotic solution [12] is a triangle which has a shock on one side and a diffuse boundary on the other, the latter ending (or beginning) at $C = 0$ and at a time $t_{R,0}$ given by the equation:

$$t_{R,0} = t_p + xQ'(0) \tag{V-75}$$

Equation V-75 is not valid if $Q'(0)$ is infinite, $e.g.$, in the case of an isotherm that is tangential to the concentration axis. The band profile can be on either side of $t = t_{R,0}$, depending on the sign of the initial curvature of the isotherm, $Q''(0)$, $i.e.$, on whether the concentration shock moves faster or more slowly than the concentration $C = 0$ (see Figure V-9). If we neglect the value of t_p, the retention time of the shock is given by:

$$t_{R,s} = xQ'(0) \pm \sqrt{|2A_p xQ''(0)|} \tag{V-76}$$

where A_p is the area of the injection profile, $e.g.$, the product $C_0 t_p$ of the concentration and duration of a wide rectangular profile. If $Q''(0) > 0$, the isotherm is convex upward, the sign in eqn. V-76 is $-$, $t_{R,s} < t_{R,0}$, and the shock elutes first, followed by a rear diffuse boundary. This is the case of the Langmuir isotherm (see profile 1 in Figure V-9). In the converse case of $e.g.$, an S-shape isotherm, the shock elutes last after a front diffuse boundary (see profile 2 in Figure V-9). Note that if $Q''(0) = 0$, the isotherm is linear and the boundary condition propagates unchanged, even along an infinitely long column.

Equation V-75 is the asymptotic limit of eqn. V-49. When bC_0 is small, we derive from this last equation:

$$t_R = t_p + t_0 \left[1 + Fa \left(1 - 2\sqrt{\frac{bC_0 t_p}{Fa t_0}} \right) \right] \tag{V-77}$$

which identical to eqn. V-76 in the case of a Langmuir isotherm.

Between times $t_{R,0}$ and $t_{R,s}$, the concentration profile of the diffuse boundary is given by the equation

$$C(x,t) = \frac{1}{Q''(0)} \left(\frac{t}{x} - Q'(0) \right) \tag{V-78}$$

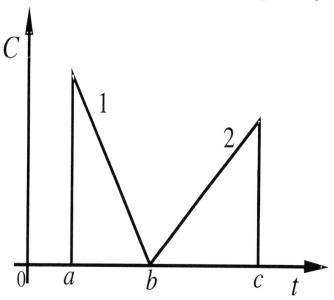

Figure V-9   Asymptotic Profile of the Ideal Model. First case: t_p is neglected. The profile 1 corresponds to a convex upward isotherm (*e.g.*, a Langmuir isotherm), the profile 2 to a convex downward isotherm. The retention times of the concentration discontinuity is given by eqn. V-76 (profile 1, point a) or eqn. V-77 (profile 2, point c). The retention time of the concentration $C = 0$ is given by eqn. V-75 (point b). The concentrations C_M and $C(x,t)$ are given by eqns. V-79, and V-78, respectively.

This is a straight line. So, the asymptotic profile is a rectangular triangle, whatever the isotherm model (see Figure V-9). This model has influence only on the relative position of the shock and the diffuse boundary, through the sign of its initial curvature, and on the numerical values of the two time boundaries ($t_{R,0}$ and $t_{R,s}$) and of the slope of the diffuse boundary ($1/Q''(0)$).

The maximum concentration of the asymptotic profile is

$$C_M = \sqrt{\frac{2A_p}{xQ''(0)}} \qquad \text{(V-79)}$$

The maximum concentration of the band increases as the square root of the amount injected and decreases as the square root of the migration distance. Figure V-9 shows two profiles, one for the compounds having a convex upward (*i.e.*, Langmuirian) isotherms (see Chapter II, section 1) the band of which have a front shock and a rear diffuse boundary, the other for the less common case of the compounds with a convex downward isotherm, the bands of which have, conversely, a rear shock and a front diffuse boundary.

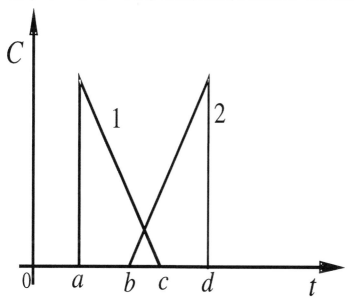

Figure V-10   Asymptotic Profile of the Ideal Model. Second case: t_p is not neglected. Same as in Figure V-9, except for a finite t_p. The retention times of a, b, c, and d are given by eqns. V-76a, V-75b, V-75a, and V-76b, respectively. Peak 1, convex upward isotherm; peak 2, convex downward isotherm.

Because the shock originates from the front discontinuity of the rectangular injection in the case of a convex upward isotherm but from the rear discontinuity of this injection in the case of a convex downward isotherm and because these two discontinuities are injected a time t_p apart, the exact equation for the retention times $t_{R,0}$ and $t_{R,s}$ when t_p cannot be neglected are:

$$t_{R,0} = t_p + xQ'(0) \tag{V-75a}$$

$$t_{R,s} = t_p + xQ'(0) - \sqrt{|2A_p xQ''(0)|} \tag{V-76a}$$

in the case of a convex upward isotherm and

$$t_{R,0} = xQ'(0) \tag{V-75b}$$

$$t_{R,s} = xQ'(0) - \sqrt{|2A_p xQ''(0)|} \tag{V-76b}$$

in the case of a convex downward isotherm.

This situation is illustrated in Figure V-10 that shows the asymptotic profiles obtained in the same two cases as illustrated above, in Figure V-9, of two compounds, one with a convex upward and the other with a convex downward isotherm.

APPENDIX A
DERIVATION OF THE LAX SOLUTION

This derivation is presented here only for the case of isotherms that are convex functions. Should the isotherm have an inflection point, the derivation is far more complex and beyond the level and scope of this book. The reader is referred to the relevant mathematical publications.

First, we need to transform eqn. V-58 into an equation involving $Q(C)$[5] but not $\partial Q/\partial C$. For this purpose, let

$$\phi(x,t) = \int_0^t C(x,t')dt' \tag{V-A-1}$$

$$\phi(0,t) = \int_0^t C(0,t')dt' \tag{V-A-2}$$

where t' is a dummy variable. Thus,

$$C(x,t) = \frac{\partial \phi(x,t)}{\partial t} = \phi_t(x,t), \qquad\qquad C_t(x,t) = \frac{\partial C(x,t)}{\partial t} = \phi_{tt}(x,t)$$

Introducing these equations into eqn. V-71 gives

$$\phi_{xt} + Q'(\phi_t)\phi_{tt} = 0$$

integrating this equation with respect to t gives

$$\phi_x + Q(\phi_t) = 0 \tag{V-A-3}$$

As desired, this equation involves Q directly and not $\partial Q/\partial C$.

Using eqns. V-A-1 and V-A-2, the equation defining a convex function (eqn. V-72) and the equation of its tangent (eqn. V-73) can be rewritten

$$Q(\phi_t) = \min_b \; [Q'(b)\phi_t + Q(b) - bQ'(b)] \tag{V-A-4}$$

$$\tilde{Q}(\phi_t) = Q'(b)\phi_t + Q(b) - bQ'(b) \tag{V-A-5}$$

The differential equation for the tangent corresponds to eqn. V-A-3 and is

$$\tilde{\phi}_x + \tilde{Q}(\tilde{\phi}_t) = 0 \tag{V-A-6}$$

This is a linear equation whose solution is $\tilde{\phi}$. Combining eqns. V-A- 5 and V-A-6 gives

$$Q'(b)\tilde{\phi}_t + \tilde{\phi}_x = bQ'(b) - Q(b) \tag{V-A-7}$$

---

[5]Where $Q(C) = [C + Ff(C)]/u$ and $f(C)$ is the equilibrium isotherm.

If we let $Q'(b) = dt/dx$, the LHS of eqn. V-A-7 is equal to $d\tilde{\phi}/dx$. Then, eqn. V-A-7 can be rewritten

$$\frac{d\tilde{\phi}}{dx} = bQ'(b) - Q(b) \qquad \text{(V-A-8)}$$
$$\frac{dt}{dx} = Q'(b)$$

On the other hand, eqn. V-A-4 can be written

$$Q(\phi_t) = Q'(b)\phi_t + Q(b) - bQ'(b) - r \qquad \text{(V-A-9)}$$

where $r > 0$ is the difference between the value of the curve $Q(C)$ and its tangent, $\tilde{Q}(C)$. Introducing eqn. V-A-9 into eqn. V-A-3 and considering that $d\phi/dx = \partial\phi/\partial x + \partial\phi/\partial t \cdot dt/dx$, we have

$$\frac{d\phi}{dx} = bQ'(b) - Q(b) + r \qquad \text{(V-A-10)}$$
$$\frac{dt}{dx} = Q'(b)$$

Comparing eqns. V-A-8 and V-A-10 shows that

$$\phi \geq \tilde{\phi} \qquad \text{(V-A-11)}$$

Which is the relationship between ϕ and $\tilde{\phi}$ that we wanted.

Now, we need to derive $\tilde{\phi}$. Integrating eqn. V-A-8 gives

$$\tilde{\phi}(x, t) = [bQ'(b) - Q(b)]x + \tilde{\phi}(o, t_b) \qquad \text{(V-A-12)}$$
$$t = Q'(b)x + t_b$$

where t_b is the time that corresponds to the location $x = 0$. Equation V-A-12 is the solution of the linear equation. Since $\tilde{\phi}(0, t) = \phi(0, t)$, $\phi(x, t)$ can be derived from eqns. V-A-11 and V-A-12. from the second equation V-A-12, we have

$$b = P\left(\frac{t - t_b}{x}\right) \qquad \text{(V-A-13)}$$

where P is the inverse function of Q', i.e., $Q'(P(\alpha)) = \alpha$. Combining eqns. V-A-12 and V-A-13 gives

$$\tilde{\phi}(x, t, t_b) = \tilde{\phi}(0, t_b) + \left\{P\left(\frac{t - t_b}{x}\right)\frac{t - t_b}{x} - Q\left[P\left(\frac{t - t_b}{x}\right)\right]\right\}x \qquad \text{(V-A-14)}$$

Combining eqns. V-A-11 and V-A-14 gives

$$\phi(x, t) = \max_{t_b} \tilde{\phi}(x, t, t_b) \qquad \text{(V-A-15)}$$

Suppose that $\tilde{\phi}(x, t, t_b)$ is maximum when $t_b = \overline{t_b}$, *i.e.*, that $\phi(x, t) = \tilde{\phi}(x, t, t_b)$. Then, $\overline{t_b}$ can be derived from the equation

$$\frac{\partial \tilde{\phi}(x, t, t_b)}{\partial t_b} = 0$$

So, we have

$$\phi(x, t) = \tilde{\phi}(x, t, \overline{t_b}) = \tilde{\phi}(0, \overline{t_b}) + \{P\left(\frac{t - \overline{t_b}}{x}\right)\frac{t - \overline{t_b}}{x} - Q\left[P\left(\frac{t - \overline{t_b}}{x}\right)\right]\}x \qquad \text{(V-A-16)}$$

The solution of the chromatographic equation (eqn. V-71) can be derived from $C(x, t) = \phi_t$. Now

$$\frac{\partial \tilde{\phi}(x, t, t_b)}{\partial t_b} = \frac{\partial \tilde{\phi}(0, t_b)}{\partial t_b} + \{P'\left(\frac{t - t_b}{x}\right)\frac{t_b - t}{x^2} + P\left(\frac{t - t_b}{x}\right)\frac{-1}{x} - Q'\left[P\left(\frac{t - t_b}{x}\right)\right] \times$$
$$P'\left(\frac{t - t_b}{x}\right)\frac{-1}{x}\}x = C_0(t_b) - P\left(\frac{t - t_b}{x}\right)$$

When $\partial \tilde{\phi}(x, t, t_b)/\partial x = 0$, we have

$$P\left(\frac{t - \overline{t_b}}{x}\right) = C_0(\overline{t_b}) \qquad \text{(V-A-17)}$$

We can derive $\overline{t_b}$ from eqn. V-A-17 but for this, we need to calculate the inverse function of Q', *i.e.*, $P\left(\frac{t - \overline{t_b}}{x}\right)$. Combining eqns. V-A-16 and V-A-17, gives

$$\phi(x, t) = \tilde{\phi}(x, t, t_b) = \tilde{\phi}(0, \overline{t_b}) + \{C_0(\overline{t_b})\frac{t - \overline{t_b}}{x} - Q\left[C_0(\overline{t_b})\right]\}x \qquad \text{(V-A-18)}$$

From eqns. V-A-17 and V-A-18, we derive

$$C(x, t) = \frac{\partial \phi_t(x, t)}{\partial t} = \frac{\partial \tilde{\phi}(x, t, t_b)}{\partial t} = C_0(\overline{t_b}) \qquad \text{(V-A-19)}$$

A simple example illustrates that this solution is reasonable. Suppose that

$$Q(C) = \frac{C + Ff(C)}{u} = \frac{(1 + Fa)}{u}C$$

this means that the equilibrium isotherm is linear. Then, $Q(C) = \tilde{Q}(C)$, $C = b$, and $t_b = \overline{t_b}$. From the second equation under V-A-12, we have

$$\overline{t_b} = t_b = t - Q'(b)x = t - Q'(C)x = t - \frac{1 + Fa}{u}x$$

Combining this result with V-A-18 gives

$$C(x, t) = C_0\left(t - \frac{1 + Fa}{u}x\right)$$

which is obviously a satisfactory result.

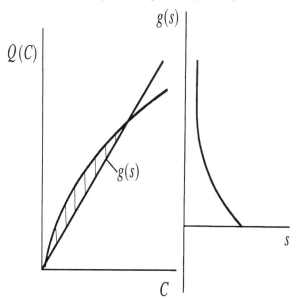

Figure V-A-1   The Conjugate Concave Function of $Q(C)$, $g(s)$.

In order to use this method in the general case, we need to derive t_b from eqn. V-90 which means that we need to find the inverse function, Q'^{-1}, of the function Q'. In other words, Lax method transform the difficulty in solving the nonlinear chromatography equation with a nonlinear isotherm into the difficulty in calculating the inverse function of Q'. Aris and Amundson [7,8] introduced a method which consists of expressing Q'^{-1} with a graph of the conjugate function which is concave. The conjugate function is defined as

$$g(s) = \max_c \; [Q(C) - sC] \qquad\qquad \text{(V-A-20)}$$

where s is the slope of the tangent. A graph illustrating this function is shown in Figure V-A-1.

The height of the hatched region in Figure V-A-1 is $Q(C) - sC$. $g(s)$ is the maximum height. It corresponds to

$$\frac{d}{dC}[Q(C) - sC] = 0$$

Accordingly, C is given by

$$C = Q'^{-1}(s) = P(s)$$

Combining this relationship and eqn. V-A-20 gives

$$g(s) = [Q(p(s)) - sP(s)]$$

Moreover, we have

$$g'(s) = Q'(P(s))P'(s) - sP'(s) - P(s) = sP'(s) - sP'(s) - P(s)$$
$$= -P(s) = -Q'^{-1}(s)$$

using the relationship $Q'(P(s)) = Q'(Q'^{-1}(s)) = s$. This last equation shows that P, the inverse function of Q', can be expressed using g, the conjugated, concave function of Q. Therefore, the problem of calculating $C(x, t)$ is replaced by a search for $g(s)$, using a graphic method. Then, g' is derived from g, which gives P and t_b, and, finally, allows the calculation of $C(x, t)$.

EXERCISES

1) Calculate the area of the elution profile predicted by the reaction model of chromatography discussed in section IV of this chapter. Compare the results obtained in the two cases when (a) $\beta \neq 0$ and (b) $\beta = 0$.

2) Calculate the elution profile predicted by the ideal, non-linear model of chromatography for a single component with an anti-Langmuir isotherm. Determine the position and the height of the shock, the function giving the diffuse profile of the band, and draw the profile.

3) Derive the Lax condition of the shock in the ideal, non-linear model of chromatography for a single component, with an anti-Langmuir isotherm.

4) Give the retention time predicted by the ideal, nonlinear model of chromatography for a single component, for a Freundlich isotherm, $q = kC^\alpha$.

5) Compare the three following methods of determination of u_s:
    (i) From eqns. V-10a to V-12.
    (ii) From eqns. V-14 to V-16.
    (iii) From eqns. V-65 to V-70.
Show the differences and similarities among these methods.

6) Demonstrate that, under linear conditions, the front of the raffinate band and the rear of the extract band in SMB both have a shock. (Tip: calculate the concentration dependence of the velocity associated with a concentration on these fronts as a function of the fractional time spent by each concentration of the profile inside the two columns between which these profiles move during a cycle, columns I and II for the extract, III and IV for the raffinate). See ref. [25].

7) Assume a compound with a Langmuir isotherm. Assume that the mobile phase used has a finite concentration, C_0 of this compound and that the breakthrough of this concentration has been achieved. Hence, the initial condition is:

$$C(x, 0) = C_0$$
$$q(x, 0) = f(C_0)$$

Now, on this concentration plateau, the injection of a rectangular pulse of concentration $C = C_1$ and duration $t = t_p$ is carried out.
    a- What is the elution profile of this pulse if $C_1 > C_0$?
    b- What is it when $C_1 < C_0$ (i.e., the injection is that of a vacancy)?
    Hint: Note that the signal observed is proportional to the difference $C_1 - C_0$. It can be positive (injection of a band) or negative (injection of a vacancy). Determine the

relationship between a concentration and the velocity that is associated with it. When does it increase with increasing signal? When does it decrease with decreasing signal? See ref. [26].

LITERATURE CITED

[1] G. Guiochon, S. Golshan-Shirazi and A. M. Katti, *"Fundamentals of Preparative and Nonlinear Chromatography "*, Academic Press, Boston, MA, 1994.

[2] F. James, *"Sur la modélisation mathématique des équilibres diphasiques et des colonnes de chromatographie,"* Ph.D. dissertation, École Polytechnique, Palaiseau, France, 1990.

[3] J.N. Wilson, *J. Amer. Chem. Soc.*, **62** (1940) 1583.

[4] D. DeVault, *J. Amer. Chem. Soc.*, **65** (1943) 532.

[5] H.-K. Rhee, B.F. Bodin and N.R. Amundson, *Chem. Eng. Sci.*, **26** (1971) 1571.

[6] H.-K. Rhee and N.R. Amundson, *Chem. Eng. Sci.*, **27** (1972) 199.

[7] R. Aris and N. R. Amundson, *"Mathematical Methods in Chemical Engineering,"* Prentice- Hall, Englewood Cliffs, NY, 1973.

[8] H.-K. Rhee, R. Aris and N.R. Amundson, *"First-Order Partial Differential Equations. II Theory and Application of Hyperbolic Systems of Quasilinear Equations,"* Prentice Hall, Englewood Cliffs, NJ, 1989.

[9] L. Jacob, P. Valentin and G. Guiochon, *J. Chim. Phys. (France)*, **66** (1969) 1097.

[10] L. Jacob and G. Guiochon, *J. Chim. Phys. (France)*, **67** (1970) 185.

[11] G. Guiochon and L. Jacob, *Chromatogr. Rev.*, **14** (1971) 77.

[12] S. Golshan-Shirazi and G. Guiochon, *J. Phys. Chem.*, **94** (1990) 495.

[13] A. Fick, *Ann. Phys. (Leipzig, Germany)*, **170** (1855) 59.

[14] J. C. Giddings, *Unified Separation Science*, Wiley, New York, NY, 1991.

[15] L. Jacob, P. Valentin and G. Guiochon, *Chromatographia*, **4** (1971) 6.

[16] F. Helfferich and D. L. Peterson, *Science*, **142** (1963) 661.

[17] M. Czok and G. Guiochon, *Anal. Chem.*, **62** (1990) 189.

[18] T. Fornstedt, PREP-2002, Washington, D.C., June 16-19, 2002.

[19] I. Quiñones, C. M. Grill, L. Miller and G. Guiochon, *J. Chromatogr. A*, **867** (2000) 1.

[20] E. Cremer and H. F. Huber, *Angew. Chem.*, **73** (1961) 461.

[21] B.Lin, S.Golshan-Shirazi, Z.Ma and G.Guiochon, *Anal. Chem.*, **60** (1989) 2647

[22] P. Lax, *Commun. Pure Applied Math.*, **7** (1954) 159.

[23] P. Lax, *Commun. Pure Applied Math.*, **10** (1957) 537.

[24] P. Lax, *"Contributions to Non-Linear Functional Analysis,"* E.H. Zarantello Ed., University of Wisconsin, Madison, WI, 1971, 603.

[25] G. Zhong and G. Guiochon, *Chem. Eng. Sci.*, **51** (1996) 4307.

[26a] G. Zhong, T. Fornstedt and G. Guiochon, *J. Chromatogr. A*, **734** (1996) 63.

[26b] P. Sajonz, T. Yun, G. Zhong, T. Fornstedt and G. Guiochon, *J. Chromatogr. A*, **734** (1996) 75.

CHAPTER VI
IDEAL, NONLINEAR,
REACTION CHROMATOGRAPHY

There can be chromatographic processes in which the solute molecules are not only adsorbed on a surface (or, more generally, retained through some molecular interactions with the stationary phase) but can also react in the stationary phase, by decomposing or by reacting with some other solute molecules, with the solvent or with an additive. Catalytic chromatographic processes belong to this group. In this last case, the column is packed with a catalyst or a mixture of a catalyst and an adsorbent, the latter helping in separating the products from the reagents and from each other, which enhances the conversion yield. Reaction chromatography has also become a method of measurement of reaction rate constants. Under linear conditions, reaction chromatography does not raise any significant problems that are specific of chromatography and need be addressed here. Important reviews of the work done in reaction gas and liquid chromatography have been published by Langer [1,2]. The use of reaction chromatography in simulated moving bed has been investigated by Carr and his associates [3]. The models of reaction chromatography were analyzed by Rhee *et al.* [4].

Under nonlinear conditions, when the adsorption isotherm of at least the reagent is nonlinear and the concentration of this reagent is high, the band profiles of reagent and product reflect both the adsorption and the reaction processes and particularly their kinetics. Although at least two compounds are always simultaneously present in reaction chromatography, we will here consider this process as a single-component problem because we will assume that the process of reaction chromatography is conducted under such experimental conditions that the reaction is irreversible and that its rate is small. So, the concentration of the product of the reaction will be small and the chromatographic process will be linear as far as this compound is concerned. This is obviously not the case for the reagent, for which it will be nonlinear. However, we will further neglect the competition between reagent and product, assuming only weakly nonlinear behavior for the former compound, hence neglecting band interference. Under these simplifying assumptions, the reagent problem is a single-component nonlinear one.

I. MATHEMATICAL MODEL OF REACTION CHROMATOGRAPHY

In Chapter III, we derived the mass balance equation of a single component in chromatography, in the ideal, nonlinear case (eqn. III-2a). Still assuming no axial dispersion and an infinite rate of radial mass transfer, we follow the same rationale to write the mass conservation equations of the reagent and the product of the following catalyzed, irreversible reaction

$$A_1 \rightarrow A_2 \tag{VI-1}$$

These equations are

$$u\frac{\partial C_1}{\partial x} + \frac{\partial C_1}{\partial t} + F\frac{\partial q_1}{\partial t} = 0 \tag{VI-2}$$

$$u\frac{\partial C_2}{\partial x} + \frac{\partial C_2}{\partial t} + F\frac{\partial q_2}{\partial t} = 0 \tag{VI-3}$$

where C_1, C_2, q_1, and q_2 are the concentrations of the reagent and the product in the mobile and the stationary phase, respectively, u is the mobile phase velocity, and F is the phase ratio.

The difference between conventional and reaction chromatography is that the relationship giving the rates of variations of the concentrations in the stationary phase $(\partial q_i/\partial t)$ is no longer related only to the isotherms of the adsorption equilibrium. They involve also the reaction rate. If an irreversible reaction takes place in the adsorbed state, e.g., at the surface of a catalyst, we have

$$\frac{\partial q_1}{\partial t} = \frac{\partial f_1(C_1, C_2)}{\partial t} + k_r f_1(C_1, C_2) \tag{VI-4}$$

$$\frac{\partial q_2}{\partial t} = \frac{\partial f_2(C_1, C_2)}{\partial t} - k_r f_1(C_1, C_2) \tag{VI-5}$$

where k_r is the rate constant of the reaction while $f_1(C_1, C_2)$ and $f_2(C_1, C_2)$ are, respectively, the reagent and the product equilibrium concentrations in the adsorbed phase in the presence of the concentrations C_1 and C_2 of these two compounds in the mobile phase. These two concentrations are given by the respective equilibrium isotherms of these products. Since the concentrations q_1 and q_2 are the actual concentrations of the reagent and the product, respectively, in the stationary phase, we have $q_1 = f_1(C_1, C_2)$ and $q_2 = f_2(C_1, C_2)$ *only* when $k_r = 0$. The signs of the second terms in the RHS of eqns. VI-4 and VI-5 are important. To understand them, observe that the mass balances of the two compounds in the combinations of eqns. VI-2 and VI-4, and of eqns. VI-3 and VI-5 for the reagent and the product, respectively, express that the variation of the actual concentration of these two compounds in the adsorbed phase arise from the sum of the effects of the mass transfer kinetics and the reaction kinetics (to simplify the presentation, we assume that the reaction takes place only in the adsorbed phase).

When the isotherm equations, $f_1(C_1, C_2)$ and $f_2(C_1, C_2)$, are both linear functions, the system of eqns. VI-4 and VI-5 can be solved using the method of the Laplace transform. Peng [5] discussed this approach in 1984. Usually, however, the adsorption isotherm of the reagent is not linear. Therefore, the discussion of the problem of reaction chromatography under nonlinear conditions is not only of theoretical importance, it is also of practical relevance. This problem is complex because of the presence of the second terms in the RHS of eqns. VI-4 and VI-5. These terms are called damped terms. A perturbation method has the advantages of being easier to solve and of illustrating more clearly the influence of the nonlinear behavior of the isotherm in reaction chromatography [6].

If the adsorption isotherms are given by the Langmuir model, we have

$$f_1(C_1, C_2) = \frac{a_1 C_1}{1 + b_1 C_1 + b_2 C_2} \tag{VI-6a}$$

$$f_2(C_1, C_2) = \frac{a_2 C_2}{1 + b_1 C_1 + b_2 C_2} \tag{VI-6b}$$

where a_1, a_2, b_1 and b_2 are numerical parameters. If we assume that the reagent concentration is moderate and that the reaction is relatively slow at the scale of the retention time of the reagent, the isotherms exhibit only moderate deviations from linear behavior. Then, the Langmuir isotherm of the first component (reagent) can be replaced by a parabolic isotherm and that of the second one (product) by a linear isotherm

$$f_1(C_1, C_2) = a_1 C_1 - a_1' C_1^2 \tag{VI-7a}$$

with $\qquad a_1' C_1 \ll a_1$

$$f_2(C_1, C_2) = a_2 C_2 \tag{VI-7b}$$

where $a_1' = a_1 b_1$. In writing the Taylor expansion of eqn. VI-6a, we neglected in eqn. VI-7a the other second order term, $a_1 b_2 C_1 C_2$, because we assumed k_r to be small and, accordingly, the concentration of the reaction product, C_2, to be negligible and its adsorption behavior to be linear (*i.e.*, $b_2 C_2 \approx 0$). Finally, we also assumed a small deviation from linear behavior for the reagent, *i.e.*, that $b_1 C_1$ also is small.

Combining eqns. VI-3 and VI-5 and eliminating the small terms of higher order gives

$$\frac{\partial q_1}{\partial t} = (a_1 - 2a_1' C_1)\frac{\partial C_1}{\partial t} + k_r a_1 C_1 \qquad \text{(VI-8)}$$

Combining this equation with eqn. VI-1 gives the new system of PDE equations

$$\frac{\partial C_1}{\partial x} + \frac{1 + Fa_1 - 2Fa_1' C_1}{u}\frac{\partial C_1}{\partial t} = -\frac{\beta C_1}{u} \qquad \text{(VI-9a)}$$

$$\frac{\partial C_2}{\partial x} + \frac{1 + Fa_2}{u}\frac{\partial C_2}{\partial t} = \frac{\beta C_1}{u} \qquad \text{(VI-9b)}$$

where $\beta = Fa_1 k_r$. The first one of these two equations does not contain C_2, so it is not coupled with the second one and it can be solved separately. The converse is not true, eqn. VI-9b contains C_1, hence is coupled with eqn. VI-9a. Equations VI-9a,b are the approximate equations that are valid when the system deviates only slightly from linear behavior (perturbation approach).

We analyze first the properties of this mathematical model (section II). Then, we discuss the solutions in two possible cases (see section III). In both of them, the initial conditions are $C_1 = 0$ and $C_2 = 0$. This corresponds to a column containing initially only the mobile and the stationary phases in equilibrium, with no reagent. In the first case, $C_1(0, t) = C_{1,0}$ for $x = 0$, *i.e.*, we do a breakthrough injection of the reagent and we have a Riemann boundary condition. In the second case, the boundary condition corresponds to a rectangular injection of the reagent, with a concentration $C_{1,0}$ and a width t_p.

II. MATHEMATICAL ANALYSIS OF THE MODEL

Along the characteristic lines, we can write the following equations

$$\frac{dC_1}{dx} = -\beta \frac{C_1}{u} \qquad \text{(VI-10)}$$

$$\frac{dt}{dx} = \frac{1 + Fa_1 - 2Fa_1' C_1}{u} \qquad \text{(VI-11)}$$

Integration of the first of these two equations gives

$$C_1 = C_1^*(\tau)\, e^{-\frac{\beta}{u}x} \qquad\qquad (\text{VI-12})$$

where $C_1^*(\tau)$ is the boundary condition, i.e., the concentration profile of the pulse of reagent injected into the column, at $x = 0$, and τ is the time given by the intersection of the characteristic line considered and the time axis. Integration of the second characteristic equation gives

$$t = \tau + \frac{1 + Fa_1}{u}x - \frac{2Fa_1'C_1^*(\tau)(1 - e^{-\frac{\beta}{u}x})}{\beta} \qquad\qquad (\text{VI-13})$$

This family of characteristic lines corresponds to eqn. VI-10. Since C_1^* is a function of τ, the characteristic lines defined by eqn. VI-13 and corresponding to different values of τ (i.e., starting from different points of the boundaty condition) have different slopes. In certain cases, these characteristic lines may intersect. If they do, the solution has a discontinuity. The coordinates of these intersection points can be calculated from the set of characteristics equations originating from different values of τ. Using eqn. VI-13, the equation of the characteristic line originating from the point at $t = \tau + d\tau$ is given by

$$t = (\tau + d\tau) + \frac{1 + Fa_1}{u}x - \frac{2Fa_1'C_1^*(\tau + d\tau)(1 - e^{-\frac{\beta}{u}x})}{\beta} \qquad\qquad (\text{VI-14})$$

Combining eqns. VI-13 and VI-14 and letting $d\tau$ tend toward 0 gives

$$0 = 1 - \frac{2Fa_1'(1 - e^{-\frac{\beta x}{u}})}{\beta}\frac{dC_1^*(\tau)}{d\tau} \qquad\qquad (\text{VI-15})$$

Since we always have $0 < 1 - e^{-\frac{\beta}{u}x} < 1$, this equation may have a solution provided that the following condition is satisfied

$$\frac{dC_1^*(\tau)}{d\tau} > \frac{\beta}{2Fa_1'} \qquad\qquad (\text{VI-16})$$

When there is a solution, the breakthrough curve exhibits a discontinuity. In the two cases listed above and discussed in section III, the step and the pulse injection, we have $dC_1^*(\tau)/d\tau = \infty$ at the front. Then the condition VI-16 is verified and the front of the injection band is always a discontinuity.

To facilitate the analysis of the discontinuity, we rewrite eqn. VI-9a under the classical conservation form

$$\frac{\partial Q_1}{\partial t} + \frac{\partial C_1}{\partial x} = -\frac{\beta C_1}{u}$$

where $Q_1 = \frac{1}{u}(C_1 + Fa_1C_1 - Fa'_1C_1^2)$. Assume that the velocity of the discontinuity is $u_s = (dx/dt)_s$. From the Rankine–Hugoniot condition (*i.e.*, from the velocity of the shock, eqn. V-12), we have

$$u_s = \frac{u[C_1]}{[C_1 + Fa_1C_1 - Fa'_1C_1^2]}$$

where $[A]$ denotes the amplitude of the jump in A associated with the discontinuity or difference between the values of A on each side of the shock. Since we assume the column to be initially empty of reagent, the initial value of the concentration is 0, so

$$u_s = \frac{u}{1 + Fa_1 - Fa'_1C_1} \qquad \text{(VI-17)}$$

This equation satisfies the entropy condition

$$u_{Z_+} < u_s < u_{Z_-} \qquad \text{(VI-18)}$$

where $u_Z = \frac{u}{1 + Fa_1 - 2Fa'_1C_1}$ is the characteristic velocity. The symbols $+$ and $-$ correspond to the two sides of the discontinuity, u_{Z_+} corresponding to the side where $C_1 = 0$. So, eqn. VI-17 is equivalent to $u_{s,left} > u_s > u_{s,right}$. In this case, the discontinuity is stable. Hence, it is a concentration shock, like in nonlinear, ideal chromatography.

III. THE BREAKTHROUGH TIME
AND THE ELUTION PROFILE

We can now discuss successively the two basic problems of chromatography, the jump in the reagent concentration from 0 to a finite value (the Riemann problem) at the column inlet and the injection of a rectangular pulse of reagent (wide rectangular injection and, at the limit, the Dirac problem).

1 - Case I. The Riemann Problem

At time $t = 0$, the concentration of the stream entering the column ($x = 0$) is raised from $C_1 = 0$ to $C_1 = C_{1,0}$, a concentration that is kept constant thereafter. The previous mathematical analysis of the problem shows that the breakthrough profile has two parts, a discontinuity originating from the injection profile that is stable and moves as a shock, at the velocity $u_s =$

$(dx/dt)_s$ followed by the plateau concentration which connects to the top of the shock and on which the concentration moves at the velocity u_z, along a characteristic line. The reason for this profile, which is the same as the one obtained when there is no chemical reaction (see Chapter V) but has a different height and retention time, is that, except for the shock, all the parts of this concentration wave are controlled by eqns. VI-10 to VI-12 that have continuous solutions. Since $C_1^*(\tau) = C_{1,0}$ in the present case, the equation of the characteristic line is $C_1 = C_{1,0}e^{-\beta x/u}$. The point at the top of the discontinuity is also at the end of the plateau. Introducing this relationship into the equation giving u_s and integrating it gives

$$t = (1 + Fa_1)\frac{x}{u} - \frac{Fa_1'C_{1,0}(1 - e^{-\frac{\beta x}{u}})}{\beta} \qquad \text{(VI-19)}$$

Compared with the result obtained under linear conditions, eqn. VI-19 contains an additional term which represents the effect of the nonlinear behavior of the isotherm. This term is a function of $\beta = Fa_1k_r$. The reason for the dependence of this additional term on the rate constant of the reaction is that, under nonlinear conditions, the velocity of the shock is related to its maximum concentration and that this concentration depends on the rate of the reaction. It decreases with increasing rate constant. When the deviation from linear behavior, i.e., $a_1'C_{1,0}$, becomes small, eqn. VI-19 tends toward the result obtained under linear conditions. Under linear and nonlinear conditions, however, the steady state concentration is $C_1 = C_{1,0}e^{-\beta x/u}$, independent of the step concentration. This concentration may affect the breakthrough retention time, not its relative height, $C_{1,\infty}/C_{1,0}$.

Figure VI-1 compares the breakthrough curves for the reagent and the product of the reaction.

2 - Case II. Rectangular pulse injection problem

In this case, the injection of a rectangular pulse of reagent of width t_p and height $C_{1,0}$ is performed at the initial time, at the inlet of the column. The front of the pulse is a stable shock (in the case of a convex upward isotherm, as assumed here) and it moves at the velocity u_s (see eqn. VI-17). The maximum concentration of the band, C_1, depends now on the reaction rate, on the width of the injection pulse, and on the deviation of the isotherm from linear behavior. As in nonlinear chromatography ($k_r = 0$), the retention time of the front depends on the concentration of the band, which is now controlled both by the reaction rate constant and by $C_1^* = C_{1,0}$, since the boundary condition is a rectangular injection. The profile of the rear of the band depends on the velocity associated with each concentration. The low

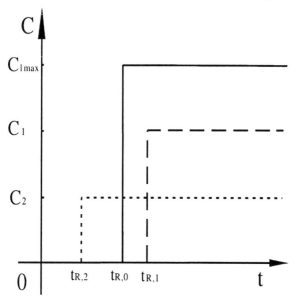

Figure VI-1   Breakthrough Curves of the Reagent and the Product of the Reaction in Reaction Chromatography.   Linear conditions for the product. Nonlinear conditions for the reagent, with a small deviation from linear behavior (parabolic isotherm).   Experimental condition, $k_r = 0.5$.   Solid line, pure reagent, no reaction, $C_{1,max} = 0.4182$.   Long- dashed line, reagent when the reaction takes place, $C_1 = 0.2800$.   Short-dashed line, reaction product, $C_2 = 0.1383$.

concentrations at the rear of the band move more slowly than the plateau concentration, hence the rear profile is a diffuse boundary and spreads during its migration. The velocity associated with the plateau concentration, however, is higher that the shock velocity for this same concentration, so the rear of the plateau moves faster than the front shock. While the band rear spreads, its top erodes and, eventually, the band profile becomes a curvilinear triangle, after which the band height decreases constantly. Each point of the rear profile of the band migrates along a characteristic line determined by an equation similar to eqn. VI-13. The rear of the band is injected at time $t = t_p$, hence the rear profile is

$$t = t_p + (1 + Fa_1)\frac{x}{u} - \frac{2Fa_1'C_1^*(1 - e^{-\frac{\beta x}{u}})}{\beta} \tag{VI-20a}$$

Since $C_1 = C_1^* e^{-\beta x/u}$, this equation may be rewritten

$$t = t_p + (1 + Fa_1)\frac{x}{u} - \frac{2Fa_1'C_1(1 - e^{-\frac{\beta x}{u}})}{\beta e^{-\frac{\beta x}{u}}} \tag{VI-20b}$$

Once the concentration plateau of the injection band has been eroded, the point at the top of the discontinuity is also at the top end of the continuous rear profile or diffuse boundary. The concentration C_1 is the same and the elution time given by eqn. VI-20 is also that of the shock. Combining eqn. VI-20 and the equation giving u_s and eliminating C_1 gives the differential equation accounting for the shock migration

$$\frac{dt}{dx} - \frac{\beta t e^{-\frac{\beta x}{u}}}{2u(1 - e^{-\frac{\beta x}{u}})} = \frac{1 + Fa_1}{u} - \frac{\beta e^{-\frac{\beta x}{u}}(t_p + \frac{1+Fa_1}{u}x)}{2u(1 - e^{-\frac{\beta x}{u}})} \qquad \text{(VI-21)}$$

This equation is an ordinary differential equation of the type $dt/dx + p(x)t = q(x)$. It can be integrated (see Appendix A) into

$$t = t_p + \frac{1 + Fa_1}{u}x + \text{constant}\sqrt{1 - e^{-\frac{\beta x}{u}}} \qquad \text{(VI- 22)}$$

This result can be verified by calculating the derivative of t, porting function and derivative into the differential equation VI-21, and obtaining an identity. We now need to derive the value of this integration constant.

3 - Determination of the integration constant

At the beginning of its migration along the column, the rectangular pulse includes a front shock, a rear diffuse boundary and a flat top at $C_{1,0}e^{-\beta x/u}$. During the elution of this band, the length of the concentration plateau decreases progressively toward zero, then the front shock erodes constantly because this front shock moves more slowly than the point at the same concentration at the top of the rear diffuse boundary (see mathematical analysis, previous section). As long as the plateau is not completely eroded, the retention time of the shock (at the front of the plateau) is given by eqn. VI-19 above, $i.e.$,

$$t_s = (1 + Fa_1)\frac{x}{u} - \frac{Fa_1' C_{1,0}(1 - e^{-\frac{\beta x}{u}})}{\beta}$$

The retention time of the point at the rear of the plateau is given by the characteristic equation VI-20, which may be rewritten as

$$t_z = t_p + (1 + Fa_1)\frac{x}{u} - \frac{2Fa_1' C_{1,0}(1 - e^{-\frac{\beta x}{u}})}{\beta}$$

The plateau is completely eroded when $t_s = t_z$. From this condition, we derive the coordinates, x_d, t_d of the shock at this time. They are:

$$x_d = -\frac{u}{\beta} \ln \left(1 - \frac{\beta t_p}{Fa_1' C_{1,0}} \right) \tag{VI-23}$$

$$t_d = -\frac{1 + Fa_1}{\beta} \ln \left(1 - \frac{\beta t_p}{Fa_1' C_{1,0}} \right) - t_p \tag{VI-24}$$

Combining eqns. VI-22 to VI-24 gives

$$\text{constant} = -2 \sqrt{\frac{Fa_1' C_{1,0} t_p}{\beta}} \tag{VI-25}$$

Since the argument of a logarithm must be positive, we derive from eqn. VI-23 that a triangle wave is formed only if $t_p \leq Fa_1' C_{1,0}/\beta$ or $t_p \leq a_1' C_{1,0}/(a_1 k_r)$. The plateau just vanishes when there is equality. If t_p is large enough, case II is equivalent to case I, the front of the band and a segment of the plateau being eluted before the plateau has had time to erode away entirely.

If we combine eqns. VI-23 and VI-25, we obtain the migration equation of the shock or, in this case, of the peak, relating its position, x, and the time:

$$t = t_p + \frac{1 + Fa_1}{u} x - 2 \sqrt{\frac{Fa_1' C_{1,0} t_p (1 - e^{-\frac{\beta x}{u}})}{\beta}} \tag{VI-26}$$

When $x = L$, eqn. VI-26 gives the retention time

$$t_R = t_p + t_0(1 + Fa_1) - 2 \sqrt{\frac{Fa_1' C_{1,0} t_p (1 - e^{-\beta t_0})}{\beta}} \tag{VI-27}$$

where $t_0 = L/u$ is the hold-up time. This expression is different from the one obtained in the linear case. The additional term, introduced by the nonlinear behavior of the isotherm, is now a function of the width of the rectangular pulse. Obviously, when the isotherm becomes linear ($b_1 = 0$, $a_1' = 0$), eqn. VI-27 tends toward the classical expression of linear chromatography.

To obtain the profile of the rear diffuse boundary, we solve eqn. VI-20 for C_1:

$$C_1(x,t) = \frac{\beta e^{-\frac{\beta x}{u}} \left[t_p + (1 + Fa_1)\frac{x}{u} - t \right]}{2Fa_1' (1 - e^{-\frac{\beta x}{u}})} \tag{VI-28}$$

At column outlet, the elution profile is given by

$$C_1(L,t) = \frac{\beta e^{-\beta t_0} \left[t_p + (1 + Fa_1)t_0 - t \right]}{2Fa_1' (1 - e^{-\beta t_0})} \tag{VI-29}$$

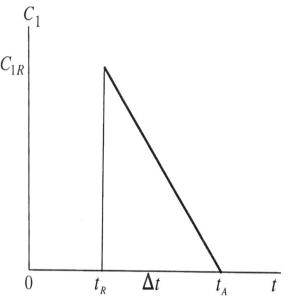

Figure VI-2   Elution Profile of a Rectangular Pulse of the Reagent in Reaction Chromatography. Nonlinear conditions but with a small deviation from linear behavior (parabolic isotherm).

This equation is that of a straight line in the (C_1, t) plane, as illustrated in Figure VI-2. The reason why we obtain such a simple profile is that we have assumed for the equilibrium isotherm a parabolic equation, $f_1 = a_1 C_1 - a_1' C_1^2$. In the general case, the expansion of the isotherm equation around the origin has higher order terms and the profile of the rear boundary would not be linear. Equation VI-29 shows also that the end of the peak, which corresponds to the elution of the concentration $C_1 = 0$, takes place at the retention time that is observed under linear conditions for the end of the elution of a rectangular pulse $(t_{R,0} = t_p + (1 + Fa_1)t_0)$. Combination of this result and of eqn. VI-26 shows that the peak width (see Figure VI-2) is

$$\Delta t = 2\sqrt{\frac{Fa_1' C_{1,0} t_p (1 - e^{-\beta \, t_0})}{\beta}} \qquad (\text{VI-30})$$

Combination of eqn. VI-29 with eqn. VI-27, which gives the retention time of the peak, permits the determination of the maximum concentration of the peak

$$C_{1,R} = \sqrt{\frac{\beta C_{1,0} t_p}{Fa_1' (1 - e^{-\beta t_0})}} e^{-\beta t_0} \qquad (\text{VI-31})$$

The pre-exponential factor reflects the influences of the nonlinear behavior of the isotherm and of the rate constant of the reaction.

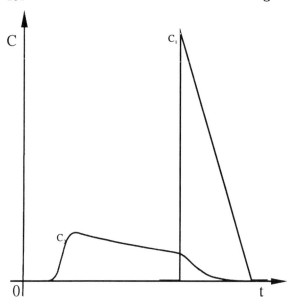

Figure VI-3   Elution profiles of the reagent and the product of the reaction in reaction chromatography.   Nonlinear conditions but with a small deviation from the linear behavior (parabolic isotherm) for the reagent.   Linear conditions for the reaction product.

Figure VI-3 shows the elution profiles of the reagent and the product of the reaction. The former profile is the same as in Figure VI-2 (where the product profile was not shown). Note that the elution profile of the product of the reaction is definitely different from any profile encountered in linear or nonlinear ideal chromatography.

In the case of a Langmuir isotherm, there are no algebraic nor closed-form solutions for the elution profiles. However, numerical solutions can be calculated as discussed later (Chapter IX). Figure VI-4 shows the elution profiles of the reagent and of the product when these compounds follow Langmuir and linear behavior, respectively.

The profiles of both the reagent and the product of the reaction depend obviously on the rate constant of the reaction. VI-5a and VI-5b show the elution profiles obtained for a reagent with Langmuir isotherm behavior (VI-5a) and for a reaction product with a linear isotherm (VI-5b), respectively. As expected, the size area of the reagent peak decreases with increasing rate constant, its height decreases, and its retention time increases since the behavior of the band is Langmuirian. By contrast, the area and the height of the band of the reaction product increase with increasing rate constant while the retention time remained unchanged since the reaction product was to have a linear isotherm.

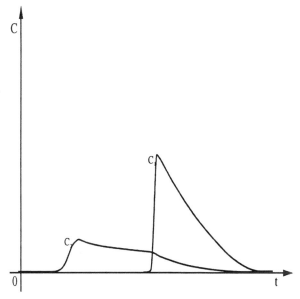

Figure VI-4   Elution profiles of the reagent and the product of the reaction in reaction chromatography. Langmuir isotherm for the reagent and linear behavior for the product of the reaction.

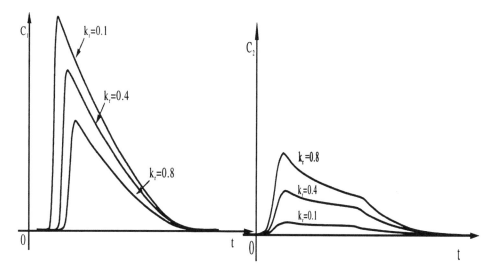

VI-5 Influence of the Rate Constant of the Reaction on the Elution Profile of the Reagent in Reaction Chromatography. Left, Profile of the Reagent. Nonlinear conditions, with a Langmuir isotherm. Right, Profile of the Reaction Product. Linear Conditions.

IV. EXPRESSION OF THE PARAMETERS

The following equations relate the parameters of the model (*i.e.*, the rate constant of the reaction and the two parameters of the parabolic isotherm) and the experimental parameters which are usually measured, the peak height ($C_{1,R}$), its retention time (t_R), the retention time of the peak end or retention time under linear conditions ($t_{R,0}$), and the hold-up time (t_0).

$$k_r = \frac{\beta}{Fa_1} = \frac{1}{t_0 + t_p - t_{R,0}} \ln \frac{(t_{R,0} - t_R)C_{1,R}}{2C_{1,0}t_p} \qquad \text{(VI-32)}$$

$$a_1 = \frac{t_{R,0} - t_0 - t_p}{Ft_0} \qquad \text{(VI-33)}$$

$$b_1 = \frac{a_1'}{a_1} = \frac{(t_{R,0} - t_R)^2 \left[\ln 2C_{1,0}t_p - \ln(t_{R,0} - t_R)C_{1,R}\right]}{2(t_{R,0} - t_0 - t_p)\left[2C_{1,0}t_p - (t_{R,0} - t_R)C_{1,R}\right]} \qquad \text{(VI-34)}$$

These equations can be useful for parameter identification. The practical problem encountered in their use, however, would be the questionable accuracy of the determination of the three parameters of the model by this method. These equations assume that the column is only moderately overloaded. Under such conditions, and unless the column efficiency is high, the peaks obtained do not have a very steep front boundary.

V. ASYMPTOTIC ANALYSIS

The analysis of reaction chromatography that we have made here is actually a perturbation analysis. It assumes that both the reaction rate and the deviation from linear isotherm behavior are moderate or small. Its main purpose was to discuss the influence of the extent of the nonlinear behavior of the isotherm and of the reaction rate on the band profile. Equations VI-20 and VI-29 show that, for an heterogeneous catalytic reaction, the main difference between a linear and a nonlinear isotherm is that in the former case the retention time of the concentration signal does not depend on the injected amount nor on the reaction rate while it depends on these two parameters in the latter case.

In case I, a first order expansion of eqn. VI-19 shows that, when β tends toward 0, the retention time of the breakthrough front is given by

$$t_R = (1 + Fa_1)t_0 - Fa_1'C_{1,0}t_0(1 - \frac{1}{2}\beta t_0) \qquad \text{(VI-35)}$$

This equation shows that increasing the reaction rate contributes to limit the decrease of the retention time due to the nonlinear behavior of the isotherm. If the reaction rate constant increases, both the concentration of the breakthrough plateau of the reagent and the corresponding shock velocity decrease. When $\beta = 0$, eqn. VI-35 becomes

$$t_R = [1 + F(a_1 - a_1' C_1)]t_0 \qquad \text{(VI-36)}$$

because, in this case, the shock velocity becomes

$$u_s = \frac{u}{1 + F(a_1 - a_1' C_1)}$$

In case II, the retention time of the pulse is

$$t_R = t_p + t_0 \left[1 + Fa_1 - 2\sqrt{Fa_1' C_{1,0} t_p (1 - \frac{\beta t_0}{2})} \right] \qquad \text{(VI-37)}$$

and we derive from eqn. VI-37 that, when β becomes equal to 0, the retention time of the pulse is

$$t_R = t_p + t_0 \left[(1 + Fa_1) - 2\sqrt{Fa_1' C_{1,0} t_p} \right]$$

$$= t_p + t_0 \left[1 + Fa_1 \left(1 - 2\sqrt{\frac{a_1' C_{1,0} t_p}{Fa_1^2 t_0}} \right) \right] \qquad \text{(VI-38)}$$

The mathematical solution of the ideal model of chromatography (Chapter V, section I-4) showed that, when the equilibrium isotherm is given by the Langmuir model ($f_1 = a_1 C_1/(1 + bC_1)$) and the injection is a rectangular pulse of height $C_{1,0}$ and width t_p, the retention time of the peak is

$$t_R = t_p + t_0 \left[1 + Fa_1 \left(1 - \sqrt{\frac{bC_{1,0} t_p}{Fa_1 t_0}} \right)^2 \right]$$

(see eqn. V-49) When $bC_{1,0} \ll a_1$, the relationship simplifies to

$$t_R = t_p + t_0 \left[1 + Fa_1 \left(1 - 2\sqrt{\frac{bC_{1,0} t_p}{Fa_1 t_0}} \right) \right] \qquad \text{(VI-39)}$$

This equation is the same as eqn. VI-38 since, in this section, we replaced the Langmuir isotherm by the parabolic isotherm having the same initial slope and curvature, hence, $a_1' = a_1 b_1$.

When β tends toward 0, the peak height given by eqn. VI-31 simplifies to

$$C_{1,R} = \sqrt{\frac{t_p}{Fa_1' C_{1,0} t_0 (1 - \frac{1}{2}\beta t_0)}} C_{1,0} e^{-\beta t_0} \qquad \text{(VI-40)}$$

When $\beta = 0$, we have

$$C_{1,R} = \sqrt{\frac{C_{1,0} t_p}{Fa_1' t_0}} = \sqrt{\frac{C_{1,0} t_p}{Fa_1 b t_0}} \qquad \text{(VI-41)}$$

This equation is the same as eqn. V-54. We have shown that when the rate constant of the reaction tends toward 0, the band profile obtained in reaction chromatography tends toward the profile given by the ideal model.

APPENDIX A
Demonstration of Equation VI-20

Equation VI-20 is an ordinary differential equation (ODE) of the form

$$\frac{dt}{dx} + p(x)t = q(x) \tag{VI-A-1a}$$

where

$$p(x) = \frac{-\beta e^{\frac{-\beta x}{u}}}{2u(1 - e^{-\frac{\beta x}{u}})} \tag{VI-A-1b}$$

$$q(x) = \frac{1 + Fa_1}{u} - \frac{\beta e^{-\frac{\beta x}{u}}\left(t_p + \frac{1+Fa_1}{u}x\right)}{2u(1 - e^{-\frac{\beta x}{u}})} \tag{VI-A-1c}$$

The solution for this form of ODE is

$$t = e^{-\int p(x)dx}\left[\int q(x) \; e^{\int p(x)dx} \; dx \; + \; Const\right] \tag{VI-A-2}$$

We need now to calculate the integrals in eqn. VI-A-2. The first one is given by

$$\int p(x)dx = \int \frac{-\beta e^{\frac{-\beta x}{u}}}{2u(1 - e^{-\frac{\beta x}{u}})}dx = \ln\frac{1}{\sqrt{1 - e^{-\frac{\beta x}{u}}}} \tag{VI-A-3}$$

In order to calculate the second integral in eqn. VI-A-2, let $q(x) = q_1(x) + q_2(x) + q_3(x)$, where

$$q_1(x) = \frac{1 + Fa_1}{u} \tag{VI-A-4a}$$

$$q_2(x) = \frac{-\beta e^{-\frac{\beta x}{u}}t_p}{2u(1 - e^{-\frac{\beta x}{u}})} \tag{VI-A-4b}$$

$$q_3(x) = \frac{-(1 + Fa_1)\beta x e^{-\frac{\beta x}{u}}}{2u^2(1 - e^{-\frac{\beta x}{u}})} \tag{VI-A-4c}$$

The calculation of the solution of eqn. VI-A-2 becomes that of three integrals. The solution of the second one is obvious

$$\int q_2(x)e^{\int p(x)dx}dx = \int \frac{\beta e^{-\frac{\beta x}{u}}t_p}{2u\left(1 - e^{-\frac{\beta x}{u}}\right)} \frac{dx}{\sqrt{1 - e^{-\frac{\beta x}{u}}}}dx = \frac{t_p}{\sqrt{1 - e^{-\frac{\beta x}{u}}}} \tag{VI-A-5}$$

The calculation of the third one gives

$$\int q_3(x)e^{\int p(x)dx}dx = \int \frac{-(1 + Fa_1)\beta x e^{-\frac{\beta x}{u}}}{2u^2\left(1 - e^{-\frac{\beta x}{u}}\right)} \frac{dx}{\sqrt{1 - e^{-\frac{\beta x}{u}}}}$$

$$= \int \frac{(1 + Fa_1)x}{u} d\left(\frac{1}{\sqrt{1 - e^{-\frac{\beta x}{u}}}}\right)$$

$$= \frac{(1 + Fa_1)x}{u\sqrt{1 - e^{-\frac{\beta x}{u}}}} - \int \frac{(1 + Fa_1)dx}{u\sqrt{1 - e^{-\frac{\beta x}{u}}}} \tag{VI-A-6}$$

The integral in this equation is also equal to

$$-\int q_1(x) \; e^{\int p(x) \; dx} \; dx$$

These two integrals cancel out in the final calculation, hence

$$t = e^{-\int p dx} \left[\int q e^{\int p dx} dx + c \right]$$

$$= t_p + \frac{1 + F a_1}{u} x + \text{constant} \sqrt{1 - e^{-\frac{\beta x}{u}}} \qquad \text{(VI-A-7)}$$

This last equation is equation VI-20.

EXERCISES

1) Derive the solution of ideal reaction chromatography for an anti-Langmuir isotherm $(q = aC/(1 - bC)$, with $b > 0)$, in the case of a Dirac pulse injection. Use the same method as described in this Chapter. Discuss the result obtained.

2) Give the numerical analysis of the equations of reaction chromatography, eqns. VI-4 to VI-6b, in the case of a Dirac pulse injection. Compare the results of the numerical calculations with those presented above.

3) Discuss the influence of the reaction rate constant on the profile of the reagent, using a computer to calculate numerical solutions.

4) Design an experimental procedure to determine the reaction rate constant using eqn. VI-32.

5) Compare the influences of the reaction rate constant, k_r, and of the diffusion coefficient on the profile of the reagent.

LITERATURE CITED

[1] C.-Y. Jeng and S. H. Langer, *J. Chromatogr.*, **589** (1992) 1.
[2] R. Thede, D. Haberland, Z. Deng and S. H. Langer, *J. Chromatogr. A*, **683** (1994) 279.
[3] M. C. Bjorklund and R. W. Carr, *Catalysis Today*, **25** (1995) 159.
[4] H.-K. Rhee, R. Aris and N. R. Amundson, *"First-Order Partial Differential Equations,"* Prentice Hall, Englewood Cliffs, NY, 1992.
[5] S. Peng, *Scientia Sinica* (Series B), **28**(1) (1985) 1.
[6] B Lin, S. Chang, S. Peng, *Scientia Sinica* (Series B), **4** (1993) 394.

CHAPTER VII

THE PROFILES OF SINGLE-COMPONENT BANDS IN NONLINEAR, NONIDEAL CHROMATOGRAPHY

In this chapter, we discuss the influence of the various kinetic phenomena that influence the band profiles under nonlinear conditions when the algebraic solutions of the mathematical models used are tractable. However, the mathematical complexity of the problem of nonlinear, nonideal chromatography is high. Accordingly, there are in general no algebraic solutions to the problem, even under the most simple assumptions of the Langmuir or the parabolic isotherm and of the equilibrium–dispersive model. There is not even a general theorem of existence of the solution, at least under certain conditions, and there is certainly no theorem regarding the conditions of convergence of a procedure of numerical calculation of a solution toward the exact solution.

There are only a few problems in nonlinear chromatography for which an algebraic solution exists. Except in the case of the Thomas model (see later, section II), such a solution is available only when the deviation of the isotherm from linear behavior is small enough for the solution of a perturbation theory to be realistic. When the concentrations are very low, the equilibrium isotherm is practically linear and the influence of the concentration on the shape of the band profile is negligible [1], as explained in Chapter IV. At high concentrations, however, this is no longer true; the equilibrium isotherms are curved, a different velocity is associated with each concentration, and the retention factor depend on the concentrations. As a consequence, as explained in Chapter V, each concentration seems to be moving at a different velocity. This phenomenon has a considerable influence on the band profiles. Due to the concentration dependence of the velocity associated with each concentration, band asymmetry arises, one side of the profile becoming self-sharpening and the other diffuse. Thus, the nonlinear behavior of the equilibrium isotherm is the most basic feature of *nonlinear chromatography*. At high concentrations, when the isotherm is strongly nonlinear, the concentration of the injected band has often more influence on the elution band profile than diffusion and the mass-transfer resistances combined.

By contrast with the influence of the solute concentration under nonlinear conditions, which tends to cause a self-sharpening of one side of the band profile (usually its front, most isotherms being convex upward, see Chapter II) and an increase of the corresponding concentration gradient, diffusion and the mass-transfer resistances are dissipative. They have a smoothing effect on both sides of the profiles. The sharper the profile, the higher the concentration gradient and the faster the diffusion. The sharper the profile, the more rapidly the concentration tends to vary when the band passes, the higher the mass-transfer resistances and the larger their dispersion effect. Thus, diffusion and mass-transfer resistances tend to counteract the influence of the nonlinear isotherm on the profile, until a balance is reached between these effects. The shock layer theory (see section V) gives profound insights into the nature and extent of this steady state effect. Finally, when the isotherm is nonlinear and the concentration is high enough, diffusion and the mass-transfer resistances may affect not only the profile but also the position of the band maximum, *i.e.*, its retention time.

In this chapter, we study the influence of the dispersive effects of axial dispersion and the mass transfer resistances under moderately nonlinear conditions, when the influence of a nonlinear isotherm begins to be observed. In a transition range, the influence of the isotherm on the band shape grows progressively with increasing concentration while the relative influence of diffusion and the mass-transfer resistances decreases. What happens in this range is of interest both for the analysts (particularly those involved in trace analyses) and for the engineers who might encounter such cases, particularly for difficult purifications, when the separation factor is low and the purity of the recovered fractions must be high.

I - INFLUENCE OF AXIAL DIFFUSION
IN NONLINEAR CHROMATOGRAPHY

There are no exact analytical solutions for the elution problem in non-ideal, nonlinear chromatography when the nonideal behavior originates from diffusion. However, there are some interesting and useful approximate algebraic solutions in cases in which the nonlinear behavior of the isotherm is small but cannot be ignored entirely, in other words, when the isotherm can be approximated by a two-term expansion

$$q = a_1 C + a_2 C^2 \qquad \text{(VII-1)}$$

and when C is small enough that $a_1 \gg a_2 C$. This approach permits a study of the onset of nonlinear behavior in chromatography. Equation VII-1 is the two-term expansion of the isotherm equation near the origin ($C = 0$). Note that if eqn. VII-1 is used in place of a Langmuir isotherm, we have $b = -a_2/a_1$ (as results from a two-term Taylor expansion of eqn. II-2). Thus the extent of nonlinear behavior of the isotherm in the concentration range sampled by the band is determined by $a_2 C_M/a_1$, where C_M is the maximum concentration of the band. This factor should be less than 0.05 for consistency with the assumption made above.

A first approximate solution of the chromatographic equation with diffusion but no mass-transfer resistances was given by Houghton [2]. Later Haarhoff and van der Linde [3] published a different and, overall, more correct approximate solution. Both solutions make simplifying assumptions on the basis of the assumed slight deviation of the isotherm from linear behavior. Because of the important differences in the nature of some of these assumptions, the two solutions are presented successively and then compared.

The problem discussed in this section is mathematically formulated by the combination of the following two equations:

$$\frac{\partial C}{\partial t} + F\frac{\partial q}{\partial t} + u\frac{\partial C}{\partial z} = D_L\frac{\partial^2 C}{\partial t^2} \qquad \text{(VII-2)}$$

$$q = a_1 C + a_2 C^2 \qquad \text{(VII-3)}$$

Equation VII-2 is the mass balance of the equilibrium–dispersive model, eqn. VII-3 the equilibrium isotherm. The system of equations must be completed by appropriate initial and boundary conditions. Note that if the actual isotherm is accounted for by the Langmuir equation, the second order expansion is

$$q = aC(1 - bC)$$

so

$$a_1 = a \qquad \text{and} \qquad a_2 = -ab$$

The two solutions that will be discussed below differ, among others things, by different choices of the initial and boundary conditions, by the use of different simplifications, and by different mathematical derivations. Still, both solutions follow a similar approach. They modify the partial differential equation (eqn. VII-2) into a Burgers equation and use the Cole–Hopf transform to solve it. In both cases, the simplifications made to this equation in order to obtain an equation that has an algebraic solution must be considered closely. An ill-considered modification of a mass balance equation may result in an equation which no longer conserves mass. This is what happens in the case of Houghton's equation [1]. This causes some difficulties in certain applications of the corresponding solution.

1 - Houghton Solution

The model of chromatography studied by Houghton [2] assumes that the column is infinite in length and that the injection is made "in the middle" of this column, at position $x = 0$. Thus, this is an "open–open" problem. Actually, the injection is part of the initial condition, not of the boundary condition (see below, eqns. VII-4 and VII-5 and Chapter III, Section I, subsections 1c,d).[1] The initial condition is

$$C(x,0) = C_0(x) = C_0 \qquad \text{if } |x| < \frac{L_0}{2}$$

$$= 0 \qquad \text{if } \frac{L_0}{2} \le |x| \qquad \text{(VII-4)}$$

The initial values of q are derived by combining eqn. VII-3 and VII-4. There is no actual boundary condition because the column is supposed to be infinite in length, so we may also write

$$C(-\infty, t) = 0 \qquad \text{(VII-5)}$$

---

[1]For a solution of a "close–open" problem, with the injection made as a conventional Dirac boundary condition, see eqn VII-69.

We have discussed earlier (Chapter IV, sections I-3c and I-3d) the lack of realism of this set of initial and boundary conditions and the difficulties that arise from their use.

Since the column length is infinite, the problem to be solved now is an initial value problem and L_0 is the injection interval. Provided that L_0 is very small compared to the length of actual columns which are used to acquire the actual elution band profiles with which the solution will eventually be compared, this initial condition is approximately equivalent to the injection of a rectangular pulse as a boundary condition.

The solution is obtained (see Appendix A) by rewriting eqn. VII-2 in a coordinate system moving along the column at the same velocity as the concentration $C = 0$ of the band. Further rearrangement and the use of second order expansions allows the transformation of eqn. VII-2 into the following Burgers equation [1,2]

$$\frac{\partial C}{\partial t} + C \frac{\partial C}{\partial \eta} = \frac{D_R}{(\lambda u_R)^2} \frac{\partial^2 C}{\partial \eta^2} \tag{VII-6}$$

with

$$\eta = -\frac{x - u_R t}{\lambda u_R}$$

$$\lambda = \frac{2 F a_2}{1 + F a_1} = \frac{2 F a_2}{1 + k_0'}$$

$$u_R = \frac{u}{1 + F a_1} = \frac{u}{1 + k_0'}$$

$$D_R = \frac{D_L}{1 + F a_1} = \frac{D_L}{1 + k_0'}$$

The Burgers' equation is solved by applying the Cole–Hopf nonlinear transform, which leads to the solution. The mathematics are detailed in Appendix A. Jaulmes et $al.$ [4] simplified the solution by assuming a Dirac injection instead of a rectangular pulse injection (the injection remains done as an initial condition, as in Houghton [2]). They obtained the following equations defining the concentration profile at elution ($i.e.$, at $x = L$)

$$X = \left| \frac{e^{-\tau^2/2}}{\sqrt{2\pi} \left[\coth m + \mathrm{erf}\left(\frac{\tau}{\sqrt{2}} \right) \right]} \right| \tag{VII-7}$$

where X, m, and τ are the dimensionless concentration, sample size, and time, respectively. They are given by the following equations

$$X = |b| \, C \frac{k'_0}{1 + k'_0} \sqrt{\frac{u^2 t}{2 D_L (1 + k'_0)}} \tag{VII-8}$$

$$\tau = \frac{k'_0 L}{(1 + k'_0) \sqrt{\frac{2 D_L t}{1 + k'_0}}} \frac{t_{R,0} - t}{t_{R,0} - t_0} \tag{VII-9}$$

$$m = \frac{L u}{2 D_L} \left[\frac{k'_0}{1 + k'_0} \right]^2 L_f \tag{VII-10}$$

where L_f is the loading factor, or ratio of the sample size to the column saturation capacity (see Chapter V, eqn. V-55). Obviously, there is no such thing as a saturation capacity with a parabolic isotherm. Because there are in fact no parabolic isotherms reported in the literature, we choose to use here the loading factor as it is defined for the Langmuir isotherm which has the parabolic isotherm (eqn. VII-3) for two-term expansion (i.e., has the same tangent and curvature at the origin),

Examples of concentration profiles along the column calculated with the Houghton equation are given in Figure VII-1. The profiles in Figure VII-1 are plotted in the band reference system of coordinates, i.e., as functions $C(\xi)$ (with $\xi = x - u_R t$, see eqn. VII-A-4), at increasing values of time. In the linear case (Figure VII-1a), the band remains symmetrical with respect to its center. Diffusion causes a broadening of the band which becomes progressively wider and shorter (the area is conserved by the solution under linear conditions). The nonlinear behavior of the isotherm causes an asymmetry of the band in the direction corresponding to that of the deviation of the isotherm from linear behavior. If $\lambda > 0$ (convex-downward or anti-Langmuir isotherm, see Appendix A), the band has a sharp rear side and a diffuse front side; its rear side moves more slowly than the band center. The converse is true if $\lambda < 0$ (convex upward or Langmuirian isotherm). The result obtained is similar to the one derived from perturbation analysis (see later, section III). If D_L tends toward 0, one side of the profile becomes steeper and steeper and tends toward a vertical line, i.e., a concentration shock, the solution of the ideal, nonlinear model (see Chapter V). Finally, we see that the position of the peak changes with the value of D_L. This is further illustrated in Figure VII-2a which shows a series of elution curves corresponding to a Dirac initial condition (eqn. VII-4) [4a]. Note that when the apparent axial dispersion coefficient increases, the peak height decreases and the retention time increases at first, since the influence of the nonlinear behavior of the isotherm on both the peak shape and the retention time decreases. At very high values of the axial dispersion coefficient, however,

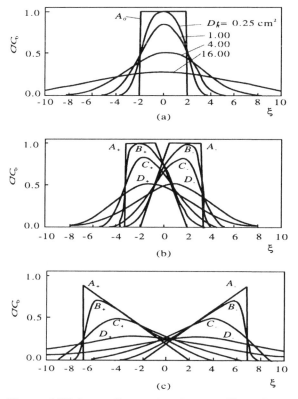

Figure VII-1.   Concentration profiles along the column derived from the Houghton equation [2]. Plots of the relative concentration, C/C_0, versus $\xi = x - Ut$.
Top.   Linear conditions, $\lambda = 0$. The product $D_L t_R$ indicates the extent of diffusive transport. Center. Nonlinear conditions, incomplete evolution of the profiles (t = 100 sec). Bottom. Nonlinear conditions, profiles fully developed (t = 400 sec).
Parabolic isotherm: $q = a_1C + a_2C^2$ $L = 25$ cm; $F = 0.25$. Values of D_R (cm^2/sec): A, 0; B, 0.0025; C, 0.10; D, 0.40. Extent of nonlinear behavior: subscript 0, $\lambda = 0$; subscript +, $\lambda C_0 = 0.05$; subscript -, $\lambda C_0 = -0.05$ [2].
Reproduced with permission. ©*1988 American Chemical Society.*

the retention time of the peak maximum decreases again [4b], as can be seen when determining the time for which the concentration distributions in eqns. IV-9 or IV-10 are maximum (see Figure VII-2b).

Jaulmes *et al.* [4a] showed that, if we differentiate eqn. VII-7, we find that the position of the peak maximum is given by

$$X_M = \frac{|\tau_M|}{2} \qquad\qquad \text{(VII-11)}$$

where τ_M is the reduced elution time (eqn. VII-9) of the peak maximum.

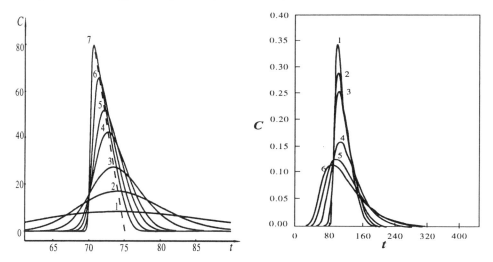

Figure VII-2.   Influence of Axial Dispersion on the Elution Profiles in Nonlinear Chromatography. Left, profiles calculated by Jaulmes *et al.* [4a]. *Reprinted with permission.* ©1988 American Chemical Society. **Right, profiles calculated with eqns. VII-7 to VII-10, a Langmuir isotherm, an injection made as a rectangular pulse boundary condition, and D_L (cm^2/s): 1, 0.004; 2, 0.008; 3, 0.012; 4, 0.040; 5, 0.080; 6, 0.120 [4b].** *Reprinted with permission by Marcel Dekker Inc.*

Combination of this relationship and eqns. VII-8 and VII-9 gives

$$\frac{t_{R,0} - t_M}{t_{R,0} - t_0} = 2bC_M \qquad \text{(VII-12)}$$

This is a convenient relationship which allows the correct prediction of the variation of the retention time (or elution time of the peak apex) with the sample size, when we know the isotherm. Indeed, experimental results confirm that the retention time of a band varies linearly with increasing maximum concentration of this band (which is not exactly proportional to the sample size) [4a]. It is important to observe that dt_M/dC_M is finite, and that, contrary to conventional wisdom, the retention time of a band remains constant at low concentrations or at low sample sizes only insofar as its variations are smaller than the precision of its measurement. Unfortunately, it is not possible to perform measurements of the peak height and the retention time that are accurate enough to allow the derivation of reasonable estimates of the curvature of the isotherm from a plot of the retention time versus the maximum concentration of the band.

Furthermore, it should be noted that the Houghton solution is approximate. The simplification made in replacing the mass balance equation (eqn. VII-9) by the Burgers' equation (eqn. IV-C-11) neglects the term λC in the denominator of the last two terms of eqn. IV-C-8 [1]. In the process, the mass balance equation was modified. The assumption made is incorrect to the first order, and the new solution does not conserve mass [1]. The profiles obtained at low values of the axial dispersion coefficient are different from those obtained as solutions of the ideal model ($D_L = 0$), showing that the solution does not give the correct limit. There is a significant loss in the peak area [1]. This loss is a function of the deviation of the isotherm from linear behavior, *i.e.*, it increases with increasing sample size. The solution is not sufficiently precise for quantitative modeling of the band profiles [5].

2 - Solution of Haarhoff and van der Linde

Haarhoff and van der Linde [3] studied the same problem as Houghton and followed the same derivation until the Burgers' equation (eqn. VII-6) is obtained. Then, they made a further simplification, which Houghton [2] did not do, but which turns out to correct the equations of the solution for the lack of mass conservation. They observed that the concentration profile of an elution band is practically always equal to zero, except around the band maximum (*i.e.*, within *ca* ± three standard deviations from the band maximum in linear chromatography). Accordingly, they suggested calculating the effect of the apparent dispersion term in eqn. VII-9 at the limit retention time, *i.e.*, at the retention time under linear conditions, $t_{R,0} = (1 + k_0')t_0$. This means replacing the term $L/\sqrt{2D_L t/(1 + k_0')}$ in eqn. VII-9 by $\sqrt{N}$ [1,3]. The solution is still given by eqn. VII-7 but with new, different definitions for the dimensionless parameters

$$X = |b| \, C \frac{k_0'}{1 + k_0'} \sqrt{N} \qquad \text{(VII-8b)}$$

$$\tau = \frac{k_0' \sqrt{N}}{1 + k_0'} \frac{t_{R,0} - t}{t_{R,0} - t_0} \qquad \text{(VII-9b)}$$

$$m = N \left[\frac{k_0'}{1 + k_0'} \right]^2 L_f \qquad \text{(VII-10-b)}$$

where $N = Lu/(2D_L)$.

The solution derived by Haarhoff and van der Linde [3] is more accurate than the one derived by Houghton [2], although it contains one more simplification and is only an approximate solution of the Burgers' equation while the Houghton solution is an exact solution of this equation. Much later, Golshan-Shirazi and Guiochon [6] showed that the equation which was really solved by Haarhoff and van der Linde can be derived from the last correct equation in Houghton's derivation (eqn. IV-A-8) by taking the concentration profile derived from the ideal model (see Chapter V) corresponding to the isotherm in eqn. VII-3 and using it to replace the factor $(1 + \lambda C)$ in the denominators of the last two terms of this equation, instead of neglecting λC and replacing the factor by 1. Then, we get

$$C = \frac{1}{2b} \frac{t_{R,0} - t}{t_{R,0} - t_0}$$

$$1 + \lambda C = \frac{1 + k'}{1 + k'_0} = \frac{t}{t_{R,0}} \qquad \text{(VII-13)}$$

This simplification is correct to the first order and conserves mass. This explains the superiority of the solution of Haarhoff and van der Linde [3] over that of Houghton [2] for all practical applications.

3 - Comparison between the two solutions

The question of this comparison, first raised in the literature by Lucy *et al.* [5], was discussed in detail by Golshan-Shirazi and Guiochon [6]. These authors showed that the Houghton equation is incorrect even at low values of the loading factor. The peak area loss is already 2.5% for a loading factor of 0.2%. The solution of Haarhoff and van der Linde also fails progressively as the sample size is increased but this is because, when the concentration increases markedly, the actual isotherm deviates more and more from its parabolic expansion at the origin. It is a general observation in chromatography that band profiles are very sensitive to the isotherm and that an excellent accuracy in their measurement is required in order to achieve a good agreement between calculated and measured band profiles [1]. So, in contrast with that of the Houghton equation, the failure of the Haarhoff and van der Linde solution at high concentrations is due to a model error in the isotherm, not to an incorrect simplification of the mass balance equation.

Still, the solution of Haarhoff and van der Linde provides very well for what it set out to do, to account for the progressive onset of column overloading and to predict band profiles at low or moderate values of the loading factor. Golshan-Shirazi and Guiochon [6] estimated that, when the equilibrium isotherm is accounted for by the Langmuir model, the solution of Haarhoff and van der Linde is valid for values of the loading factor not exceeding 0.2%.

From the mathematical point of view, the two solutions are very similar, both using the Burgers equation and the Cole–Hopf transform (a well-known approach to solve nonlinear parabolic problems). However, Houghton's solution is based on an open-open problem, the column being defined in the entire 1-D space, $(-\infty, +\infty)$, and uses as the initial condition a Dirac function in space, $C(x, 0) = A\delta(x)$, to introduce the sample. Finally, when the sample size decreases indefinitely, the profile tends toward a Gaussian distribution, which is the correct solution under linear conditions. Haarhoff and van der Linde [3] solved a closed-open problem (as explained in the appendix of their paper), used also an initial condition to define the sample injection and their solution tends toward a Gaussian distribution under linear conditions. Some of these conditions are not realistic in chromatography. Actually, the chromatographic column is close-open (i.e., semi-infinite), there is a free initial condition $(C(x, 0) = 0 \ \forall \ x)$ and, as a first approximation, the boundary condition is a Dirac function in time, $C(0, t) = B\delta(t)$. Accordingly, the limit of the solution under linear conditions is a Gaussian function multiplied by $\frac{x}{t}$.

4 - Further Comments

There are several ways to handle axial dispersion in nonlinear chromatography. The most simple is the use of numerical methods of calculations. With the powerful microcomputers available nowadays, this is the simplest and fastest method. Solutions of all but the most complex problems are available in minutes (in 1986 it used to take a fast algorithm 30 minutes to calculate a band profile for a 5000 plate column with a DEC 750; in 1999 it took a Pentium II 200 MHz one hour to calculate the band profiles after 200 cycles on an SMB with eight 2000-plate columns). However, the usefulness of numerical solutions to understand physical problems in depth is limited. We need the results of an analytical approach instead. Because we have algebraic solutions of two chromatographic models, the linear, nonideal model and the nonlinear ideal model, we may attempt to use one of the classical perturbation methods available in connection with either solution. The solutions derived by Houghton [2] or by Haarhoff and van der Linde [3] are examples of a perturbation of the solution of the linear nonideal model using a two-term expansion of the isotherm. These solutions account for what happens when the concentration of the injected band is such that the equilibrium isotherm is moderately nonlinear. Their study informs us on what happens at the onset of column overloading.

The perturbation of the ideal nonlinear model, the solution of which contains a shock or discontinuity (see Chapter V), is entirely different from that of a continuous solution, such as that of the nonideal linear model. The perturbation methods needed to account for the influence of a small amount of axial dispersion on the solution of the ideal nonlinear model are called *singular perturbations*. Initially, these methods were developed for investigations of the boundary layer in hydrodynamics. In the boundary layer, the viscosity of the fluid cannot be ignored and the thickness of this layer is very small. This makes the boundary layer a singular area. The shock layer in nonlinear chromatography can be seen as a moving boundary layer. Its limit at zero dispersion coefficient is a true concentration shock or discontinuity. In this region, the diffusion term, $\partial^2 C/\partial x^2$ is very large and the effect of diffusion cannot be entirely neglected. The physical shock that we observe is the result of a dynamic compensation between the dispersive effects of diffusion and the self-sharpening effect of thermodynamics arising from the nonlinearity of the isotherm. What makes the singular perturbation so useful is that it represents exceptionally well the properties of the boundary or shock layer. This approach will be discussed in section V. More fundamental details on the singular perturbation method can be found in the mathematical literature [7].

II - INFLUENCE OF THE MASS-TRANSFER RESISTANCES IN NONLINEAR CHROMATOGRAPHY

In chromatography, the convection caused by the mobile phase stream percolating through the column displaces any possible equilibrium which could arise locally between the two phases by constantly moving the liquid phase with respect to the solid phase. If any local equilibrium could exist, even temporarily, it does not subsist long. The incoming liquid phase, which pushes away the solution in equilibrium with the solid has a different composition. A radial concentration gradient in the two phases appears and mass transfer takes place in the direction which tends to relax this gradient. Mass transfer kinetics is finite, however, and the mass transfer resistances prevent the establishment of any equilibrium.

The rate of the mass transfer kinetics depends on the nature of the packing material used and on the characteristics of the solute considered. This rate usually decreases with increasing molecular weight of the solute and also with increasing polarity of this solute or rather with its increasing ability to give selective, high-energy interactions with the stationary phase. The models discussed here are particularly suited to solutes belonging to this last group, the one of small molecules (*i.e.*, solutes with a relatively large molecular diffusivity) that are able to interact strongly with the solid surface of an adsorbent, *e.g.*, strongly basic compounds eluted on RPLC phases, with an unbuffered liquid phase. Far from being exceptional, this situation is frequently encountered in routine separations, particularly in the purification of enantiomers in chiral separations. These models are also suitable for the investigations of the band profiles of biopolymers when their mass transfer kinetics is slow in comparison to their diffusion rate, which is not infrequent. However, in many real situations, the mass transfer kinetics is complex and cannot be solved by using any of the simple models that are presented in this work.

1 - The Thomas Model and its Solution

As explained earlier (see Chapter II, section IV-5), there are several models available to account for the mass transfer kinetics in a chromatographic column. The most important of these models are the Langmuir kinetic model, the first-order kinetic model, and the linear driving force models. Although the last of these models, and particularly the liquid-film linear driving-force model, have a simpler formulation, it is remarkable that a closed-form algebraic solution has been obtained only with the former kinetic model.

In 1944, Thomas [8] derived a solution for the Riemann problem (*i.e.*, for the breakthrough curve) of a model of chromatography combining the mass balance equation of the ideal model (no axial diffusion) and the (second-order) Langmuir kinetics. This model is written:

$$u\frac{\partial C}{\partial x} + \frac{\partial C}{\partial t} + F\frac{\partial q}{\partial t} = 0 \qquad \text{(VII-14a)}$$

$$\frac{\partial q}{\partial t} = k_a\, C\, (q_s - q)\, -\, k_d\, q$$

$$\text{(VII-14b)}$$

where k_a and k_d are the rate constants of adsorption and desorption, respectively, and q_s is the saturation capacity of the adsorbent. This kinetic model was used by Langmuir to derive the Langmuir equilibrium isotherm (obtained when $\partial q/\partial t = 0$), with $b = k_a/k_d$. The derivation of the algebraic solution of the Thomas model is complex and beyond the scope of this book. The initial and boundary conditions are:

$$C(x,0) = q(x,0) = 0$$
$$C(0,t) = C_0 \qquad \text{for} \quad t > 0 \qquad \text{(VII-14c)}$$

The solution is [8]

$$C(x,t) = C_0\frac{I_0\left(2k\sqrt{\alpha y z}\right) + \Phi\left[k(1 + \kappa C_0)y, \frac{\alpha k z}{1+\kappa C_0}\right]}{I_0\left(2k\sqrt{\alpha y z}\right) + \Phi\left[k(1 + \kappa C_0)y, \frac{\alpha k z}{1+\kappa C_0}\right] + \Phi[\alpha k z, k y]}$$

$$\text{(VII-14d)}$$

with

$$\Phi(u,v) = e^u \int_0^u e^{-s} I_0(2\sqrt{vs})ds \qquad z = \frac{x}{V} \qquad \alpha = \frac{k_a q_s}{k_b}$$

$$y = \frac{\varepsilon}{1-\varepsilon}\left(t - \frac{x}{V}\right) \qquad \kappa = \frac{k_a}{k_b} \qquad k = \frac{(1-\varepsilon)k_b}{\varepsilon}$$

It includes a zero-order modified Bessel function of the first kind, $I_0(2k\sqrt{\alpha yz})$. Accordingly, it is complicated to program and this algebraic solution would not be useful to calculate numerical solutions of the problem in specific cases, which is one of the main reasons for which algebraic solutions are looked for in engineering.

Later, Goldstein [9] derived a solution of the Thomas model that is valid in the case of a rectangular pulse injection of any width and height. This solution is too complex to be derived or even discussed here. It has been discussed elsewhere [1]. Wade et al. [10] derived a simpler solution of the Thomas model that is valid in the case of a Dirac injection. This solution is the limit toward which the solution of Goldstein tends when the width of the rectangular sample pulse tends toward 0 (and accordingly, its height toward infinity, at constant sample size). It is written

$$\frac{C}{C_{d,0}} = \frac{1 - e^{-\gamma b C_{d,0}}}{\gamma b C_{d,0}} \frac{\gamma\sqrt{\frac{k_0'}{\theta}} I_1(2\gamma\sqrt{k_0'\theta})e^{-\gamma(\theta - k_0')}}{1 - T(\gamma k_0', \gamma\theta)\left[1 - e^{-\gamma b C_{d,0}}\right]} \qquad \text{(VII-14e)}$$

with

$$T(u,v) = 1 - J(u,v) = e^{-v}\int_0^u e^{-t} I_0(2\sqrt{vt})dt$$

where I_1 is the first-order modified Bessel function of the first kind and $C_{d,0}$ is the ratio of the number of moles of solute injected to the column dead volume, V_0, with

$$C_{d,0} = \frac{n}{V_0} = \frac{n}{\varepsilon SL} = \frac{A_p}{t_0} = C_0\frac{t_p}{t_0}$$

According to Wade et al. [10], the dimensionless band profile depends on three dimensionless parameters, (1) k_0', the retention factor at infinite dilution, $\gamma = k_a t_0$, (2) the product of the rate constant and the hold-up time which characterize the mass transfer kinetics in the column, and $bC_{d,0}$, which characterizes the degree of column overloading. Wade et al. [10] found that their solution accounts well for the elution profiles of p-nitrophenyl-α-D-mannopyranoside on silica-bonded Concanavalin A with a high concentration buffer. This result is important because the separation studied belongs to the bioaffinity chromatography mode. The mass transfer kinetics of the

separation studied was very slow and it was found that this property is characteristic of this mode of chromatography [10]. Bioaffinity chromatography
has since been little used in preparative liquid chromatography. The desired product is effectively retained with great selectivity but it is extremely
difficult to recover without nearly destroying the column.

It should be emphasized that all these algebraic solutions of the Thomas
model are exact. No simplifications are introduced in their derivation. The
price to pay for that is the extreme complexity of these algebraic expressions.

2 - Solutions of other Lumped Kinetic Models

The Thomas model is often also called the reaction model, in order to dis-
tinguish it from the transport model, another kinetic model that is obtained
when using one of the linear kinetic models (e.g., the liquid film linear driving force model, see later, eqn. VII-58b) [1]. This name of reaction model
should not cause confusion with the matters discussed in the previous chapter (Chapter VI - Ideal Nonlinear Reaction Chromatography). There are no
actual chemical reactions involved in the Thomas model. The word "reaction" in the name of the reaction model refers to the adsorption–desorption
equilibrium of the adsorbate,

$$M(\text{dissolved}) \rightleftharpoons M(\text{adsorbed})$$

There are also a reaction–dispersive and a transport–dispersive model. These
two models are obtained by adding an axial dispersion term to the mass balance equation (to eqn. VII-14a in the case of the reaction dispersive model),
i.e., by using the mass balance equation of the equilibrium–dispersive model
instead of that of the ideal model. These models are not more complex to
solve numerically than the models without the axial dispersion term but
they have no analytical solutions.

The solution of the liquid-film linear driving-force model was discussed
earlier, in the linear case (Chapter IV, section II). When the isotherm is no
longer linear, the kinetic equation (see Chapter II, section IV; Chapter III,
section I-2b, eqn. IV-13) must be rewritten accordingly

$$\frac{\partial q}{\partial t} = -k_f \left[q - f(C)\right] \qquad \text{(VII-15)}$$

In this case, the method of the Laplace transform that was used to solve the
problem in the linear case cannot be used. The corresponding equations are
too complex to be solved. The only procedure available to obtain solutions
of the problem is to resort to numerical calculations. Early in 1962, Ting

and Chu [11] discussed this problem and suggested a solution using analog computing, a calculation process which was used at the time because it was still competitive with the then very slow digital computers. This process was based on electrical analogies. In practice, it was limited to nearly linear conditions. In 1974, Villermaux [12] gave the result of another numerical calculation, also performed with an analog method, under linear conditions. Now, digital computing allows the easy derivation of accurate numerical solutions (see Chapter IX).

When k_f is very large, the peak is strongly unsymmetrical and exhibits a very steep front. It much resembles the peak obtained with the same isotherm and the same loading factor but with the ideal model. It is merely smoothed by the influence of the mass-transfer resistances that rounds off the edges and transforms the concentration shock into a steep concentration front (see Figure VII-3). The profile of this front, which is called a *"shock layer"* will be discussed later (see section V). When k_f increases indefinitely, the front tends toward a discontinuity or shock (much like when D_a tends toward 0 in the equilibrium dispersive model).

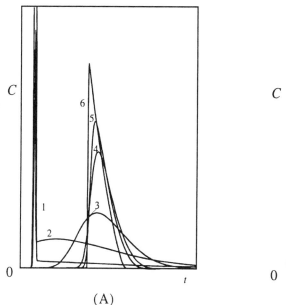
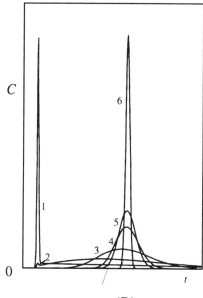

(A)

(B)

Figure VII-3   Influence of Mass Transfer Kinetics on Elution Band Profile [10]. **A, Nonlinear Chromatography,** k_f (s^{-1}): **1, 0.004; 2, 0.02; 3, 0.1; 4, 0.5; 5, 1; 6, 50. B, Linear Chromatography,** k_f (s^{-1}): **1, 0.004; 2, 0.01; 3, 0.1; 4, 0.5; 5, 1; 6, 5.** *Reproduced with permission.* ©*1988 American Chemical Society.*

When the rate coefficient decreases and the mass transfer resistance becomes high, the dispersion factor is dominant, the band profile becomes less unsymmetrical, and the process tends to become quasi-linear, see Figure VII-3 (besides, for a given loading factor, the peak is wider and shorter, hence the nonlinear effect is reduced at constant loading factor). The first moment[2] of the band remains independent of k_f in the lumped kinetic model (eqn. VII-14, first order kinetics). The second and third moments, however, are functions of k_f. The third moment increases with decreasing value of k_f and the peak becomes increasingly unsymmetrical (a symmetrical peak has a third moment equal to 0) at low values of k_f. Then, the peak profile again deviates significantly from a Gaussian profile. When k_f decreases, the first moment remains constant while the retention time of the peak maximum decreases (because of the increasing degree of asymmetry of the band). At low values of k_f, the velocity of the peak maximum tends toward u and its retention time toward t_0, the hold-up time (see Figure VII-3).

The variation of the retention time with decreasing mass-transfer rate coefficient is, to some extent, similar to the variation of the retention time with increasing dispersion coefficient derived in the previous section. This feature is common to the different forms of nonlinear, nonideal chromatography, which is independent of the exact reason for the nonideal behavior, axial dispersion or mass-transfer resistances. When the dissipative or dispersive effect increases from zero, the retention time increases first, but slightly (in the case of a convex upward isotherm), because the intensity of the nonlinear effect decreases. Then, it goes through a maximum and, eventually, decreases toward the hold-up time. At the same time, there is a progressive transition from a situation in which the nonlinear character dominates entirely toward one in which the nonideal character contributes markedly to control the profile of chromatographic bands [12]. Actually, however, whether the behavior of the peak profile is linear or nonlinear depends only on the maximum concentration of the band, on the one hand, and on the coefficients of the isotherm, on the other hand, i.e., it depends on the degree to which the isotherm behavior deviates from linear in the concentration range sampled by the band, that is from 0 to the maximum concentration of the band. Axial dispersion and the mass transfer resistances may affect this balance only insofar as it can reduce the maximum concentration of the band.

III - NONLINEAR CHROMATOGRAPHY WITH BOTH AXIAL DIFFUSION AND MASS-TRANSFER RESISTANCES (PERTURBATION SOLUTION)

When the solute concentration becomes large, the isotherm equation can-

---

[2]See definition of the moments in Chapter III, section II-3.

not be accounted for by a linear relationship, *i.e.*, the isotherm deviates from its initial tangent. Then, the nonlinear behavior of the isotherm must be taken into account. If the deviation from linear behavior is moderate and if the lumped mass-transfer coefficient, k_f, is relatively large, this problem can be linearized. Otherwise, a different approach must be selected because the nonlinearity of the isotherm becomes the major factor controlling the peak profile.

Let us write the chromatographic model as in eqns. IV-23 and IV-24 in Chapter IV, but after introducing a nonlinear isotherm equation

$$u\frac{\partial C}{\partial x} + \frac{\partial C}{\partial t} + F\frac{\partial q}{\partial t} = D_a\frac{\partial^2 C}{\partial x^2} \qquad \text{(VII-16a)}$$

$$\frac{\partial q}{\partial t} = -k_f\left[q - f(C)\right] \qquad \text{(VII-16b)}$$

where $f(C)$ is the isotherm equation. We will use here the initial and boundary conditions that correspond to the elution of a wide rectangular pulse. These conditions are

$$C(x, t = 0) = q(x, 0) = 0 \qquad \text{for} \quad x > 0 \qquad \text{(VII-17a)}$$

$$C(x = 0, t) = \begin{cases} C_0 & 0 < t < t_p \\ 0 & t_p \leq t \end{cases} \qquad \text{(VII-17b)}$$

where t_p is the duration of the rectangular injection pulse.

We can rewrite eqn. VII-16 as

$$q = f(C) - \frac{1}{k_f}\frac{\partial q}{\partial t} \qquad \text{(VII-18)}$$

When k_f increases indefinitely, the second term of the RHS of this equation becomes small and tends toward 0. The zero-order and the first-order approximations of eqn. VII-18 are, respectively

$$q = f(C) \qquad \text{(VII-19a)}$$

$$q = f(C) - \frac{1}{k_f}\frac{\partial f(C)}{\partial t} \qquad \text{(VII-19b)}$$

If the concentration range investigated is such that the deviation of $f(C)$ from a linear behavior is moderate, we can replace the isotherm by its first two-term expansion and write

$$f(C) = a_1 C + a_2 C^2 \qquad \text{(VII-20)}$$

where a_1 and a_2 are constant and are such that, in the concentration studied, $a_2C \ll a_1$.

Combining these equations gives

$$u\frac{\partial C}{\partial x} + \frac{\partial C}{\partial t} - D_a\frac{\partial^2 C}{\partial x^2} + Fa_1\frac{\partial C}{\partial t} + Fa_2\frac{\partial C^2}{\partial t} - \frac{Fa_1}{k_f}\frac{\partial^2 C}{\partial t^2} - \frac{Fa_2}{k_f}\frac{\partial^2 C^2}{\partial t^2} = 0$$
$$\text{(VII-21)}$$

Neglecting the last term, because it is a second order term, and simplifying, we obtain

$$u\frac{\partial C}{\partial x} + (1 + Fa_1)\frac{\partial C}{\partial t} - D_a\frac{\partial^2 C}{\partial x^2} + Fa_2\frac{\partial C^2}{\partial t} - \frac{Fa_1}{k_f}\frac{\partial^2 C}{\partial t^2} = 0 \quad \text{(VII-22)}$$

The solution of this equation is a distribution of the concentration $C(x,t)$ over the column length which is a function of time, *i.e.*, it migrates along the column. Let us write it as the sum of two terms

$$C = C_l + C_p \tag{VII-23}$$

where C_l is the solution of the linear, zero-order, approximation, *i.e.* of equation VII-16 with $q = a_1C$, which writes

$$u\frac{\partial C_l}{\partial x} + (1 + Fa_1)\frac{\partial C_l}{\partial t} - D_a\frac{\partial^2 C_l}{\partial x^2} = 0 \tag{VII-24}$$

and C_p is a perturbation term which tends toward 0 with a_2C^2. Combination of eqn. VII-22 with eqn. VII-24 gives

$$u\frac{\partial C_p}{\partial x} + (1 + Fa_1)\frac{\partial C_p}{\partial t} - D_a\frac{\partial^2 C_p}{\partial x^2} = -Fa_2\frac{\partial C_l}{\partial t} + \frac{Fa_1}{k_f}\frac{\partial^2 C_l}{\partial t^2} \tag{VII-25}$$

Because a_2 and $1/k_f$ are both small, a_2C_p and C_p/k_f are of the second order and small enough to be neglected in the derivation of eqn. VII-25. This is a second-order, linear, nonhomogeneous PDE. Let us assume that the solution for C_p is of the following type

$$C_p = C_{p,1}\exp(\lambda_1 x + \lambda_2 t) \tag{VII-26a}$$

with

$$\lambda_1 = \frac{u}{2D_a} \tag{VII-26b}$$

$$\lambda_2 = -\frac{u^2}{4D_a(1 + Fa_1)} \tag{VII-26c}$$

Then, we can write

$$\frac{\partial C_{p,1}}{\partial t} - D_R \frac{\partial^2 C_{p,1}}{\partial x^2} = \frac{1}{1 + Fa_1}\left(-Fa_2 \frac{\partial C_l^2}{\partial t} + \frac{Fa_1}{k_f}\frac{\partial^2 C_l}{\partial t^2}\right)$$
$$\exp\left[-\frac{u}{2D_a}x + \frac{u^2 t}{4D_a(1 + Fa_1)}\right] \quad \text{(VII-27)}$$

where $D_R = 1/(1 + Fa_1)$. This equation is a nonhomogeneous diffusion equation. It can be solved using the method of the Green function [14]. Since the initial and boundary conditions are the same for C and for C_l, these conditions are homogeneous for C_p and $C_{p,1}$, i.e.,

$$C_{p,1}(0, x) = C_p(0, x) = 0 \quad \text{(VII-28a)}$$
$$C_{p,1}(t, 0) = C_p(t, 0) = 0 \quad \text{(VII-28b)}$$

In order to obtain the Green function, we need a boundary condition for $x = L$, i.e., at the end of the column. Since

$$C_{p,1}(t, L) = C_p(t, L)\exp\left[-\frac{u}{2D_a}L + \frac{u^2}{4D_a(1 + Fa_1)}t\right] \quad \text{(VII-29)}$$

$C_{p,1}(t, L)$ tends toward 0 when L tends toward infinity. Owing to this condition, the Green function for eqn. VII-27 is

$$g(x - \xi, t - \tau) = \frac{n}{4\pi D_R(t - \tau)}\left\{\exp\left[\frac{-(x - \xi)^2}{4D_R(t - \tau)}\right] - \exp\left[\frac{-(x + \xi)^2}{4D_R(t - \tau)}\right]\right\}$$
$$\text{(VII-30)}$$

Thus

$$C_{p,1}(x, t) = \int_0^t \int_0^L g(x - \xi, t - \tau)\varphi(\xi, \tau)d\xi d\tau \quad \text{(VII-31a)}$$

where

$$\varphi(\xi, \tau) = \frac{1}{1 + Fa_1}\left(-Fa_2 \frac{\partial C_l^2}{\partial \tau} + \frac{Fa_1}{k_f}\frac{\partial^2 C_l}{\partial \tau^2}\right)\exp\left[-\frac{u\xi}{2D_a} + \frac{u^2\tau}{4D_a(1 + Fa_1)}\right]$$
$$\text{(VII-31b)}$$

In eqns. VII-31a and VII-31b, ξ and τ are dummy integration variables. So

$$C_p(L, t) = \exp\left[\frac{uL}{2D_a} - \frac{u^2\tau}{4D_a(1 + Fa_1)}\right]\int_0^t \int_0^L g(L - \xi, t - \tau)\varphi(\xi, \tau)d\xi d\tau$$
$$\text{(VII-32)}$$

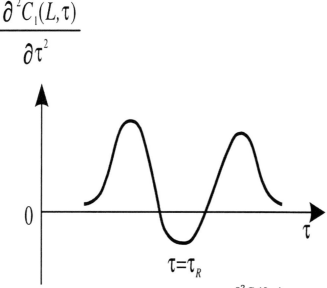

$$\frac{\partial^2 C_1(L,\tau)}{\partial \tau^2}$$

$$\tau = \tau_R$$

Figure VII-4   **Shape of the function:** $\frac{\partial^2 C_l(L,\tau)}{\partial \tau^2}$.

When D_a tends toward 0, $g(L - \xi, t - \tau)$ tends toward $\delta(L \pm \xi)$ and since the first integral is taken in the interval $(0, L)$, the equation VII-63 can be written

$$C_l = \frac{1}{1 + Fa_1} \int_0^t \left(-Fa_2 \frac{\partial C_l^2(L,\tau)}{\partial \tau} + \frac{Fa_1}{k_f} \frac{\partial^2 C_l(L,\tau)}{\partial \tau^2} \right)$$

$$\exp \left[-\frac{uL}{2D_a} + \frac{u^2 \tau}{4D_a(1 + Fa_1)} \right] d\tau \quad \text{(VII-33)}$$

or

$$C_1 = \frac{1}{1 + Fa_1} \exp \left[-\frac{uL}{2D_a} \right] \left\{ -Fa_2 I_1 + \frac{Fa_1}{k_f} I_2 \right\}$$

where

$$I_1 = \int_0^t \frac{\partial C_l^2(L,\tau)}{\partial \tau} \exp \left[\frac{u^2 \tau}{4D_a(1 + Fa_1)} \right] d\tau \tag{VII-34}$$

$$I_2 = \int_0^t \frac{\partial^2 C_l(L,\tau)}{\partial \tau^2} \exp \left[\frac{u^2 \tau}{4D_a(1 + Fa_1)} \right] d\tau \tag{VII-35}$$

Because both C_l and C_l^2 are Gaussian functions, the differential $\frac{\partial C_l^2(L,\tau)}{\partial \tau}$ is positive when $t_R < t$ and negative when $t < t_R$. A plot of the second order differential, $\frac{\partial^2 C_l(L,\tau)}{\partial^2 \tau}$, intersects the abscissa axis in two points, one at a finite value of τ and the other at infinity. The function is negative in the central part of the range (Figure VII-4) while it is positive on both sides. Finally, partial integration gives

$$I_1 = C_l^2(L, \tau) \exp\left[\frac{u^2 \tau}{4D_a(1 + Fa_1)}\right]\Big|_0^{\tau=t}$$

$$- \int_0^t \frac{u^2 C_l^2(L, \tau)}{4D_a(1 + Fa_1)} \exp\left[\frac{u^2 \tau}{4D_a(1 + Fa_1)}\right] d\tau \quad \text{(VII-36)}$$

So, the second part of I_1 is always negative while the first part tends toward 0 when $t \gg t_R$. Accordingly, I_1 is positive when t is small but becomes negative when t is sufficiently large. When a_2 is negative (Langmuirian isotherm), the band tails and the tailing increases with increasing deviation of the isotherm from linear behavior. The converse is true if $a_2 > 0$, i.e., in the case of an isotherm which is convex toward the axis of mobile phase concentrations.

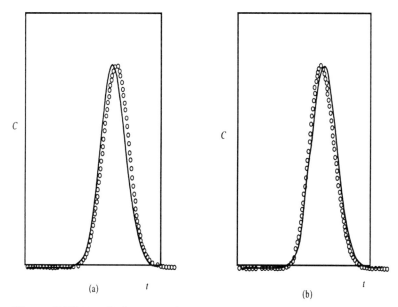

(a)

(b)

Figure VII-5   **Influence of the Sign of the Initial Curvature of the Iso therm in Nonlinear Chromatography [14]. Left, $L = 25$ cm; $u = 0.625$ cm/s; $F = 0.25$; $k_f = 10$ s^{-1}; $D_a = 0.001$ cm^{2}s^{-1}; $T = 10$; $C_0 = 0.1$. Linear isotherm, solid line, $a_1 = 12$; nonlinear isotherm, dashed line, $a_2 = +3.6$. Right, same numerical values for all parameters, except $a_2 = $ -3.6.**

The value of the integral I_2 (eqn. VII-35) depends on the second order derivative $\frac{\partial^2 C_l}{\partial \tau^2}$. The curves in Figure VII-4 show that this integral decreases with increasing τ around the central point, at $t = t_R$. On either side, far from the central point, I_2 is positive. Consequently, the band is broadened.

However, the influence of I_2 depends on $1/k_f$. It becomes negligibly small when k_f increases indefinitely. When the rate coefficient is sufficiently large the effect of the mass transfer resistance is similar to that of dispersion in linear chromatography, as shown by Giddings [15]. Figure VII-5 illustrates the influence of the initial curvature of the isotherm on the band profile in nonlinear chromatography, at low value of the loading factor.

The results of this perturbation analysis are the same as those derived from the numerical analysis of the same set of equations [16].

IV - THE ZERO-ORDER MOMENT AND THE BOUNDARY CONDITIONS OF CHROMATOGRAPHY

The zero-order moment of a chromatogram or of a concentration distribution along the column is equal to the peak area. It is an important characteristic of the peak. We need to explain here, however, that there are two different zero-order moments of a chromatographic band. The first one is the area of the elution profile. It is proportional to the integral of the detector signal and is given by $\int_0^\infty C_{out}(t)dt$, where $C_{out}(t)$ is the concentration of the solute at the column outlet, during the elution of the band. The other one is the area of the concentration profile along the column. It is the zero-order moment of the concentration function and is given by $\int_0^\infty C(L,t)dt$. Most chromatographers consider that the latter is also an invariant for a given amount injected and think that it is "obviously" proportional to the sample size. This is certainly correct for the first of these two moments, $\int_0^\infty C_{out}(t)dt$, but this is not so obvious for the second one, $\int_0^\infty C(L,t)dt$, especially in the cases in which there is axial dispersion, which is the general case in chromatography, because generally $\int_0^\infty C(L,t)dt \neq \int_0^\infty C_{out}(t)dt$.

In this section, we discuss the zero-moment of the concentration profile along the column. This zero-order moment is actually not a mass and its invariance must be demonstrated. As we will show, this invariance is achieved when the proper boundary conditions are selected for the calculation of the integral. It turns out that these conditions are also those used conventionally in chemical engineering and in chromatography [17,18]. This subsection also provides a method to discuss the mass conservation of chromatography equations that are similar to the Houghton equation.

1 - The Variation of the Zero-Order Moment of a Moving Band

The classical equation of the equilibrium dispersive model of chromatography

$$\frac{\partial C}{\partial t} + F\frac{\partial q}{\partial t} + u\frac{\partial C}{\partial x} = D_L\frac{\partial^2 C}{\partial x^2} \qquad \text{(VII-37)}$$

is a mass balance or mass conservation equation. Its integration along the column gives

$$\frac{d}{dt}\int_0^L S[C + Fq]dx = S(uC - D_L\frac{\partial C}{\partial x})_0 - S(uC - \frac{\partial C}{\partial x})_L \qquad \text{(VII-38)}$$

Obviously, the integral in the LHS of eqn. VII-38 is the mass of compound while the term $S(uC - D_L\partial C/\partial x)$ is a mass flux. Since, when $t = 0$ and $t = \infty$, we have $C = q = 0$ in the whole column, integration of equation VII-38 gives

$$S\int_0^\infty (uC - D_L\frac{\partial C}{\partial x})_0 dt = S\int_0^\infty (uC - D_L\frac{\partial C}{\partial x})_L dt \qquad \text{(VII-39)}$$

If $D_L = 0$, we have

$$\int_0^\infty C(0,t)dt = \int_0^\infty C(L,t)dt \qquad \text{(VII-40)}$$

This equation shows that, in the case of the ideal model, the zero-order moment or area of the injection profile is equal to that of the elution profile. In the more general case, when $D_L \neq 0$, a further analysis becomes necessary. The zero-order moment is not the sample mass nor is it usually directly related to it, so its conservation is still uncertain. If the zero-order moment is not a constant, it is a function of x. So, let us calculate it and search for the conditions under which its values for $x = 0$ and for $x = L$ are equal. Integrating eqn. VII-37, we have

$$\int_0^\infty \frac{\partial C}{\partial x}dt + F\int_0^\infty \frac{\partial q}{\partial t}dt + u\int_0^\infty \frac{\partial C}{\partial x}dt = D_L\int_0^\infty \frac{\partial^2 C}{\partial x^2}dt \qquad \text{(VII-41)}$$

or

$$(C + Fq)\Big|_{t=\infty} - (C + Fq)\Big|_{t=0} + u\frac{dA(x)}{dx} = D_L\frac{d^2 A(x)}{dx^2} \qquad \text{(VII-42)}$$

where $A(x) = \int_0^\infty C(x,t)dt$. Accordingly

$$\int_0^\infty \frac{\partial C}{\partial x}dt = \frac{d}{dx}\int_0^\infty C(x,t)dt = \frac{dA}{dx}$$

$$\int_0^\infty \frac{\partial^2 C}{\partial x^2}dt = \frac{d^2}{dx^2}\int_0^\infty C(x,t)dt = \frac{d^2 A}{dx^2}$$

Since

$$(C + Fq)\,|_{t=\infty} = 0 \tag{VII-43a}$$

$$(C + Fq)\,|_{t=0} = g_1(x) \tag{VII-43b}$$

equation VII-42 becomes

$$D_L \frac{d^2 A}{dx^2} - u \frac{dA}{dx} = -g_1(x) \tag{VII-44}$$

The solution of this differential equation is

$$A(x) = a_1(x) e^{\frac{ux}{D_L}} + a_2(x) \tag{VII-45}$$

and, using the method of the variable coefficients [19]

$$a_1(x) = -\frac{1}{u} \int_0^x g_1(x) e^{\frac{-ux}{D_L}} dx + a_{1,0} \tag{VII-46a}$$

$$a_2(x) = -\frac{1}{u} \int_0^x g_1(x) dx + a_{2,0} \tag{VII-46b}$$

so

$$A(x) = a_{1,0} e^{\frac{ux}{D_L}} + a_{2,0} - \frac{e^{\frac{ux}{D_L}}}{u} \int_0^x g_1(x) e^{\frac{-ux}{D_L}} dx + \frac{1}{u} \int_0^x g_1(x) dx \tag{VII-47}$$

where $a_{1,0}$ and $a_{2,0}$ are integration constants. These constants are derived from the value of the zero-order moment and its first derivative at the boundary. Hence, they depend on the boundary conditions.

Equation VII-47 is the relationship between the zero-order moment and the migration distance, x. We discuss now the conditions under which this moment will remain constant during the elution of the band.

2 - The Zero-Order Moment and the Boundary Conditions

In a well-posed problem involving a PDE, the boundary conditions belong generally to one of the following three different types. In a convection–dispersion problem like the chromatography problem, these conditions are

$$\text{Type I:} \quad C(x = 0, t) = \varphi_1(t) \tag{VII-48a}$$

$$\text{Type II:} \quad \frac{\partial C}{\partial x}\Big|_{x=0} = \varphi_2(t) \tag{VII-48b}$$

$$\text{Type III:} \quad \alpha C(x = 0, t) + \beta \frac{\partial C}{\partial x}\Big|_{x=0} = \varphi_3(t) \tag{VII-48c}$$

In chemical engineering, the boundary conditions are usually of the first or the third type. We discuss now the properties of the zero-order moment under the types I and III boundary conditions.

a - Boundary conditions of the third type

An excellent example of such boundary conditions is the classical Danckwerts conditions, which are most often used in models of nonideal chromatography and are expressed as follows

$$uC - D_L \frac{\partial C}{\partial x} = u\varphi_3(t) \qquad \text{for} \quad x = 0 \qquad \text{(VII-49a)}$$

$$\frac{\partial C}{\partial x} = 0 \qquad \text{for} \quad x = L \qquad \text{(VII-49b)}$$

In these equations, the function $\varphi_3(t)$ is determined by the process of injecting the sample into the column (see Chapter II, section II-1b, *Initial and Boundary conditions*). Integration of these equations in the present situation gives

$$D_L \frac{dA(x)}{dx} \Big|_{x=0} = uA(x) \Big|_{x=0} - uA_0 \qquad \text{(VII-50a)}$$

$$\frac{dA}{dx} \Big|_{x=L} = 0 \qquad \text{(VII-50b)}$$

where $A_0 = \int_0^\infty \varphi_3(t)dt$. Combining eqns. VII-50, VII-50b, and VII-47 gives the integration constants

$$a_{1,0} = \frac{1}{u} \int_0^L g_1(x) e^{\frac{-ux}{D_L}} dx \qquad \text{(VII-51a)}$$

$$a_{2,0} = A_0 \qquad \text{(VII-51b)}$$

And from eqn. VII-47, we obtain

$$A(L) = a_{1,0} e^{\frac{uL}{D_L}} + a_{2,0} - \frac{e^{\frac{uL}{D_L}}}{u} \int_0^L g_1(x) e^{\frac{-ux}{D_L}} dx + \frac{1}{u} \int_0^L g_1(x)dx$$

$$= \frac{e^{\frac{ux}{D_L}}}{u} \int_0^L g_1(x') e^{\frac{-ux'}{D_L}} dx' + \frac{1}{u} \int_0^L g_1(x')dx' + A_0$$

$$\text{(VII-52)}$$

In chromatography problems, $g_1(x)$, *i.e.*, the initial condition, is generally equal to 0. Then, the two integrals in eqn. VII-52 are also equal to 0 and the zero-order moment is invariant. However, there are cases in which the initial condition is different from 0 (*e.g.*, in the modeling of simulated moving bed separations) and then the zero-order moment must be used carefully. *It is usually not constant.*

b - Boundary conditions of the first type

These boundary conditions correspond either to a column of infinite length or to the ideal model (no axial dispersion, $D_L = 0$ in eqns. VII-48c and VII-49a). These conditions can be rewritten as

$$C(x = 0, t) = \varphi_1(t) \tag{VII-53a}$$

$$C(x = \infty, t) \neq 0 \tag{VII-53b}$$

Integration of these equations gives

$$A(0) = \int_0^\infty \varphi_1(t) \tag{VII-54a}$$

$$A(\infty) = \int_0^\infty C(x = \infty, t)dt \neq 0 \tag{VII-54b}$$

From eqns. VII-47 and VII-54a, we have

$$A(0) = a_{1,0} + a_{2,0} = \int_0^\infty \varphi_1(t) \tag{VII-55}$$

Since $A(\infty)$ is finite (see eqn. VII-47), $a_{1,0}$ must be equal to 0, thus

$$a_{2,0} = \int_0^\infty \varphi_1(t)dt \tag{VII-56}$$

So, we have

$$A(x) = \int_0^\infty \varphi_1(t)dt - \frac{e^{\frac{ux}{D_L}}}{u} \int_0^x g_1(x)e^{\frac{-ux'}{D_L}} dx' + \frac{1}{u} \int_0^x g_1(x')dx' \tag{VII-57}$$

where x' is an integral variable. Since, in chromatography, $g_1(x) = 0$ in almost all cases, the zero-order moment is invariant. However, this conclusion is not necessarily valid when the initial condition happens to be different from 0.

V. SHOCK LAYER ANALYSIS

The mathematical concept of a concentration shock is not physically realistic. It assumes an instantaneous change in the concentration of the solute at a given point. This concentration discontinuity would cause an infinite concentration gradient. It would also require the immediate adjustment of the stationary phase concentration to the abrupt change in the mobile phase concentration. Yet the formation of a concentration shock derives logically from a model of chromatography that assumes that there is no axial diffusion and that radial mass transfer is infinitely fast. In all actual experiments, these conditions cannot be realized. While thermodynamics tends to build up concentration shocks and to propagate them, the finite rates of axial dispersion and mass transfer kinetics tend to disperse any rapid variation of the local concentration. The resulting profile that is actually observed is called a shock layer. The shock layer theory discusses the problem of the interactions between these two sets of effects that act in opposite directions, the self-sharpening effects that are caused by the nonlinear thermodynamics of phase equilibrium and the dispersive effects that originate from axial dispersion and from the finite rate of the mass transfer kinetics. A steady-state is eventually achieved when a stable shock layer propagates and the breakthrough profile follows a constant pattern [20]. These interactions are at the heart of the phenomenon of chromatographic separation at finite concentrations. Because the shock layer theory is the most sophisticated analysis of these interactions that leads to algebraic solutions, it is most important and we are of the opinion that mastering the shock layer concept and its theory are critical for a good understanding of all the observations made in nonlinear chromatography.

The formation of a shock layer is similar to the phenomenon encountered in the study of the nonlinear effects associated with the propagation of vibrations in elastic or compressible media, *i.e.*, in the formation of shock waves. Rhee and Amundson [21,22] introduced the concept of shock layer in chromatography. It is the most fruitful solution to the nonlinear, nonideal model of chromatography. In this form, however, it is applicable only to the simple problem of the breakthrough curve. Later, Lin *et al.* [23] gave a detailed analysis of the concept in a particular case of elution.

1 - Migration of a Breakthrough Profile in Nonideal, Nonlinear Chromatography

In this section, we assume that axial dispersion proceeds at a finite rate and that the mass transfer resistances are no longer negligible. We further assume that the behavior of the band is described properly by the lumped kinetic model, using the solid film linear driving force model (see Chapter II, section IV-5c). In such a case, the basic equations of chromatography can be written in dimensionless form as follows

$$\frac{\partial C}{\partial x} + \frac{\partial C}{\partial \tau} + F\frac{\partial q}{\partial \tau} = \frac{1}{Pe}\frac{\partial^2 C}{\partial x^2} \qquad \text{(VII-58a)}$$

$$\frac{\partial q}{\partial \tau} = St\left(f(C) - q\right) \qquad \text{(VII-58b)}$$

where $f(C)$ stands for the equation of the isotherm, Pe is the Peclet number, St the Stanton number and x and τ are the space and time reduced coordinates, respectively. These parameters are related to the classical ones by

$$x = \frac{z}{L} \tag{VII-58c}$$

$$\tau = \frac{ut}{L} \tag{VII-58d}$$

$$Pe = \frac{uL}{D_L} \tag{VII-58e}$$

$$St = \frac{k_f L}{u} \tag{VII-58f}$$

We assume that the space domain is infinite and that the concentration is C_f at $x = -\infty$ and C_i at $x = +\infty$. We study the migration of the front corresponding to the concentration jump from C_i to C_f. Rhee *et al.* [21,22] showed that a constant pattern solution arises after a certain time and a stable, steep front, *i.e.*, the breakthrough profile, propagates along the column, provided that a simple condition is met. We assume here that the isotherm does not have an inflection point. Then the condition is $(C_i - C_f)f''(0) > 0$. If we further restrict the discussion to the general case of an isotherm convex upward, hence with $f''(0) < 0$, we must have $C_i < C_f$. Otherwise, we have a diffuse front, not a shock layer.

Rhee *et al.* [21,22] showed that the breakthrough profile or, in this case, the shock layer propagates at the same velocity as the shock does in the ideal model, a velocity that is proportional to the mobile phase velocity, with

$$\lambda = \frac{u_s}{u} = \frac{1}{1 + F\frac{\Delta q^*}{\Delta C}} = \frac{1}{1 + K} \tag{VII-59}$$

The last equation serves also as the definition of $K = F(f(C_f) - f(C_i))/(C_f - C_i)$. Finally, Rhee *et al.* showed that the profile of the shock layer is obtained by integrating the following nonlinear differential equation

$$\frac{-\lambda}{Pe\,St}\frac{d^2C}{d\eta^2} + \left[\frac{1}{Pe} + \frac{\lambda(1-\lambda)}{St}\right]\frac{dC}{d\eta} = F\,\lambda\,g(C; C^f; C^i) \tag{VII- 60}$$

where η is a new variable describing the position of the breakthrough curve in the column, a function of the position z and the time τ [1]. The function $g(C; C^f; C^i)$ depends both on the isotherm and on the amplitude of the concentration jump

$$g(C; C^f; C^i) = \frac{\Delta q^*}{\Delta C}(C - C^f) - f(C) + f(C^f) \tag{VII-61}$$

Rhee *et al.* [22] did not attempt to further calculate this function. Later Zhu *et al.* [24] showed that it is particularly simple to calculate $g(C; C^f; C^i)$ in the case of the Langmuir isotherm. In this case, the shock layer thickness becomes

$$\Delta\eta_z = \left(\frac{D_a(1+K)}{Ku} + \frac{u}{(1+K)k_f}\right) \frac{bC_0+2}{bC_0} \ln\left|\frac{1-\theta}{\theta}\right| \qquad \text{(VII-62)}$$

In this equation, θ is a fraction of the concentration jump, so the shock layer is assumed to begin at the relative concentration θ and to end at $1 - \theta$. Usually, a value of 0.05 is taken for θ [1].

The shock layer thickness is a most useful tool to study the influence of axial dispersion and of the mass transfer resistances on the steepness of the breakthrough curves [1,24,25]. It is also useful because it allows the definition of a measure of the column efficiency under nonlinear conditions.

2 - Shock Layer Analysis in Elution Chromatography

In elution chromatography, the height of the elution band decreases constantly during its migration whence the injection plateau has been eroded away (see Chapter V, section I-4a and, *e.g.*, the derivation of eqn. VII-49). Accordingly, since the shock layer propagates at the same velocity as the shock in the equilibrium model, the propagation velocity of the shock layer of an elution band varies during its migration. It decreases continuously in the case of a convex-upward isotherm. For this reason, it is difficult to derive an analytical solution for the profile of the shock layer under elution condition in the general case. This is possible only in some particular cases. We discuss here one such case, when (1) the degree of nonlinear behavior is weak (*i.e.*, for a Langmuir isotherm, when the maximum value of bC is moderate compared to unity); (2) only the mass-transfer resistances are considered among the possible dispersion factors (*i.e.*, axial and eddy diffusions are neglected); and (3) the mass-transfer resistances are small. Because of this combination of assumptions, the nonlinear, ideal, near-equilibrium model can be used [23]. Then, the profile and the thickness of the shock layer can be derived as analytical functions.

The mathematical model of the problem just defined is simpler than the one used in the previous section (eqn. VII-58a,b). It is written as follows:

$$F\frac{\partial q}{\partial t} + \frac{\partial C}{\partial t} + u\frac{\partial C}{\partial x} = 0 \qquad \text{(VII-63a)}$$

$$\frac{\partial q}{\partial t} = k_f(q^* - q) \qquad \text{(VII-63b)}$$

$$q^* = \frac{aC}{1+bC} \qquad \text{(VII-63c)}$$

and the initial and boundary conditions of the rectangular injection are

$$C(x,0) = 0 \tag{VII-63d}$$

$$C(0,t) = \begin{cases} C_0 & 0 < t \leq t_p \\ 0 & t > t_p \end{cases} \tag{VII-63e}$$

where C_0 is the injection concentration and t_p the injection duration.

Because of the restrictive assumptions made, bC is small and the mass-transfer coefficient k_f is large. Since k_f is large, we can write the following approximation

$$q = q^* - \frac{1}{k_f}\frac{\partial q}{\partial t} \approx q^* - \frac{1}{k_f}\frac{\partial q^*}{\partial t}$$

Because of the isotherm selected and introduced into eqn. VII-61, and since bC is small, we have

$$q = q^* - \frac{a}{k_f}\frac{\partial C}{\partial t} \tag{VII-64}$$

Finally, with all these simplifications, the mass balance equation can be written

$$\frac{\partial C}{\partial x} + \frac{1 + Fa(1 - bC)}{u}\frac{\partial C}{\partial t} = \frac{Fa}{uk_f}\frac{\partial^2 C}{\partial t^2} \tag{VII-65}$$

In the case of the breakthrough curve, a shock layer analysis could be made by following the method described in the previous section. It would just be simpler. The reader could perform the calculations as an exercise.

a - Profile of the Shock Layer in Elution Chromatography

We define a new coordinate system with

$$\tau = \frac{1 + Fa}{2Fab}\left(x - \frac{u}{1 + Fa}t\right) \tag{VII-66}$$

In this system, eqn. VII-65 can be rewritten as

$$\frac{\partial C}{\partial x} + C\frac{\partial C}{\partial \tau} = D_f\frac{\partial^2 C}{\partial \tau^2} \tag{VII-67}$$

with $D_f = u/(4Fab^2k_f)$. The initial and boundary conditions in eqns. VII-17a and VII-17b become

$$C(x,0) = 0$$

$$C(0,\tau) = \begin{cases} C_0 & -\frac{u}{2Fab}t_p < \tau < 0 \\ 0 & \tau \leq -\frac{u}{2Fab}t_p \end{cases} \tag{VII-68}$$

Equation VII-67 is a Burgers' equation. The solution of this equation is obtained by using a Cole–Hopf transform (see Appendix A).

For a Dirac boundary condition ($t_p \to 0$), the solution of eqn. VII-67 is

$$C(x,t) = \sqrt{\frac{S_k}{2R_k x}} \; \frac{(e^{R_k} - 1)e^{-W^2}}{\sqrt{\pi} + \frac{\sqrt{\pi}}{2}(e^{R_k} - 1)\mathrm{erfc}(W)} \qquad \text{(VII-69)}$$

with

$$W = \frac{\tau}{\sqrt{\frac{2S_k x}{R_k}}} = \frac{\tau}{2\sqrt{D_f x}}$$

$$S_k = \frac{uC_0 t_p}{2Fab}$$

$$R_k = \frac{S_k}{2D_f} = C_0 t_p bk_f$$

The difference between the solutions in eqns. VII-7 and VII-69 comes from the difference in the problems that they solve. The Houghton problem uses the initial condition to perform the injection while, in this problem, it is a boundary condition.

The front part of the profile is a shock layer, the profile of which is represented by eqn. VII-69. Equation VII-69 is the solution described by Haarhoff and Van der Linde [3]. It was also derived by Whitham [26]. A similar solution is easily derived for the boundary condition of a wide injection, following the same procedure.

b - *Thickness of the Shock Layer in Elution Chromatography*

The thickness of the shock layer can easily be derived from its profile (eqn. VII-69). Obviously, the problem of the shock layer thickness is relevant only at high column efficiency, when the peak front is steep. This means that k_f, hence R_k is large. When R_k increases indefinitely, $e^{R_k} - 1$ is equivalent to e^{R_k}. We introduce, for the sake of convenience, the following factor

$$t_s = \frac{W}{\sqrt{R_k}} = \frac{\tau}{\sqrt{2S_k x}}$$

Then, the numerator of the second factor in eqn. VII-69 becomes equal to $e^{R_k(1-t_s^2)}$. At the same time, $\mathrm{erfc}(W)$ becomes equivalent to $e^{-W^2}/W\sqrt{\pi}$. The shock layer profile ends at the peak maximum ($t_s = 1$). It begins at the origin of the peak front ($C = 0$, for $|t_s| \gg 1$ or $|\tau| \gg \sqrt{2S_k x}$). Obviously, however, the thickness of the shock layer cannot be expressed as the distance between this indeterminate origin and the peak maximum.

Similar to what was done in the previous subsection for the thickness of the shock layer of the breakthrough curve, we can take as a reasonable estimate of the shock layer thickness of the elution band the expression $|1 - t_s(\theta)|$, where θ is a fraction, to be defined for any practical application, of the maximum peak concentration, C_M. We observe that the concentration is significantly different from 0 only near the peak maximum, *i.e.*, for t_s close to unity. Then, $t_s^2 - 1$ is equivalent to $2(t_s - 1)$ and eqn. VII-69 can be approximated by

$$C(x, \tau) = \frac{\tau}{x} \frac{1}{1 + 2\sqrt{\pi R_k} e^{2R_k(t_s - 1)}} \qquad \text{(VII-70)}$$

From eqn. VII-70, it is easy to derive the shock layer thickness, $|1 - t_s|$, as a function of C. Let $\theta = \frac{C}{C_{max}}$ with C_{max} maximum concentration of the peak, achieved for $t_s = 1$. We have $C_{max} = \frac{\tau}{x}|_{t_s=1}$. Accordingly,

$$|\Delta t_s| = |1 - t_s| = \frac{1}{2R_k} log(\frac{1}{2\sqrt{\pi R_k}} \frac{1 - \theta}{\theta}) \qquad \text{(VII-71)}$$

Finally, replacing t_s by its definition as a function of time, we obtain the thickness of the shock layer

$$|\Delta t| = \frac{1}{k_f \sqrt{L_f}} log(\frac{1}{2\sqrt{\pi L_f St k_0'}} \frac{1 - \theta}{\theta}) \qquad \text{(VII-72)}$$

Where $R_k = L_f St k_0'$, $St = k_f L/u$ and $L_f = u C_0 t_p/(LFq_s)$.
   Equation VII-72 shows that, since θ is finite, so is $|1 - t_s|$. It is of the same order of magnitude as R_k^{-1}, *i.e.*, $|1 - t_s| = O(R_k^{-1})$.

c - Asymptotic Analysis

When $R_k \rightarrow \infty$, $e^{R_z(t_s - 1)}$ tends toward 0 if t_s is smaller than 1 and towards infinity if it is larger than 1. Then, eqn. VII-71 can be rewritten

$$C(x, \tau) = \frac{\tau}{x} \qquad \text{for} \quad 0 < t_s < 1 \qquad \text{(VII-73)}$$

Combining this expression with the definition of τ gives

$$C(x, t) = \frac{1 + Fa}{4Fab}(x - \frac{u}{1 + Fa}t) \qquad 0 < \frac{1 + Fa(x - \frac{u}{1+Fa}t)}{2\sqrt{uC_0 t_p x Fab}} < 1 \qquad \text{(VII-74)}$$

For the peak, $C(x, t) = C_{max} = \frac{\tau}{x}|_{t_s=1} = \sqrt{\frac{2S_k}{x}} = \sqrt{\frac{uC_0 t_p}{Fabx}}$, and, accordingly, we have

$$\sqrt{\frac{uC_0 t_p}{Fabx}} = \frac{1 + Fa}{2Fab}(x - \frac{u}{1 + Fa}t)$$

or

$$t = \frac{x}{u}[1 + Fa(1 - 2\sqrt{\frac{buC_0t_p}{Fax}})] \tag{VII-75}$$

And for $x = L$, we have

$$t_R = \frac{L}{u}[1 + Fa(1 - 2\sqrt{\frac{bC_0t_p}{Fat_0}})] \tag{VII-76}$$

This is the retention time of the peak given by the solution of the ideal, equilibrium model at low concentrations, when the second-order terms can be neglected and the isotherm equation is parabolic. In this case, it is in agreement with eqn. V-49.

The results obtained in this section are illustrated in Figure VII-6.

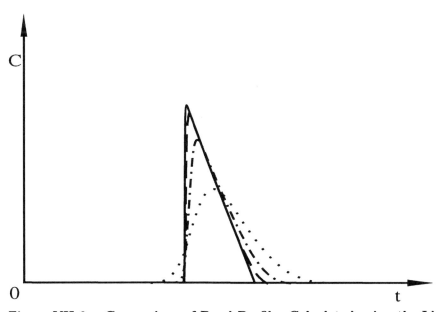

Figure VII-6   Comparison of Band Profiles Calculated using the Ideal and the Equilibrium–Transport Models [23]. Parabolic isotherm, $q^* = 12\ C$ (1 - 0.024 C) (C in mg/ml). $u = 0.19$ cm/s, $F = 0.45$, $L = 25$ cm, $C_0 = 10$ mg/ml, $t_p = 10$ s. Solid line: ideal model, $k_f = \infty$. Dashed line, $k_f = 10$ s^{-1}. Chain-dotted line, $k_f = 2$ s^{-1}. Dotted line, $k_f = 0.2$ s^{-1}. *Reproduced with permission by Elsevier.*

VI - TRANSITION BETWEEN LINEAR
AND NONLINEAR CHROMATOGRAPHY

Strictly speaking, there is no concentration range within which chromatography is linear because all isotherms have a finite curvature at the origin. The only possible exception would be an isotherm having an inflection point right at the origin, a chance encounter that, to the extent of our knowledge, has not been made yet. However, there are many cases in which no deviations from linear behavior can be observed because these deviations are smaller than the error made on the measurements of the property used to assess deviations from linear behavior. Therefore, the deviations are not significant and the behavior can be considered as linear. However, if the measurements are extended toward higher concentrations, nonlinear behavior will become obvious.

For example, even at low concentrations, the retention time of the maximum of an elution band, t_M, is a function of the sample size as shown by eqn. VII-12. Jaulmes et $al.$ [4] derived an equation that they validated with appropriate experimental data in a companion paper

$$t_M = t_{R,0} \left(1 - 2b \frac{k_0'}{1 + k_0'} C_M - \frac{1}{2N} \frac{k_0'}{1 + k_0'} \right) \qquad \text{(VII-77)}$$

where C_M is the maximum concentration of the band, the column efficiency, N, is equal to $u^2 t_R / [2D_L(1 + k_0')] = uL/[2D_L]$, and the other symbols are the same as in eqns. VII-12 [1,4]. If the second term in eqn. VII-77 is lower than the error made on the measurement of the retention time, the variation of the retention time in the range of peak height studied is not significant. The product bC accounts for the nonlinear behavior of the Langmuir isotherm (whether the actual equilibrium isotherm is described by the Langmuir model is irrelevant here because we are dealing now with a perturbation solution and eqn. VII-77 uses a second order expansion which can be applied to any isotherm). So, if the value of bC_M is less than 0.01, the precision on the retention time measurements must be of the order of 0.1% for nonlinear behavior to be measurable. In most cases, the short-term reproducibility of the retention time measurements in HPLC is such that the relative standard deviation is of the order of 0.1% for moderately polar compounds [27]. Because the chromatographic process involves dilution during elution, the solute concentration in the elution band is much lower than in the sample, as long as the sample volume is small compared to the volume occupied by the band, so the actual concentration of the band is usually low and measurements can easily be carried out for values of bC_M much lower than 0.01. Then, the chromatographic behavior may be considered as linear. On the other hand, by using large sample sizes, through proper calibration of the detector, it could be possible, in principle, to determine an approximate value of the isotherm curvature. Unfortunately, this method cannot be developed further into an acceptable procedure for the determination of isotherms. It involves too many sources of model and random errors, hence it gives inaccurate results.

Another approach to investigate the transition between apparent linear and true nonlinear chromatography is to consider the loading factor, L_f, or ratio of the sample size to the monolayer capacity of the amount of adsorbent contained in the column. By definition, the loading factor is

$$L_f = \frac{nb}{\varepsilon S L k'_0} \qquad \text{(VII-78)}$$

where $n \ (= C_0 V_0)$ is the sample size, $k'_0 = aF$, and S is the cross-sectional area of the column. Golshan-Shirazi and Guiochon [1,28,29] have shown that the degree of agreement between the ideal and the equilibrium–dispersive models depends on the value of an effective loading factor defined as

$$m = N \left(\frac{k'_0}{1 + k'_0} \right)^2 L_f \qquad \text{(VII-79)}$$

This equation is most important because it combines three parameters of fundamentally different origins, (1) a parameter that depends only on diffusion, dispersion and mass transfer kinetics, the plate number at infinite dilution, N; and two parameters that depend only on the thermodynamics of the phase system used, (2) the retention factor at infinite dilution that is proportional to the initial slope of the isotherm, and (3) the loading factor, L_f. The plate number, $N = uL/2D_L$, is a measure of the intensity of the dispersive effects in the system studied. N increases with decreasing D_L, that is with decreasing intensity of the dispersive effects. The product bC is a measure of the intensity of the nonlinear effects, which are of thermodynamic origin. Note, however, that bC_M varies during elution of the band and may decrease by one to two orders of magnitude (or even more) between injection and elution. Like the loading factor, L_f, it increases with increasing sample size. The influence of the retention factor in eqn. VII-79 is moderate or even negligible, unless this factor is small. In the practically useful range of k'_0, between 1 and 10, $[k'_0/(1 + k'_0)]^2$ increases from 0.25 to 0.83. Accordingly, eqn. VII-79 expresses the balance between the influence of thermodynamics and that of dispersion on the result of a chromatographic experiment. The parameter m is a dimensionless number. We call it the Shirazi number.

The degree of agreement between the profiles predicted by the ideal and the equilibrium–dispersive models increases with increasing column efficiency and with increasing loading factor. It has been shown [1] that the agreement between these profiles is excellent when the effective loading factor, m, is larger than about 35. This corresponds approximately to values of the product NL_f larger than 50, *i.e.*, to an efficiency larger than 5000 theoretical plates for a loading factor of 1%, which is still only a moderate degree of overloading. Equation VII-79 can be used to ascertain the effective boundary region between apparently linear chromatography ($m < 2$) and truly nonlinear chromatography, with little axial dispersion ($m > 35$). The importance of the value of m in the description of the shape of chromatographic peaks is illustrated by the following values corresponding to the different peaks in Figures VII-2 and VII-3.

APPENDIX A

Derivation of the Houghton Solution
of the Nonlinear Model

The equations of the equilibrium–dispersive model for a parabolic isotherm are written

$$\frac{\partial C}{\partial t} + F\frac{\partial q}{\partial t} + u\frac{\partial C}{\partial z} = D_L\frac{\partial^2 C}{\partial t^2} \qquad \text{(VII-A-1)}$$

$$q = a_1 C + a_2 C^2 \qquad \text{(VII-A-2)}$$

where D_L is the apparent axial dispersion coefficient.

The boundary condition is $C(0,t) = 0$ and the initial conditions for elution are

$$C(x,0) = C_0(x) = C_0 \qquad \text{if } |x| < \frac{L_0}{2}$$

$$= 0 \qquad \text{if } \frac{L_0}{2} \leq |x| \qquad \text{(VII-A-3)}$$

The initial values of q are derived by combining eqn. VII-10 and VII-11. Note that the injection is part of the initial conditions, not of the boundary condition as is conventional in chromatography. The initial conditions assume that the column is infinite in length and that the injection was made "in the middle" of it, at position $x = 0$, instantaneously, at time $t = 0$.

The coordinates are first transformed so that the following calculations will be carried out no longer in the fixed reference system of the column but in a reference coordinate system moving at the same velocity, $u_R = u/(1 + k'_0)$, as that of the band propagating under analytical *i.e.*, linear conditions

$$\xi = x - \frac{u}{(1 + Fa_1)}t = x - \frac{u}{1 + k'_0}t = x - u_R t \qquad \text{(VII-A-4)}$$

If we consider any function, $\varphi(t, \xi)$ of the variables t and ξ, we have

$$\varphi(t, \xi) = \varphi(t(t, x),\ \xi(t, x)) \qquad \text{(VII-A-5a)}$$

hence
$$d\varphi = \left(\frac{\partial \varphi}{\partial t}\right)_\xi dt + \left(\frac{\partial \varphi}{\partial \xi}\right)_t d\xi$$

$$= \left(\frac{\partial \varphi}{\partial t}\right)_x dt + \left(\frac{\partial \varphi}{\partial x}\right)_t dx \qquad \text{(VII-A-5b)}$$

So
$$\left(\frac{\partial \varphi}{\partial t}\right)_x = \left(\frac{\partial \varphi}{\partial t}\right)_\xi \frac{dt}{dt} + \left(\frac{\partial \varphi}{\partial \xi}\right)_t \left(\frac{\partial \xi}{\partial t}\right)_x$$

$$= \left(\frac{\partial \varphi}{\partial t}\right)_\xi - u_R \left(\frac{\partial \varphi}{\partial \xi}\right)_t \qquad \text{(VII-A-6a)}$$

$$\left(\frac{\partial \varphi}{\partial x}\right)_t = \left(\frac{\partial \varphi}{\partial \xi}\right)_t \qquad \text{(VII-A-6b)}$$

Using these relationships, we can rewrite eqn. VII-A-1 into

$$\frac{\partial C}{\partial t} - u_R\frac{\partial C}{\partial \xi} + u\frac{\partial C}{\partial \xi} + F\frac{\partial q}{\partial t} - Fu_R\frac{\partial q}{\partial \xi} = D_L\frac{\partial^2 C}{\partial \xi} \qquad \text{(VII-A-7)}$$

Substituting the isotherm equation (eqn. VII-A-2) into eqn. VII-A-7 gives, after dividing both sides of the intermediate equation by $1 + Fa_1 + 2Fa_2C$

$$\frac{\partial C}{\partial t} - \left(u_R - \frac{u}{1 + Fa_1 + 2Fa_2C} \right) \frac{\partial C}{\partial \xi} = D_R' \frac{\partial^2 C}{\partial \xi^2} \qquad \text{(VII-A-8)}$$

with

$$D_R' = \frac{D_L}{1 + Fa_1 + 2Fa_2C} \simeq \frac{D_L}{1 + Fa_1} = D_R \qquad \text{(VII-A-9)}$$

We consider than the deviation of the isotherm from linear behavior is small. Accordingly, we assume that

$$\frac{u}{1 + Fa_1 + 2Fa_2C} = \frac{u}{(1 + Fa_1)(1 + \lambda C)} \simeq u_R - \lambda u_R C \qquad \text{(VII-A-10)}$$

where $\lambda = 2Fa_2/(1 + Fa_1) = 2Fa_2/(1 + k_0')$ is related to the initial curvature of the isotherm. Introducing this simplifying assumption in eqn. VII-C-8 and letting $\eta = -\xi/(\lambda u_R)$ gives

$$\frac{\partial C}{\partial t} + C \frac{\partial C}{\partial \eta} = \frac{D_R}{(\lambda u_R)^2} \frac{\partial^2 C}{\partial \eta^2} = D_u \frac{\partial^2 C}{\partial \eta^2} \qquad \text{(VII-A-11)}$$

The last one of these two equations is the definition of $D_u = D_R/(\lambda u_R)^2$, a symbol introduced merely for the sake of simplicity in the following derivations.

Equation VII-A-11 is a Burgers' equation, a nonlinear convection–diffusion equation. Its solution can be derived using the Cole–Hopf transform, which replaces the function $C(\eta, t)$ by a function $S_c(\eta, t)$:

$$C = -2D_u \frac{1}{S_c} \frac{\partial S_c}{\partial \eta} \qquad \text{(VII-A-12)}$$

This transform and its effects are better understood if it is subdivided into two successive transforms

$$C = \frac{\partial \phi}{\partial \eta} \qquad \text{(VII-A-13a)}$$

$$\phi = -2D_u \log S_c \qquad \text{(VII-A-13b)}$$

Introducing eqn. VII-A-13a into eqn. VII-A-11 and integrating the differential equation obtained gives

$$\frac{\partial \phi}{\partial t} + \frac{1}{2} \left(\frac{\partial \phi}{\partial \eta} \right)^2 = D_u \frac{\partial^2 \phi}{\partial \eta^2} \qquad \text{(VII-A-14)}$$

Then, doing the second step of the transform (eqn. VII-A-13b) gives

$$\frac{\partial S_c}{\partial t} = D_u \frac{\partial^2 S_c}{\partial \eta^2} \qquad \text{(VII-A-15)}$$

Since the column length is infinite, only the initial condition should be considered. There are no conditions at $x = L$. So, from eqn. VII-A-12, VII-A-13 and VII-A-14, we obtain

$$S_c(\eta, 0) = S_c(\eta) = \exp\left[-\frac{1}{2D_u} \int_0^\eta C_l(\eta')d\eta' \right] \qquad \text{(VII-A-16)}$$

Equations VII-A-15 and VII-A-16 form an initial value problem. Its solution is

$$S_c(t, \eta) = \frac{1}{\sqrt{4\pi D_u t}} \int_{-\infty}^{\infty} S_c(\eta') \exp\left[-\frac{(\eta - \eta')^2}{4D_u t} \right] d\eta' \qquad \text{(VII-A-17)}$$

From eqn. VII-A-17, it is easy to derive $\partial S_c / \partial \eta$, hence

$$C(\eta, t) = -2D_u \frac{1}{S_c} \frac{\partial S_c}{\partial \eta} = \frac{\int_{-\infty}^{\infty} \frac{\eta - \eta'}{t} S_c(\eta') \exp\left[-\frac{(\eta - \eta')^2}{4D_u t}\right] d\eta'}{\int_{-\infty}^{\infty} S_c(\eta') \exp\left[-\frac{(\eta - \eta')}{4D_u t}\right] d\eta'} \qquad \text{(VII-A-18)}$$

If we let $G_m = \int_0^{\eta} C_l(\eta') d\eta' + \frac{(\eta - \eta')^2}{2t}$, where η' is an integration variable, $C(\eta, t)$ can be written as

$$C(\eta, t) = \frac{\int_{-\infty}^{+\infty} \frac{\eta - \eta'}{t} e^{-\frac{G_m}{2D_u}} d\eta'}{\int_{-\infty}^{+\infty} e^{-\frac{G_m}{2D_u}} d\eta'}$$

Eventually, with the Dirac injection condition, the solution is given by the following set of equations [4].

$$X = \left| \frac{e^{-\tau^2/2}}{\sqrt{2\pi}(\coth m + \text{erf} \frac{\tau}{\sqrt{2}})} \right| \qquad \text{(VII-A-19)}$$

where X, m, and τ are the dimensionless concentration, sample size, and time, respectively. These parameters are defined by the following equations

$$X = |b| \, C \frac{k_0'}{1 + k_0'} \sqrt{\frac{u^2 t}{2D_L (1 + k_0')}} \qquad \text{(VII-A-20a)}$$

$$\tau = \frac{k_0' L}{(1 + k_0') \sqrt{\frac{2D_L t}{1 + (1 + k_0')}}} \frac{t_{R,0} - t}{t_{R,0} - t_0} \qquad \text{(VII-A-20b)}$$

$$m = \frac{Lu}{2D_L} \left[\frac{k_0'}{1 + k_0'}\right]^2 L_f \qquad \text{(VII-A-20c)}$$

where L_f is the loading factor, $L_f = \frac{bC_0 t_p}{t_R - t_0}$, and $t_p = \frac{L_0(1 + k_0')}{u}$. The loading factor is the ratio of the sample size to the column saturation capacity as defined for the Langmuir isotherm which has the same two-term expansion as the parabolic isotherm used here, hence has the same initial slope and curvature (i.e., $q_s = -a_1^2/a_2$). See also eqn. VII-10. If we define $L_f = \frac{L_0 C_0}{LF q_s} = \frac{L_0 C_0 a_2}{LF a_1^2}$, eqn. VII-A-20c can be transformed into

$$m = \frac{Lu}{2D_L} \frac{k_0'^2}{1 + k_0'} L_f \qquad \text{(VII-A-20d)}$$

EXERCISES

1) Derive the equation for the equilibrium–dispersive model of chromatography, for a single component. Write the set of equations for two components.

2) Derive the solution of equations VII-2b and VII-3 using Laplace transform.

3) Derive from equations VII-2b and VII-3 the first, second, and third-order moments of a chromatographic band, using the Laplace transform. Explain why the third moment, μ_3, is different from 0 for a Gaussian profile.

4) Derive the solution of the equilibrium–dispersive model of chromatography for a breakthrough profile (boundary conditions of frontal analysis; first one component, then two components) and draw profiles corresponding to different values of the experimental conditions, using a computer.

5) Derive from eqns. VII-21 and VII-22 a first order approximation of the expression giving q when $k_f \rightarrow \infty$. Show that the effect of k_f is the same when $D_L = \frac{u^2 Fa}{(1+Fa)^2 k_f}$.

6) Compare the areas of the injection profile and that of elution profiles calculated with Houghton's equation for different values of the parameters. Use a computer to calculate the integral of Houghton's profiles.

LITERATURE CITED

[1] G. Guiochon, S. Golshan-Shirazi and A. M. Katti, *"Fundamentals of Preparative and Nonlinear Chromatography "*, Academic Press, Boston, MA, 1994.
[2] G. J. Houghton, *J. Phys. Chem.*, **67** (1963) 84.
[3] P. C. Haarhoff and H. J. Van der Linde, *Anal. Chem.*, **38** (1966) 573.
[4a] A. Jaulmes, C. Vidal-Madjar, A. Ladurelli and G. Guiochon, *J. Phys. Chem.*, **88** (1984) 5379.
[4b] B. Lin, Z. Ma and G. Guiochon, *Separat. Sci. Technol.*, **24** (1989) 809.
[5] C. A. Lucy, J. L. Wade and P. W. Carr, *J. Chromatogr.*, **484** (1989) 61.
[6] S. Golshan-Shirazi and G. Guiochon, *J. Chromatogr.*, **506** (1989) 495.
[7] M. van Dyke, *"Perturbation Methods in Fluid Mechanics"* (annoted edition), The Parabolic Press, 1975.
[8] Ting Chin-Chun and Chu Pao-Lin, *Scientia Sinica*, 6(9) (1962) 1269.
[9] J. Villermaux, *J. Chromatogr. Sci.*, **12** (1974) 822.
[10] B. Lin, S. Golshan-Shirazi and G. Guiochon, *J. Phys. Chem.*, **93** (1989) 3363.
[11] H.C. Thomas, *J. Amer. Chem. Soc.*, **66** (1944) 1644.
[12] S. Goldstein, *Proc. Roy. Soc. (London)*, **A219** (1953) 151.
[13] J.L. Wade, A.F. Bergold and P.W. Carr, *Anal. Chem.*, **59** (1987) 1286.
[14] R. Courant and D. Hilbert, *"Methods of Mathematical Physics"*, Vol. 1 1962.
[15] J. C. Giddings, *"Dynamics of Chromatography "*, M. Dekker, New York, NY, 1965.
[16] Bingchang Lin, J. Wang, Bingcheng Lin, *J. Chromatogr.*, **438** (1988) 171.
[17] R. Aris and N. R. Amundson, *"Mathematical Methods in Chemical Engineering"*, Prentice-Hall, Englewood Cliffs, NJ, 19.

[18] B. Lin, Z. Ma, G. Guiochon, *J. Chromatogr.*, **542** (1992) 1.
[19] M. Braun, *"Differential Equations and their Applications"*, Springer- Verlag, New York, NY, 1983.
[20] D.O. Cooney and E.N. Lightfoot, *Ind. Eng. Chem. (Fundam.)*, **4** (1965) 233.
[21] H.-K. Rhee, B.F. Bodin and N.R. Amundson, *Chem. Eng. Sci.*, **26** (1971) 1571.
[22] H.-K. Rhee and N.R. Amundson, *Chem. Eng. Sci.*, **27** (1972) 199.
[23] B. Lin, T. Yun, G. Zhong and G. Guiochon, *J. Chromatogr. A*, **708** (1995) 1.
[24] J. Zhu, Z. Ma and G. Guiochon, *Biotechnol. Progr.*, **9** (1993) 421.
[25] J. Zhu and G. Guiochon, *AIChE J.*, **41**(1995) 45.
[26] G. B. Whitham, *"Linear and Nonlinear Waves"*, Wiley, New York, NY, <u>1974</u>.
[27] M. Kele and G. Guiochon, *J. Chromatogr.*, **830** (1999) 55.
[28] S. Golshan-Shirazi and G. Guiochon, *J. Chromatogr.*, **506** (1990) 495.
[29] S. Golshan-Shirazi and G. Guiochon, *Anal. Chem.*, **61** (1989) 462.

CHAPTER VIII
TWO-COMPONENT
IDEAL, NONLINEAR CHROMATOGRAPHY

The basic function of chromatography is the separation and/or the purification of compounds. Therefore investigation of multicomponent chromatography is the essential part of the theory of chromatography, the one
that stands to have useful consequences. The studies of the band profiles
of single compounds that were discussed in the two previous chapters were
necessary for this critical discussion. They are insufficient alone. While the
superposition principle applies to all linear chromatography experiments,
it does not hold true in the general, nonlinear, multicomponent case. Because of the competition between the different components of the system,
the presence of the other components modifies the equilibrium concentration of each compound to an extent which depends on the concentrations
of these other components. Thus, the whole composition of the feed influences the band profiles of each of the components and their separation [1].
By contrast with single-component chromatography, the detailed study of
two-component problems is both fundamental and sufficient for an understanding of all practical problems. It introduces all the important concepts
and affords all the tools required for the discussion of the separation of

more complex mixtures. Detailed investigations of the behavior of three-component mixtures under nonlinear conditions have not shown new effects but only different combinations of the effects seen with binary mixtures [2,3].

In this chapter, we discuss ideal, nonlinear, two-component chromatography [1]. The influence of diffusion and of the mass transfer resistances will be ignored. The discussion will be essentially based on the use of the coherent condition [4-7] which is introduced after a detailed discussion of the mathematical problem and of the characteristics method. The main features of this discussion include the determination of the characteristic directions and of the Riemann invariant, $R(C_1, C_2)$ [8], of the two-component system along the characteristic direction, the "simple wave" solution and the "shock" solution. Then, we discuss the development of the bands and the calculation of the retention times in the two-component case, under Langmuir-isotherm conditions. We examine in Appendix A the Hodograph transform, another method of calculation of the solution which applies in the two-component case.

I. THE MATHEMATICAL FOUNDATIONS OF THE THEORY OF IDEAL, NONLINEAR, MULTICOMPONENT CHROMATOGRAPHY

In Chapter III, we discussed the properties of the ideal, nonlinear model of chromatography in the case of a single component and we presented its solution. We introduced the characteristic lines and the characteristic form [8]. In this chapter, we give the mathematical definition for both of them in the case of a binary feed mixture and we apply them to the study of the properties of the ideal model in the separation of two components.

1 - Characteristic Lines

The characteristic lines of first-order PDEs are curves along which the function $C(x, t)$ is weakly discontinuous, i.e., its derivatives may be discontinuous in some points but the function itself is continuous everywhere [8]. Mathematically, this means that the derivative of $C(x, t)$ is not uniquely defined everywhere on the characteristic lines.

This fact has important consequences. Let us consider the following equations

$$\sum_j a_{ij} \frac{\partial C_j}{\partial t} + b_{ij} \frac{\partial C_j}{\partial x} + d_{ij} = 0 \qquad\qquad i, j = 1, 2, \cdots \qquad \text{(VIII-1)}$$

where the functions $a_{ij} = a_{ij}(x, t, C_1, C_2, \cdots)$, $b_{ij} = b_{ij}(x, t, C_1, C_2, \cdots)$, and $d_{ij} = d_{ij}(x, t, C_1, C_2, \cdots)$ are known and assume that the equation of the characteristic lines are

$$x = x(\eta) \tag{VIII-2a}$$

$$t = t(\eta) \tag{VIII-2b}$$

where η is a parameter.

Then, we can derive from eqns. VIII-1 and VIII-2 the differentials of $C(x, t)$ along the characteristic lines:

$$C'_j \equiv \frac{dC_j}{d\eta} = \frac{\partial C_j}{\partial x}\frac{dx}{d\eta} + \frac{\partial C_j}{\partial t}\frac{dt}{d\eta} = \frac{\partial C_j}{\partial x}x' + \frac{\partial C_j}{\partial t}t' \tag{VIII-3}$$

From eqn. VIII-3, we derive

$$\frac{\partial C_j}{\partial t} = \left(C'_j - \frac{\partial C_j}{\partial x}x'\right)t'^{-1} \tag{VIII-4}$$

Combining eqns. VIII-1 and VIII-4 gives

$$\sum_j \left\{ a_{ij}\left[C'_j - x'\frac{\partial C_j}{\partial x}\right] + b_{ij}t'\frac{\partial C_j}{\partial x} + d_{ij}t' \right\} = 0$$

or

$$\sum_j (a_{ij}x' - b_{ij}t')\frac{\partial C_j}{\partial x} = \sum_j a_{ij}C'_j + d_{ij}t' \tag{VIII-5}$$

In order for the partial derivatives of C_j not to be given by a unique solution, it is necessary that the determinant of the coefficients of the nonhomogeneous equation VIII-5 be equal to zero, *i.e.* that

$$\det\lfloor b_{ij}t' - a_{ij}x' \rfloor = 0 \tag{VIII-6a}$$

$$\text{or} \quad \det\lfloor b_{ij} - a_{ij}\frac{dx}{dt} \rfloor = \det\lfloor b_{ij} - a_{ij}\rho \rfloor = 0 \tag{VIII-6b}$$

In the system of equations of ideal, nonlinear, multi-component chromatography, the coefficients d_{ij} are all equal to 0 and $b_{ij} = \delta_{ij}$ (*i.e.*, it is equal to the Kronecker's δ). Accordingly, we have

$$\det\lfloor \delta_{ij}t' - a_{ij}x' \rfloor = 0$$

$$\text{or} \quad \det\lfloor \delta_{ij} - a_{ij}\rho \rfloor = 0 \tag{VIII-7}$$

Equations VIII-6 and VIII-7 give the secular equation which defines the direction of the characteristic lines.

2 - Characteristic form

In Chapter III, we showed that the partial differential equation of ideal nonlinear chromatography leads to the following equations for the characteristic lines

$$\frac{dC}{dt} = 0 \qquad\qquad\text{(VIII-8)}$$

$$\frac{dx}{dt} = \rho \qquad\qquad\text{(VIII-9)}$$

Equation VIII-8 is the time derivative of the concentration along the characteristic lines, which are themselves determined by eqn. VIII-9. Equation VIII-8 under the condition VIII-9 is called the characteristic form. We must now find the corresponding characteristic form in the multi-component case. To do that, we make a linear combination of the eqns. VIII-1. This gives the following new equation:

$$\sum_i \sum_j \left(L_i a_{ij} \frac{\partial C_j}{\partial t} + L_i b_{ij} \frac{\partial C_j}{\partial x} \right) = 0 \qquad\qquad\text{(VIII-10)}$$

or

$$\sum_j \left\{ \left(\sum_i L_i a_{ij} \right) \left[\frac{\partial C_j}{\partial t} + \frac{\sum_i L_i b_{ij}}{\sum_i L_i a_{ij}} \frac{\partial C_j}{\partial x} \right] \right\} = 0 \qquad\qquad\text{(VIII-11)}$$

It is clear that if we let $dx/dt = \sum_i L_i b_{ij} / \sum_i L_i a_{ij}$, we have [8]:

$$\sum_j \sum_i \left(L_i a_{ij} \frac{dC_j}{dt} \right) = 0 \qquad\qquad\text{(VIII-12)}$$

where the derivative dC_j/dt is calculated along the characteristic line of the multi-component chromatography problem. Equation VIII-12 is a characteristic form. It can be integrated and written as $dR(C_1, C_2,)/dt = 0$. It is easy to show that the ratio $\sum_i L_i b_{ij} / \sum_i L_i a_{ij}$ is equal to the characteristic velocity. This is easily demonstrated in the case of a binary mixture, with $i, j = 1, 2$. We have

$$g = \frac{dx}{dt} = \frac{L_1 b_{11} + L_2 b_{21}}{L_1 a_{11} + L_2 a_{21}} = \frac{L_1 b_{12} + L_2 b_{22}}{L_1 a_{12} + L_2 a_{22}} \qquad\qquad\text{(VIII-14)}$$

from which we derive

$$(b_{11} - a_{11}g)L_1 + (b_{21} - a_{21}g)L_2 = 0 \qquad\qquad\text{(VIII-15)}$$
$$(b_{12} - a_{12}g)L_1 + (b_{22} - a_{22}g)L_2 = 0 \qquad\qquad\text{(VIII-16)}$$

In order for this system of two homogeneous linear equations of the two variables L_1 and L_2 to have a nontrivial solution, *i.e.*, to have roots that are different from zero, it is necessary that the determinant of the coefficients be equal to zero, *i.e.*

$$det[b_{ij} - a_{ij}g] = 0 \qquad i,j = 1,2 \qquad \text{(VIII-17)}$$

Comparing eqns. VIII-6 and VIII-17 shows clearly that g is equal to the characteristic velocity, ρ. It would be easy to extend this result to the multicomponent case.

II. TWO-COMPONENT NONLINEAR CHROMATOGRAPHY

In this section we discuss the ideal model of chromatography for two components and its essential properties [9,10]. In the next section, the general results obtained in this discussion will be applied to solving the particular case of two components following Langmuir competitive isotherm behavior. Note that in many practical cases, the influence of the nonlinear behavior of the isotherm is considerable while that of the dispersion factors is relatively moderate, modern columns having a high efficiency. In such cases, the ideal model is also a very good approximation as we show in Chapter X.

1 - The mathematical model

The mass-balance equations of the ideal model in two-component chromatography are

$$\frac{\partial C_1}{\partial t} + F\frac{\partial q_1}{\partial t} + u\frac{\partial C_1}{\partial x} = 0 \qquad \text{(VIII-18a)}$$

$$\frac{\partial C_2}{\partial t} + F\frac{\partial q_2}{\partial t} + u\frac{\partial C_2}{\partial x} = 0 \qquad \text{(VIII-18b)}$$

The initial and boundary conditions are

$$\begin{cases} C_i(x,0) = 0, & i = 1,2 \\ C_i(0,t) = C_i(t), & i = 1,2 \end{cases} \qquad \text{(VIII-20)}$$

where C_1, C_2, q_1, and q_2 are the concentrations of the components 1 and 2 in the mobile and the stationary phase, respectively. In the nonlinear case, $q_i = q_i(C_1, C_2)$ in equations VIII-18a and VIII-18b, so the local concentrations C_1 and C_2 are coupled to each other. This means that the behavior of the two component bands in multicomponent chromatography must be discussed at the same time. This is allowed by the characteristics method, the same method used in the single-component case. We will use this method here.

In the single-component case, the equation of the ideal model of chromatography is

$$\frac{\partial C}{\partial t} + F\frac{\partial q}{\partial t} + u\frac{\partial C}{\partial x} = 0 \qquad \text{(VIII-21)}$$

and its characteristic form is

$$\frac{dC}{dt} = 0 \qquad \text{(VIII-22a)}$$

$$\frac{dx}{dt} = \frac{u}{1 + F\frac{dq}{dC}} \qquad \text{(VIII-22b)}$$

where dx/dt is the characteristic velocity. From these equations it is clear that the characteristic velocity depends on the concentration, C. In the two-component case, since the two concentrations are coupled together, we cannot obtain separate results for them or for their properties. The following mathematical analysis will demonstrate that, for the system of eqns. VIII-1 and VIII-2, the characteristic form is given by

$$\frac{dR(C_1, C_2)}{dt} = 0 \qquad \text{(VIII-23a)}$$

$$\frac{dx}{dt} = \rho(C_1, C_2) \qquad \text{(VIII-23b)}$$

Equation VIII-23a shows that $R(C_1, C_2)$ is constant. Thus, along a characteristic direction, $\rho(C_1, C_2)$, there is a relationship, $R(C_1, C_2) = \text{constant}$, between the two concentrations C_1 and C_2. $R(C_1, C_2)$ is called the Riemann invariant.

In a two-component system, there are generally two characteristic directions, $\rho_{+(C_1,C_2)}$ and $\rho_{-(C_1,C_2)}$. Accordingly, there are two Riemann invariants, $R_{+(C_1,C_2)}$ and $R_{-(C_1,C_2)}$. Under certain conditions, one of the invariants, R_+ or R_-, can be a constant, not only along a certain characteristic direction (ρ_+ or ρ_-), but also in any point of the (x, t) plane. In this case, the discussion will be simplified, since then only one direction propagates the "information" contained in the concentration wave. The solution under this condition is called a "simple wave". In the simple wave case, the Riemann invariant which remains constant gives the relationship between C_1 and C_2. In the following discussion, it will be shown that the above condition is often satisfied, and that the simple wave solution is often observed.

In fact, the main purpose of this discussion is to calculate the retention times in nonlinear two-component chromatography. This solution is derived as in nonlinear single-component chromatography, using the migration equations of the shock and of the wavelet. The retention times $t_{R,1}$ and $t_{R,2}$ are those of the peak maxima. In the two-component case also, the migration of each peak is determined by these migration equations.

2 - Analysis of the mathematical model

For the sake of a convenient general discussion, the equation of two-component ideal nonlinear chromatography can be written as:

$$\alpha_1 \frac{\partial C_1}{\partial x} + \beta_1 \frac{\partial C_1}{\partial t} + \gamma_1 \frac{\partial C_2}{\partial x} + \delta_1 \frac{\partial C_2}{\partial t} = 0 \tag{VIII-24a}$$

$$\alpha_1 \frac{\partial C_1}{\partial x} + \beta_1 \frac{\partial C_1}{\partial t} + \gamma_1 \frac{\partial C_2}{\partial x} + \delta_1 \frac{\partial C_2}{\partial t} = 0 \tag{VIII-24b}$$

Comparing with the general form of multi-component PDE (eqn. VIII-1), we observe that $\alpha_1 = b_{11}$, $\beta_1 = a_{11}$, $\gamma_1 = b_{12}$, $\delta_1 = a_{12}$, $\alpha_2 = b_{21}$, $\beta_2 = a_{21}$, $\gamma_2 = b_{22}$, and $\delta_2 = a_{22}$. Comparing with the equations of the ideal model of chromatography (eqns. VIII-18 and VIII- 19), we have

$$\alpha_1 = u, \quad \beta_1 = 1 + F\frac{\partial q_1}{\partial C_1}, \qquad \gamma_1 = 0, \quad \delta_1 = F\frac{\partial q_1}{\partial C_2}, \tag{VIII-25a}$$

$$\alpha_2 = 0, \quad \beta_2 = F\frac{\partial q_2}{\partial C_1}, \qquad \gamma_2 = u, \quad \delta_2 = 1 + F\frac{\partial q_2}{\partial C_2}, \tag{VIII-25b}$$

To obtain the characteristic lines of the system of eqns. VIII-24a and VIII-24b, we multiply these equations by λ_1 and λ_2, respectively, and add the results. We obtain

$$(\lambda_1\alpha_1 + \lambda_2\alpha_2)\frac{\partial C_1}{\partial x} + (\lambda_1\beta_1 + \lambda_2\beta_2)\frac{\partial C_1}{\partial t} +$$
$$(\lambda_1\gamma_1 + \lambda_2\gamma_2)\frac{\partial C_2}{\partial x} + (\lambda_1\delta_1 + \lambda_2\delta_2)\frac{\partial C_2}{\partial t} = 0 \tag{VIII-26}$$

The characteristic equations of equation VIII-26 are

$$(\lambda_1\alpha_1 + \lambda_2\alpha_2)\frac{dC_1}{dx} + (\lambda_1\gamma_1 + \lambda_2\gamma_2)\frac{dC_2}{dx} = 0 \tag{VIII-27a}$$

$$\frac{dt}{dx} = \frac{\lambda_1\beta_1 + \lambda_2\beta_2}{\lambda_1\alpha_1 + \lambda_2\alpha_2} = \frac{\lambda_1\delta_1 + \lambda_2\delta_2}{\lambda_1\gamma_1 + \lambda_2\gamma_2} \tag{VIII-27b}$$

λ_1 and λ_2 are obtained as solutions of the two eqns. VIII-27b. Then, $\rho_{\pm(C_1,C_2)}$ can be obtained. Porting λ_1 and λ_2 into eqn. VIII-27a and integrating the result gives the Riemann invariant, $R(C_1, C_2)$.

The solution of eqn. VIII-27b is that of an eigenvalue problem, $dt/dx = \rho^{-1}$, where ρ is the direction of the characteristic line. For the sake of convenience, we replace ρ^{-1} by σ. The eqns. VIII-27b become

$$\lambda_1 \alpha_1 \sigma + \lambda_2 \alpha_2 \sigma = \lambda_1 \beta_1 + \lambda_2 \beta_2$$
$$\lambda_1 \gamma_1 \sigma + \lambda_2 \gamma_2 \sigma = \lambda_1 \delta_1 + \lambda_2 \delta_2$$

These equations can be written in matrix form as

$$\begin{bmatrix} \alpha_1 \sigma - \beta_1 & \alpha_2 \sigma - \beta_2 \\ \gamma_1 \sigma - \delta_1 & \gamma_2 \sigma - \delta_2 \end{bmatrix} \begin{bmatrix} \lambda_1 \\ \lambda_2 \end{bmatrix} = 0 \qquad \text{(VIII-28)}$$

Equation VIII-28 is an eigenequation. The condition under which it has a nontrivial solution is that the determinant of the coefficients be equal to zero, $i.e.$

$$\begin{vmatrix} \alpha_1 \sigma - \beta_1 & \alpha_2 \sigma - \beta_2 \\ \gamma_1 \sigma - \delta_1 & \gamma_2 \sigma - \delta_2 \end{vmatrix} = 0 \qquad \text{(VIII-29)}$$

or

$$(\alpha_1 \gamma_2 - \alpha_2 \gamma_1)\sigma^2 + [\gamma_1 \beta_2 - \gamma_2 \beta_1 + \alpha_2 \delta_1 - \alpha_1 \delta_2]\sigma + \beta_1 \delta_2 - \beta_2 \delta_1 = 0 \quad \text{(VIII- 30)}$$

Introducing the values of the parameters of this equation in chromatography, as given in eqn. VIII-25, into eqn. VIII-30, we obtain

$$u^2 \sigma^2 - u[2 + F(q_{11} + q_{22})]\sigma + (1 + Fq_{11})(1 + Fq_{22}) - F^2 q_{12} q_{21} = 0 \quad \text{(VIII-31)}$$

where $q_{ij} = \partial q_i / \partial C_j$, with $i, j = 1, 2$. The solutions of eqn. VIII-31 are

$$\sigma_\pm = \frac{2 + F(q_{11} + q_{22})}{2u} \pm$$
$$\frac{\sqrt{[2 + F(q_{11} + q_{22})]^2 - 4[(1 + Fq_{11})(1 + Fq_{22}) - F^2 q_{12} q_{21}]}}{2u} \qquad \text{(VIII-32)}$$

Knowing the two characteristic velocities, we can solve eqn. VIII-30 for λ_1 and λ_2. Porting them into eqn. VIII-27 and integrating it gives the Riemann invariant, $R(C_1, C_2)$. This method is complex. When $n = 2$ ($i.e.$, in the two-component case), there is another method, the Hodograph transform (see Appendix A). Here, however, we introduce another, simpler method for the case $n = 2$.

From eqn. VIII-26, we could also derive another set of characteristic equations:

$$(\lambda_1\beta_1 + \lambda_2\beta_2)\frac{dC_1}{dt} + (\lambda_1\delta_1 + \lambda_2\delta_2)\frac{dC_2}{dt} = 0 \qquad \text{(VIII-27'a)}$$

$$\frac{dx}{dt} = \frac{\lambda_1\gamma_1 + \lambda_2\gamma_2}{\lambda_1\alpha_1 + \lambda_2\alpha_2} = \frac{\lambda_1\delta_1 + \lambda_2\delta_2}{\lambda_1\beta_1 + \lambda_2\beta_2} \qquad \text{(VIII-27'b)}$$

Since when $n = 2$ $R(C_1, C_2)$ exists, we derive from eqn. VIII-27'a and b that

$$\frac{dC_1}{dC_2} = \frac{\lambda_1\gamma_1 + \lambda_2\gamma_2}{\lambda_1\alpha_1 + \lambda_2\alpha_2} = \frac{\lambda_1\delta_1 + \lambda_2\delta_2}{\lambda_1\beta_1 + \lambda_2\beta_2} = \xi \qquad \text{(VIII-33)}$$

the last equation being the definition of ξ. The matrix equation of the eigenvalue problem becomes

$$\begin{bmatrix} \alpha_1\xi + \gamma_1 & \alpha_2\xi + \gamma_2 \\ \beta_1\xi + \delta_1 & \beta_2\xi + \delta_2 \end{bmatrix} \begin{bmatrix} \lambda_1 \\ \lambda_2 \end{bmatrix} = 0 \qquad \text{(VIII-29')}$$

The necessary condition for a nontrivial solution of this equation is that the determinant of the coefficients of eqn. VIII-29' be equal to zero. The secular equation is as follow:

$$[\alpha_1\beta_2 - \alpha_2\beta_1]\xi^2 - [\alpha_2\delta_1 + \beta_1\gamma_2 - \alpha_1\delta_2 - \beta_2\gamma_1]\xi + \gamma_1\delta_2 - \gamma_2\delta_1 = 0 \quad \text{(VIII-30')}$$

The solution of this equation gives the values of dC_1/dC_2, hence the Riemann invariant. It is clear that the values of $\sigma_\pm$ are functions of C_1 and C_2. If there is no coupling, $i.e.$ if $q_{12} = q_{21} = 0$, we derive from eqn. VIII-32

$$\sigma_\pm = \frac{[2 + F(q_{11} + q_{22})] \pm F(q_{11} - q_{22})}{2u}$$

or

$$\begin{cases} \sigma_+ = \frac{1 + Fq_{11}}{u} \\ \sigma_- = \frac{1 + Fq_{22}}{u} \end{cases} \qquad \text{(VIII-34)}$$

This is the classical result obtained in linear two-component chromatography. In this case ($q_1 = a_1C_1, q_{11} = a_1, q_2 = a_2C_2, q_{22} = a_2$), the bands of the each component migrate independently and the resulting chromatogram is the superimposition of the bands obtained independently for each component alone, under the same experimental conditions. In this case, the Riemann invariants, $R_{+(C_1,C_2)}$ and $R_{-(C_1,C_2)}$, are equal to C_1 and C_2, respectively.

The characteristic velocities, ρ_+ and ρ_-, are $u/(1+Fq_{11})$ and $u/(1+Fq_{22})$, respectively. In this case, eqns. VIII-23a and VIII-23b become

$$\frac{dC_1}{dt} = 0$$

$$\frac{dx}{dt} = \frac{u}{1+Fq_{11}}$$

and

$$\frac{dC_2}{dt} = 0$$

$$\frac{dx}{dt} = \frac{u}{1+Fq_{22}}$$

Actually, in ideal two-component chromatography, because of the nature of the equations considered, the Riemann invariant can be obtained through a more direct method and the character of the band migration determined more easily than following the approach described in this section. The coefficients of the equations of two-component ideal chromatography, eqns. VIII-18 and VIII-19, depend only on the concentrations C_1 and C_2. They do not depend on t nor x. Both equations are homogeneous. Equations of this type are generally called reducible and reducible equations can be linearized through the Hodograph transform (see Appendix A). This allows the calculation of the Riemann invariant and of the direction of the characteristic lines.

III. MULTIPLE-COMPONENT PROBLEM

When there are three or more components in the feed mixture, the mathematical problem becomes impossible to solve rigorously. There are no theorems proving the existence of the Riemann invariant for the system of equations VIII-18 when $n > 2$ [4-7]. However, we may postulate its existence and, from the chemical engineering viewpoint, derive an acceptable solution assuming the validity of the coherence condition. This condition is written:

$$\frac{Dq_1}{DC_1} = \frac{Dq_2}{DC_2} = \cdots = \frac{Dq_n}{DC_n} \tag{VIII-35}$$

The validity of this condition is easily demonstrated in the case of a Riemann problem, in which case the initial and boundary conditions are constant concentrations (*i.e.*, frontal analysis).

$$C(x,0) = C_j{}^i$$
$$C(0,t) = C_j{}^e, \qquad j = 1, 2, \cdots, n$$

where $C_j{}^i \neq C_j{}^e$, but both are constants. The superscripts i and e denote the initial and boundary data respectively, and the subscript j corresponds to the jth component of the feed or the system. The validity of the coherence condition, however, is only an assumption in the general case of multiple-component mixtures.

For the Riemann problem, it was proved that the solution is a function of $\frac{x}{t}$, i.e.,

$$C_i(x,t) = C_i(\frac{x}{t}), \qquad i = 1, 2, \cdots, n$$

Let I_i be the inverse function of the concentration, C_i,

$$\frac{x}{t} = I_i(C_i) = I_j(C_j), \qquad i,j = 1, 2, \cdots, n \qquad \text{(VIII-36)}$$

It is obvious that

$$C(\frac{x}{t}) = C_i(I_i(C_i)) = C_i(I_j(C_j)) = q_{ij}(C_j), \qquad i,j = 1, 2, \cdots, n$$
$$\text{(VIII-37)}$$

This means that there is a relationship between $C_1, C_2, \cdots$, and C_n and that there exists a single curve, Γ, in the concentration space, with x/t as a parameter, along which lie all the points representing the composition of the system. Along this curve, we calculate the directional derivative

$$\frac{Dq_i}{DC_i} = \sum_j \frac{\partial q_i}{\partial C_j} \frac{dC_j}{dC_i} \qquad \text{(VIII-38)}$$

Then, the multi-component equation:

$$\frac{\partial C_i}{\partial t} + F\frac{\partial q_i}{\partial t} + u\frac{\partial C_i}{\partial x} = 0$$

can be rewritten as

$$(1 + F\frac{Dq_i}{DC_i})\frac{\partial C_i}{\partial t} + u\frac{\partial C_i}{\partial x} = 0 \qquad \text{(VIII-39)}$$

or

$$\frac{\partial C_i}{\partial t} + \sigma\frac{\partial C_i}{\partial x} = 0 \qquad \text{(VIII-40)}$$

where

$$\sigma = (\frac{dt}{dx})_\omega \equiv \frac{F\frac{Dq_i}{DC_i}}{u} \qquad \text{(VIII-41)}$$

and ω is the parameter describing the curve Γ. It is important to understand the difference between eqns. VIII-32 and VIII-40. In eqn. VIII-32, σ does not depend on the value of dC_1/dC_2. In eqn. VIII-40, σ depends on the ratio dC_i/dC_j. In the general case, in which $n > 2$, this ratio cannot be calculated. Since σ must be the same for every component, i, the coherent condition follows automatically [4-7].

IV. SOLUTION WITH THE COMPETITIVE
LANGMUIR ISOTHERM [9]

When the adsorption (or partition) isotherm is given by the competitive Langmuir equation, the relationships between mobile and stationary phase concentrations are given by

$$q_1 = \frac{q_{s,1}b_1C_1}{1 + b_1C_1 + b_2C_2}$$

$$q_2 = \frac{q_{s,2}b_2C_2}{1 + b_1C_1 + b_2C_2}$$

Thus

$$q_{11} = \frac{\partial q_1}{\partial C_1} = \frac{q_{s1}b_1(1 + b_2C_2)}{\Delta^2} \qquad \text{(VIII-42a)}$$

$$q_{12} = \frac{\partial q_1}{\partial C_2} = -\frac{q_{s1}b_1b_2C_1}{\Delta^2} \qquad \text{(VIII-42b)}$$

$$q_{22} = \frac{\partial q_2}{\partial C_2} = \frac{q_{s2}b_2(1 + b_1C_1)}{\Delta^2} \qquad \text{(VIII-42c)}$$

$$q_{21} = \frac{\partial q_2}{\partial C_1} = -\frac{q_{s2}b_1b_2C_2}{\Delta^2} \qquad \text{(VIII-42d)}$$

where $\Delta = 1 + b_1C_1 + b_2C_2$. We will assume that the isotherm model is thermodynamically consistent [1,11], hence that $q_{s1} = q_{s2} = q_s$.

Porting all these equations into the coherent condition, eqn. VIII-35, then multiplying by $\Delta^2/q_sb_1b_2$, we have

$$C_2\xi^2 + \left[\left(\frac{1}{b_2} - \frac{1}{b_1}\right) + (C_2 - C_1)\right]\xi - C_1 = 0 \qquad \text{(VIII-43)}$$

Equation VIII-43 can be rewritten as

$$C_1 = C_2\xi + \left(\frac{1}{b_2} - \frac{1}{b_1}\right)\frac{\xi}{1 + \xi} \qquad \text{(VIII-44)}$$

Note that, if $q_{s1} \neq q_{s2}$, the selectivity factor at infinite dilution is $\alpha = q_{s2}b_2/(q_{s1}b_1)$. Then, eqn. VIII-43 becomes $C_1 = C_2\xi + (1 - \alpha)/(\alpha b_1 + b_2/\xi)$. We shall not pursue the discussion of this assumption on the ground that the Langmuir isotherm is no longer thermodynamically consistent and that a more complex formulation should probably be used instead [1,11]. Then, however, the equations become much too complicated. Note that, in this case, a so-called thermodynamically consistent Langmuir isotherm can be considered [12]. Although arbitrary and purely empirical, this approach offers an elegant way out. The equation might be easier to solve with this assumption.

Equation VIII-44 belongs obviously to the following general type of differential equations

$$C_1 = C_2 \xi + \phi(\xi) \tag{VIII-45}$$

This type is known as the Clairaut type of differential equation. Its solution has the general form $\xi_\pm(C_1, C_2) = \text{constant}$. The characteristic lines are two families of straight lines in the (C_1, C_2)-plane with the equations

$$C_1 \Gamma_+ - C_2 \xi_+ - \left(\frac{1}{b_2} - \frac{1}{b_1} \right) \frac{\xi_+}{1 + \xi_+} = 0 \tag{VIII-46a}$$

$$C_1 \Gamma_- - C_2 \xi_- - \left(\frac{1}{b_2} - \frac{1}{b_1} \right) \frac{\xi_-}{1 + \xi_-} = 0 \tag{VIII-46b}$$

From eqn. VIII-43, $\xi_\pm$ can be expressed as

$$\xi_\pm(C_1, C_2) = -\frac{1}{2C_2} \left[\left(\frac{1}{b_2} - \frac{1}{b_1} \right) + (C_2 - C_1) \right]$$

$$\pm \frac{1}{2C_2} \sqrt{ \left[\left(\frac{1}{b_2} - \frac{1}{b_1} \right) + (C_2 - C_1) \right]^2 + 4 C_1 C_2 } \tag{VIII-47}$$

The range of variations of $\xi_\pm$ can be determined by rewriting eqn. VIII-43 as:

$$(\xi + 1) \left(\xi - \frac{C_1}{C_2} \right) = \frac{\xi}{C_2} \left(\frac{1}{b_1} - \frac{1}{b_2} \right) \tag{VIII-48}$$

When C_2 becomes very large, this equation becomes

$$(\xi + 1) \left(\xi - \frac{C_1}{C_2} \right) \sim 0 \tag{VIII-49}$$

The limits when $C_2 \to \infty$ are

$$\xi_- \sim -1$$

$$\xi_+ \sim \frac{C_1}{C_2} \tag{VIII-50}$$

$\xi_- > -1$ because $b_2 > b_1$ (since, by convention, the second component is more retained than the first one and we have a thermodynamically consistent

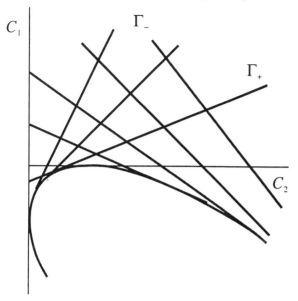

C_1

Γ_-

Γ_+

C_2

Figure VIII-1 The two Families of Characteristic Lines, Γ_+ and Γ_-.

Langmuir isotherm). The two families of straight lines, Γ_- and Γ_+ are shown in Figure VIII-1. These two families of straight lines have a common envelope in the (C_1, C_2)-plane. It is the strange solution of the Clairaut-type equation, eqn. VIII-44, and it can be derived following the conventional method. Since eqns. VIII-44 and VIII-45 are the same equation, it is more convenient to begin with the latter form. Let $\xi = p$. The equation becomes

$$a'p^2 + b'p + c' = 0 \qquad \text{(VIII-51)}$$

where $a' = C_2$, $b' = (1/b_2 - 1/b_1) + (C_2 - C_1)$, and $c' = -C_1$. Differentiating this equation with respect to p gives

$$2a'p + b' = 0$$

or

$$p = \frac{-b'}{2a'} = \frac{\frac{1}{b_1} - \frac{1}{b_2} + C_2 - C_1}{2C_2} \qquad \text{(VIII-52)}$$

Porting eqn. VIII-52 into eqn. VIII-51 gives

$$\left[\left(\frac{1}{b_1} - \frac{1}{b_2} \right) + (C_2 - C_1) \right]^2 + 4C_1 C_2 = 0 \qquad \text{(VIII-53)}$$

This equation is of the second order with respect to C_1 and C_2. It does not contain any other constant than the two parameters of the Langmuir isotherm. It is the strange solution of the Clairaut equation VIII-45 and it is the envelope of the two straight line families, Γ_+ and Γ_-. It is a parabolic curve in the (C_1, C_2)-plane.

The previous analysis showed the character of the migration of the bands of the two components of a binary mixture in nonlinear chromatography under Langmuir isotherm behavior. It demonstrates that, under nonlinear conditions, it is impossible to consider separately the behavior of the two compounds. Information regarding the system propagates along the characteristic lines, *i.e.* the functions $\xi_\pm(C_1, C_2)$ remain constant along the lines $\sigma_\pm(C_1, C_2)$, respectively. Since the values of the functions $\xi_\pm$ are different along the different curves σ, the previous results are general. But, as was pointed out in the previous section, if one family of characteristics remains absolutely constant during a chromatography experiment, *i.e.*, if it is constant not only along some characteristic direction but at any point of the (x, t) plane, there is only one propagation direction (σ_+ or σ_-). This case is named "simple wave". The function $\xi(C_1, C_2)$ which is constant everywhere, all the time, gives a functional relationship between C_1 and C_2. The practical consequences are important for the investigation of the properties of the nonlinear, two-component concentration waves.

V. SIMPLE WAVE

When C_1 and C_2 are independent of each other, the curves $\Sigma_\pm$ (i.e., $\sigma_\pm$ in the (x, t)-plane) and the curves $\Gamma_\pm$ (i.e., $\xi_\pm$ in the (C_1, C_2)-plane) correspond to each other through the Hodograph transform (see Appendix A). If C_1 and C_2 are constant, the mapping of a characteristic line, $\Sigma_\pm$ in the (x, t)-plane is a single point in the (C_1, C_2) plane. If we have a relationship between the two concentrations, *i.e.*, $C_1 = C_1(C_2)$, the mapping of one family, Σ_+ or Σ_-, degenerates into a single curve in the (C_1, C_2)-plane and the mapping of the other family degenerates into a series of points on this curve (Figure VIII-2). If the relationship $C_1 = C_1(C_2)$ corresponds to Γ_- in the (C_1, C_2)-plane, all the Σ_- in the (x, t) plane correspond to the curve Γ_- in the (C_1, C_2)-plane (Figure VIII-2) [1,9,10,13,14].

Suppose that a wide rectangular pulse of a binary mixture is injected into a column. The corresponding boundary condition is the simultaneous injection of rectangular pulses of compounds 1 and 2, both having a width t_p, and C_{10} and C_{20} being their respective heights. During the migration, once the top plateau of the rectangular injection has been eroded away, the heights of the two bands decrease during their migration. In the (C_1, C_2)-plane, the point representing the composition of the eluent at a given time migrates from point P (C_{10}, C_{20}) to point O(0,0) (Figure VIII-3). There

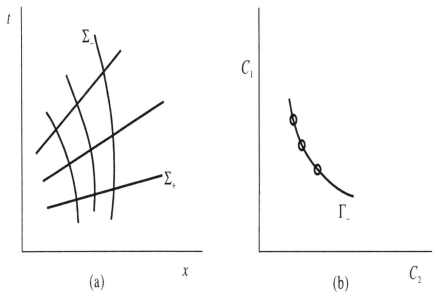

Figure VIII-2   Characteristic lines in the (x, t)- and the (C_1, C_2)-planes.

are two possible pathways under the condition of a competitive Langmuir isotherm. One is PBO, the other is PEO (Figure VIII-4). It can be proved that, since the mapping of the points of Σ_- in the (x, t)-plane is on the line PB(ξ_+) and that of the points of Σ_+ in the same plane is on the line PE (ξ_-), the direction of the migration of the simple wave is σ_- along the path PB.

In summary, under Langmuir isotherm behavior, the concentrations will change continuously along the pathway PBO. Accordingly, a two-component simple wave corresponds to the segment PB. It migrates along the direction σ_-. A single component simple wave, corresponding to BO, migrates along the direction σ_+. The corresponding rear profile of the band is shown in Figure VIII-4. Note that, in this figure, the concentration profiles along the column length are illustrated. The elution profiles, $C_i(t)$ at $x = L$ are of course different.

From this discussion, it is obvious that a constant state zone is followed by a simple wave zone (at least in the case of a Langmuir isotherm). This fact, which is well known in hydrodynamics [15], is important in chromatography. Most injections are rectangular profiles, so their elution includes a constant state that is, in most cases, followed by a simple wave. In the rare case of an anti-Langmuirian competitive isotherm, the simple wave would precede the constant state.

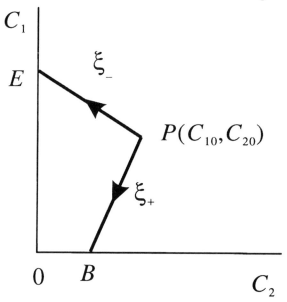

Figure VIII-3   Characteristic lines of ideal, nonlinear, two-component chromatography in the $(C_1, C_2$-plane).

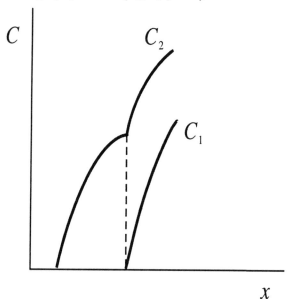

Figure VIII-4   Rear profile of the band of a binary mixture in nonlinear chromatography.

VI. SHOCK WAVE

Shock waves or concentration discontinuities form spontaneously during the elution of two-component bands as they do in single component chromatography [1,9, 10,13,14]. The fundamental reason for this is the relationship between the characteristic velocities and the concentrations with which these velocities are associated. In the Langmuir case, the function $\sigma(C_1, C_2)$ decreases with increasing C_i, thus, in two-component as in single component chromatography, the "pursuit" of low concentrations by higher ones and the intersection of characteristic lines will take place. As high concentrations cannot pass low ones, this process will eventually generate the formation of concentration discontinuities.

The mathematical analysis shows that there will be discontinuities of the concentration C_1 between both P and E and between E and O (Figure VIII-3), while there will be a discontinuity of the concentration C_2 only between P and E. In point E, C_2 becomes equal to zero.

A derivation similar to the one made in section I-1 of Chapter V, leading to eqn. V-16, allows the derivation of the velocity of the discontinuity from the law of mass conservation. The method uses an integration step, similar to the one deriving eqn. V-12 from eqn. V-9. We can derive from eqn. VIII-18a that the net mass flow of component 1 in the mobile phase which migrates into the shock during the time Δt is

$$S(u - u_s)[C_1]\Delta t \qquad \text{(VIII-54)}$$

During the same time interval, Δt, the net mass flow of component 1 in the stationary phase, which flows out from the shock because the shock migrates at velocity u_s while the molecules of the first component that are in the stationary phase remain behind, is

$$Su_s\Delta t F[q_1] \qquad \text{(VIII-55)}$$

From the law of mass conservation, the terms in eqns. VIII-54 and 55 should be equal. Thus, we must have

$$(u - u_s)[C_1] = Fu_s[q_1]$$

or

$$u_s = \frac{u}{1 + F\frac{[q_1]}{[C_1]}} \qquad \text{(VIII-56)}$$

For the second component, we have similarly

$$u_s = \frac{u}{1 + F\frac{[q_2]}{[C_2]}} \qquad \text{(VIII-57)}$$

In eqns. VIII-56 and VIII-57, $[q_i]$ and $[C_i]$ represent the jump in the corresponding concentrations from one side of the discontinuity to the other, *i.e.* $[q_i] = q_{i,+} - q_{i,-}$, where the subscripts "+" and "-" stand for the front and the rear of the concentration wave, respectively. Since $q_1 = q_1(C_1, C_2)$ and $q_2 = q_2(C_1, C_2)$, the concentration discontinuities of the two components should take place simultaneously and the values of u_s given by eqns. VIII-56 and VIII-57 should be the same, *i.e.*

$$\frac{u}{1 + F\frac{[q_1]}{[C_1]}} = \frac{u}{1 + F\frac{[q_2]}{[C_2]}} \qquad \text{(VIII-58)}$$

or

$$\frac{[q_1]}{[C_1]} = \frac{[q_2]}{[C_2]} \qquad \text{(VIII-59)}$$

This equation is similar to eqn. VIII-41:

$$\frac{Dq_1}{DC_1} = \frac{Dq_2}{DC_2}$$

which expresses the relationship between the concentrations of the two compounds along the continuous part of the profile. When the two compounds follow Langmuir competitive equilibrium behavior, we derive from eqn. VIII-59

$$\frac{b_1}{b_2}\left(1 + \frac{b_2\,C_{1+}\,C_{2-} - b_2\,C_{2+}\,C_{1-}}{C_{1+} - C_{1-}}\right) = \left(1 + \frac{b_1\,C_{2+}\,C_{1-} - b_1\,C_{1+}\,C_{2-}}{C_{2+} - C_{2-}}\right) \qquad \text{(VIII-60)}$$

since, $C_{1+} = C_{1-} + [C_1]$ and $C_{2+} = C_{2-} + [C_2]$, we have

$$\frac{b_1}{b_2}\left(1 + b_2 C_{2-} - b_2 C_{1-}\frac{[C_2]}{[C_1]}\right) = \left(1 + b_1 C_{1-} - b_1 C_{2-}\frac{[C_1]}{[C_2]}\right) \qquad \text{(VIII-61)}$$

Multiplying both sides of eqn. VIII-61 by $[C_2]/[C_1]$ gives

$$C_{1-}\left(\frac{[C_2]}{[C_1]}\right)^2 - \left[\left(\frac{1}{b_2} - \frac{1}{b_1}\right) + (C_{2-} - C_{1-})\right]\frac{[C_2]}{[C_1]} - C_{2-} = 0 \quad \text{(VIII- 62)}$$

Equation VIII-62 gives $[C_2]/[C_1]$ in the same way as eqn. VIII-A-16 gave $\xi = dC_2/dC_1$. Since eqn. VIII-62 is valid at point C_{1-}, C_{2-}, we have

$$\frac{[C_2]}{[C_1]} = \xi_\pm(C_{1-}, C_{2-}) \qquad \text{(VIII-63a)}$$

or

$$C_{2+} = C_{2-} + (C_{1+} - C_{1-})\xi_{\pm}(C_{1-}, C_{2-}) \qquad \text{(VIII-63b)}$$

In the Langmuir case, $\xi_{\pm}(C_{1-}, C_{2-}) = $ constant, so eqn. VIII-63a represents a straight line with slope $\xi_{\pm}$.

It was demonstrated in the previous section that, in the case of Langmuir isotherm behavior, the rear concentration profile of the band was continuous and that it corresponds to a simple wave along the pathway PBO in the (C_1, C_2)-plane. It was also shown that there was a discontinuity along the pathway PEO. If the point $P(C_{10}, C_{20})$ corresponds to the injection composition, the discontinuity takes place between P and E, E corresponding to the front of the discontinuity $i.e.$, $C_{1E} = C_{1+}, C_{2E} = C_{2+}$ and P to the continuous rear of the profile, $i.e.$, $C_{1P} = C_{1-}, C_{2P} = C_{2-}$. Thus, $\xi_-(C_{1-}, C_{2-}) = \xi_-(C_{1P}, C_{2P}) = \xi_-(C_{10}, C_{20})$ is the slope of PE in Figure VIII-3.

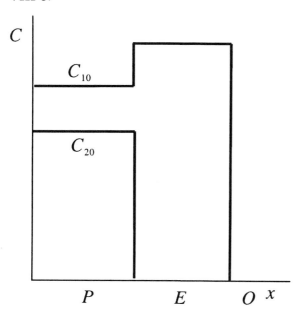

Figure VIII-5   Band Profile along the Pathway PEO.

The band profile corresponding to the pathway PEO in Figure VIII-3 is shown in Figure VIII-5. Since the passage from (C_{1P}, C_{2P}) to (C_{1E}, C_{2E}) is a jump and is not continuous, the line PE in Figure VIII-3 should not be shown as a solid line. It is an imaginary line.

VII. DEVELOPMENT OF THE BANDS OF THE TWO COMPONENTS OF A BINARY MIXTURE FOLLOWING LANGMUIR ISOTHERM BEHAVIOR. CALCULATION OF THEIR RETENTION TIMES

To sum up the results of the sections VIII-III and VIII-IV, if the equilibrium isotherm follows competitive Langmuir isotherm behavior and if the injection profile is a wide rectangular plug, the front of the bands of each of the two components of a binary mixture in nonlinear, ideal chromatography remains a concentration discontinuity or shock while its rear profile becomes continuous and is a simple wave [1,16]. In the general case, the successive profiles of the bands of the two components obtained during the development of the chromatogram are shown in Figure VIII-6 (Note that the bands spread rapidly and, accordingly, their heights decrease markedly; in order to account for this effect, the concentration scale for the figures is five times larger for those in the second column than for those in the first column). The figure illustrates the progressive separation of the two bands. The retention times of the two parts of these profiles can be calculated following the development process of each of the two bands. We see in Figure VIII-6 that, under the conditions detailed above, there are two shock waves resulting from the injection shock. The retention times of the bands are the migration times of these shocks along the column. Both shock waves are connected to the rear, diffuse profiles of the two components by a plateau. During the development process, this plateau shrinks, then fades away and, when it has disappeared (which takes place between Figures VIII-6c and VIII-6d), the height of the shock decreases. Figure VIII-6 shows also that the migration of the second shock (*i.e.* of the discontinuity of C_2) is similar to that of the single-component shock, discussed in paragraph III-1. The development of the second band is somewhat different from that of a band of the pure second component, however, because it migrates in the presence of a certain concentration of first component, hence at a faster velocity. This concentration is constant in Figures VIII-6a to VIII-6c and decreases in further figures. For the shock of the first component band, however, the migration problem is more profoundly different because the rear of the plateau between the two bands is the second shock (see Figure VIII-6). The first plateau does not collapse until the second shock begins to erode.

The calculation of the retention time of the shock of the second component of a binary mixture is carried out as follows. First, consider the migration process of the second shock after the plateau of the second band has been totally eroded away (Figure VIII-6d). In this case, the band has the shape of a curvilinear triangle. The shock migrates at the velocity u_s. It is at the

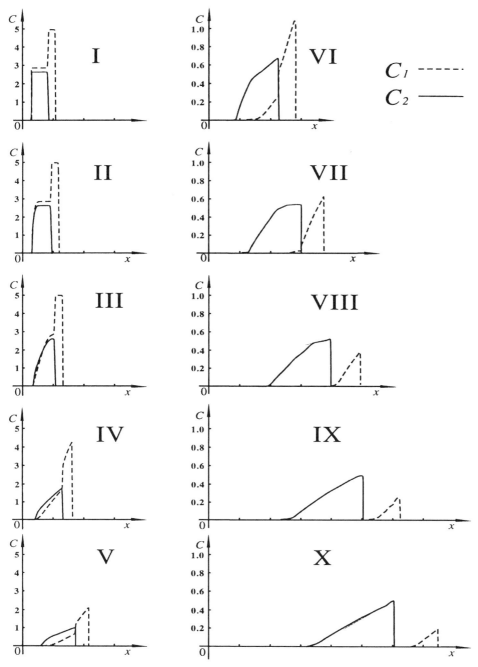

Figure VIII-6   Development of the bands of the components of a binary mixture in ideal, nonlinear chromatography. Retention times of the first shock (s): I, 90; II, 105; III, 125; IV, 175; V, 300; VI, 500; VII, 850; VIII, 1250; IX, 1750; X, 2150.

front of a diffuse band. Each concentration on this diffuse boundary migrates
at the velocity of the corresponding characteristic line, a function of the local
concentration. The migration equation corresponds with the characteristic
equation. From this analysis, the migration of the shock should obey the
following equations

$$\frac{dt}{dx} = \frac{1}{u_s} \qquad\qquad \text{(VIII-64a)}$$

$$t = t_p + x\sigma_- \qquad\qquad \text{(VIII-64b)}$$

In eqn. VIII-64b, we use σ_- rather than σ_+ because, as shown by the analysis above, the continuous boundary of the profile migrates along path PB
in Figure VIII-3, following the direction of σ_-. The equations VIII-64 are
similar to those in Section 3.1. The first equation corresponds to the migration velocity of the shock and the second is the characteristic equation,
which depends on the concentration of the point that is at the top of the
shock, on the continuous side of the profile. If the concentration of the first
component is still different from zero when the second shock reaches the
end of the column, then we can combine eqns. VIII-57 (which gives u_s) and
VIII-64a and we obtain

$$\frac{dt}{dx} = \frac{1}{u}\left(1 + F\frac{q_s b_2}{1 + b_1 C_1 + b_2 C_2}\right) \qquad\qquad \text{(VIII-65)}$$

From eqn. VIII-40, we have:

$$\sigma = \frac{1 + F\frac{Dq_i}{DC_i}}{u} \qquad \text{and} \qquad \frac{Dq_1}{DC_1} = q_{11} + q_{12}\xi$$

since $q_{12} < 0$, we have $\sigma_- = q_{11} + q_{12}\xi_+$ and

$$\sigma_- = \frac{q_s b_1(1 + b_2 C_2) - q_s b_1 b_2 C_1 \xi_+}{(1 + b_1 C_1 + b_2 C_2)^2}$$

So, finally, eqn. VIII-64b can be rewritten as

$$t = t_p + \frac{x}{u}\left[1 + \frac{F q_s b_1(1 + b_2 C_{2B})}{(1 + b_1 C_1 + b_2 C_2)^2}\right] \qquad\qquad \text{(VIII-66)}$$

Combining eqns. VIII-65 and VIII-66 and eliminating C_1 and C_2 (*i.e.*, eliminating $1 + b_1 C_1 + b_2 C_2$), we have

$$\frac{d(t - t_p - \frac{x}{u})}{d(\frac{x}{u})} = \sqrt{\frac{b_2}{b_1}\frac{F q_s b_2}{1 + b_2 C_{2B}}\frac{t - t_p - \frac{x}{u}}{\frac{x}{u}}} \qquad\qquad \text{(VIII-67)}$$

This equation is similar to eqn. III-36 in Section III-1. Its solution is :

$$\sqrt{\gamma F q_s b_2 \frac{x}{u}} - \sqrt{t - t_p - \frac{x}{u}} = const \qquad \text{(VIII-68)}$$

where $\gamma = b_2/(b_1(1 + b_2 C_{2B}))$. Equation VIII-68 may also be written

$$t = t_p + \frac{x}{u}\left[1 + \gamma F q_s b_2 \left(1 - \frac{const}{\sqrt{\gamma F q_s b_2 \frac{x}{u}}}\right)^2\right] \qquad \text{(VIII-69)}$$

The integration constant can be determined using the same method as was used in Section III-1 *i.e.*, through the condition that the plateau is just eroded away, because this is just the moment when the band profile begins to decay.

When the plateau has just been eroded away and the band begins to decay, we have

$$u_s = \frac{u}{1 + F\frac{q_s b_2}{1 + b_1 C_{10} + b_2 C_{20}}} \qquad \text{(VIII-70)}$$

In this case, the migration equation of the shock is:

$$t = \frac{x}{u}\left(1 + \frac{q_s b_2}{1 + b_1 C_{10} + b_2 C_{20}}\right) \qquad \text{(VIII- 71)}$$

and the migration equation of the simple wave connected to the shock is

$$t = t_p + \frac{x}{u}\left[1 + F\frac{q_s b_1 (1 + b_2 C_{2B})}{(1 + b_1 C_{10} + b_2 C_{20})^2}\right] \qquad \text{(VIII-72)}$$

Combining eqns. VIII-71 and VIII-72, we have

$$x_d = \frac{u t_p \Delta_0^2}{F q_s b_2 (\Delta_0 - \frac{1}{\gamma})} \qquad \text{(VIII- 73a)}$$

$$t_d = t_p + \frac{x_d}{u}\left(1 + \frac{F q_s b_2}{\gamma \Delta_0^2}\right) \qquad \text{(VIII-73b)}$$

where $\Delta_0 = 1 + b_1 C_{10} + b_2 C_{20}$. Porting x_d and t_d into eqn. VIII-68, we have

$$const = \sqrt{\gamma F q_s b_2 \frac{x_d}{u}} - \sqrt{t_d - t_p - \frac{x_d}{u}} \qquad \text{(VIII- 74a)}$$

$$= \sqrt{\gamma F q_s b_2 \frac{x_d}{u}} \left(1 - \sqrt{\frac{t_d - t_p - \frac{x_d}{u}}{\gamma F q_s b_2 \frac{x_d}{u}}} \right) \qquad \text{(VIII-74b)}$$

$$= \sqrt{\frac{\gamma t_p \Delta_0^2}{\Delta_0 - \frac{1}{\gamma}}} \left(1 - \sqrt{\frac{1}{\Delta_0^2 \gamma^2}} \right) \qquad \text{(VIII-74c)}$$

$$= \sqrt{\frac{\gamma t_p \Delta_0^2}{\Delta_0 - \frac{1}{\gamma}} \frac{\gamma \Delta_0 - 1}{\gamma \Delta_0}} \qquad \text{(VIII-74d)}$$

$$= \sqrt{t_p (\Delta_0 \gamma - 1)}$$

This result is similar to the one obtained in Section III-1. In the single-component case, $b_1 = b_2 = b$, $\gamma = 1$, and $\Delta_0 = 1 + b C_0$, i.e., $const = \sqrt{b C_0 t_p}$. Combining eqns. VIII-68 and VIII-74 gives

$$t = t_p + \frac{x}{u} \left[1 + \gamma F q_s b_2 \left(1 - \sqrt{\frac{(\Delta_0 \gamma - 1) t_t}{\gamma F q_s b_2 \frac{x}{u}}} \right)^2 \right] \qquad \text{(VIII-75)}$$

When $x = L$, we have

$$t_R = t_p + t_0 \left[1 + \gamma F q_s b_2 \left(1 - \sqrt{\frac{(\Delta_0 \gamma - 1) t_d}{\gamma F q_s b_2 t_0}} \right)^2 \right] \qquad \text{(VIII-76)}$$

Since $\Delta_0 = 1 + b_1 C_{10} + b_2 C_{20}$, eqn. VIII-76 shows that, under nonlinear conditions, the retention time of the second component depends also on the concentration of the first component in the injection pulse. This is a direct consequence of the nonlinear behavior of the isotherm. Accordingly, this result also shows that the superimposition principle does not hold under nonlinear conditions while it does so under linear conditions.

Obviously, in the single component case, since $b_1 = b_2 = b$, $\gamma = 1$, and $\Delta_0 = 1 + b C_0$, and $a = b q_s$, eqn. VIII-76 simplifies and we have

$$t_R = t_p + t_0 \left[1 + F a \left(1 - \sqrt{\frac{b C_0 t_p}{F a t_0}} \right)^2 \right] \qquad \text{(VIII-77)}$$

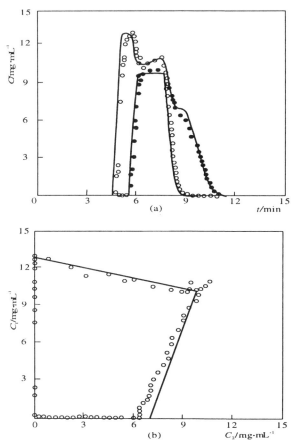

Figure VIII-7   Elution of a wide rectangular pulse of a binary mixture [17].
Solid lines, profiles of the two components calculated with the lumped kinetic
model. Symbols, concentrations of the two components measured experimentally (fraction collection and analysis). Top, concentration profiles of the two
compounds obtained by analysis of the collected fractions. Bottom, plot of
the concentration of the first component versus that of the second (hodograph
transform). *Reproduced with permission.* ©*1988 American Chemical Society.*

This equation is the same as that giving the retention time of a pure compound in nonlinear chromatography, eqn. V-49.

As we showed in the analysis above (derivation of eqn. VIII-64), the essential condition of the validity of eqn. VIII-76 is that the two components
are not yet separated, *i.e.* that $C_1 \neq 0$ when the shock of the second component exits from the column. If the two components are resolved during
their migration along the column(which takes a finite time in the framework
of the ideal model since the chromatographic problem is then hyperbolic),
we must calculate the coordinates of the point where this separation takes

place, the abscissa in the column and the time. This means that we need to know also the migration process of the first component. In two-component nonlinear chromatography, the migration process of the less retained component is far more complicated than that of the more retained one because of the interference between the two shocks. Since the velocity of the first shock is larger than that of the second one, the two shocks move apart and separate. The diffuse rear of the first profile moves under the influence of the second shock and the velocity associated with a concentration of the first component is higher if there is a second component shock behind it. The reason is that the presence of the second shock prevents the spreading of the first component band. This is the origin of the displacement effect. As a consequence, the calculation of the retention time of the first component becomes more complicated. This calculation will not be pursued here. The reader can find its results in the literature [1,18].

Figure VIII-7 illustrates the separation of a binary mixture in which the two solutes have the same concentration [17,19]. It compares the results of calculations made with the ideal model, using the equations derived above (solid lines), and experimental results (symbols). In general, as long as the column efficiency is significant (more than a few hundred plates), there is an excellent agreement between the results of the theoretical calculations based on the theory of nonlinear chromatography, the numerical calculations performed with computers, and the experimental results, as illustrated in Figure VIII-7.

APPENDIX A
The Hodograph transform

The Hodograph transform provides another method for solving the equations of nonlinear, ideal two-component chromatography when the equations are reducible (see Chapter III, section II-2f). It uses a mathematical transform under which the role of the unknown functions (C_1 and C_2) and of the variables (x and t) are exchanged. It means that x and t become functions of C_1 and C_2. The transform is written as follows

$$x = x(C_1, C_2) \qquad \text{(VIII-A-1a)}$$
$$t = t(C_1, C_2) \qquad \text{(VIII-A-1b)}$$

From these definitions, it follows that

$$\frac{dx}{dx} = \frac{\partial x}{\partial C_1}\frac{\partial C_1}{\partial x} + \frac{\partial x}{\partial C_2}\frac{\partial C_2}{\partial x} = 1 \qquad \text{(VIII-A-2)}$$

$$\frac{dt}{dx} = \frac{\partial t}{\partial C_1}\frac{\partial C_1}{\partial x} + \frac{\partial t}{\partial C_2}\frac{\partial C_2}{\partial x} = 0 \qquad \text{(VIII-A-3)}$$

These two equations can also be written as

$$\begin{bmatrix} \frac{\partial x}{\partial C_1} & \frac{\partial x}{\partial C_2} \\ \frac{\partial t}{\partial C_1} & \frac{\partial t}{\partial C_2} \end{bmatrix} \begin{bmatrix} \frac{\partial C_1}{\partial x} \\ \frac{\partial C_2}{\partial x} \end{bmatrix} = \begin{bmatrix} 1 \\ 0 \end{bmatrix} \qquad \text{(VIII-A-4)}$$

Solving this equation gives

$$\begin{bmatrix} \frac{\partial C_1}{\partial x} \\ \frac{\partial C_2}{\partial x} \end{bmatrix} = \begin{bmatrix} \frac{\partial x}{\partial C_1} & \frac{\partial x}{\partial C_2} \\ \frac{\partial t}{\partial C_1} & \frac{\partial t}{\partial C_2} \end{bmatrix}^{-1} \begin{bmatrix} 1 \\ 0 \end{bmatrix} = J \begin{bmatrix} \frac{\partial t}{\partial C_2} \\ -\frac{\partial t}{\partial C_1} \end{bmatrix} \qquad \text{(VIII-A-5)}$$

In eqn. VIII-A-5, J is defined as

$$J = \frac{\partial(C_1, C_2)}{\partial(x, t)} = \begin{bmatrix} \frac{\partial C_1}{\partial x} & \frac{\partial C_1}{\partial t} \\ \frac{\partial C_2}{\partial x} & \frac{\partial C_2}{\partial t} \end{bmatrix} \qquad \text{(VIII-A-6)}$$

and we used the following relationship

$$\begin{bmatrix} \frac{\partial x}{\partial C_1} & \frac{\partial x}{\partial C_2} \\ \frac{\partial t}{\partial C_1} & \frac{\partial t}{\partial C_2} \end{bmatrix}^{-1} = \begin{bmatrix} \frac{\partial C_1}{\partial x} & \frac{\partial C_1}{\partial t} \\ \frac{\partial C_2}{\partial x} & \frac{\partial C_2}{\partial t} \end{bmatrix}$$

which can be easily proven from eqns. VIII-A-2 and VIII-A-3. J is the transform determinant. It is clear from eqn. VIII-6 that only when $J \neq 0$, the Hodograph transform can be taken.

We may derive another pair of equations similar to eqns. VIII-A-2 and VIII-A-3:

$$\frac{dx}{dt} = \frac{\partial x}{\partial C_1}\frac{\partial C_1}{\partial t} + \frac{\partial x}{\partial C_2}\frac{\partial C_2}{\partial t} = 0 \qquad \text{(VIII-A-7a)}$$

$$\frac{dt}{dt} = \frac{\partial t}{\partial C_1}\frac{\partial C_1}{\partial t} + \frac{\partial t}{\partial C_2}\frac{\partial C_2}{\partial t} = 1 \qquad \text{(VIII-A-7b)}$$

Following a derivation similar to the one given above, we obtain for the solution of eqn. VIII-A- 7

$$\begin{bmatrix} \frac{\partial C_1}{\partial t} \\ \frac{\partial C_2}{\partial t} \end{bmatrix} = J \begin{bmatrix} -\frac{\partial x}{\partial C_2} \\ \frac{\partial x}{\partial C_1} \end{bmatrix} \qquad \text{(VIII-A-8)}$$

Porting eqns. VIII-A-6 and VIII-A-8 into eqns. VIII-24a and VIII-24b, gives

$$\alpha_1 \frac{\partial t}{\partial C_2} - \gamma_1 \frac{\partial t}{\partial C_1} - \beta_1 \frac{\partial x}{\partial C_2} + \delta_1 \frac{\partial x}{\partial C_1} = 0 \qquad \text{(VIII-A-9a)}$$

$$\alpha_2 \frac{\partial t}{\partial C_2} - \gamma_2 \frac{\partial t}{\partial C_1} - \beta_2 \frac{\partial x}{\partial C_2} + \delta_2 \frac{\partial x}{\partial C_1} = 0 \qquad \text{(VIII-A-9b)}$$

To obtain the characteristic lines of this system of equations, we follow the same process as the one with which we derived eqns. VIII-24a and VIII-24b in section II-2. Combination of eqns. VIII-A-9a and VIII-A-9b, using the factors L_1 and L_2, respectively gives the equation

$$(L_1\alpha_1 + L_2\alpha_2)\frac{\partial t}{\partial C_2} - (L_1\gamma_1 + L_2\gamma_2)\frac{\partial t}{\partial C_1} - (L_1\beta_1 + L_2\beta_2)\frac{\partial x}{\partial C_2} + (L_1\delta_1 + L_2\delta_2)\frac{\partial x}{\partial C_1} = 0$$
$$\text{(VIII-A-10)}$$

This equation can also be written

$$(L_1\alpha_1 + L_2\alpha_2)\frac{dt}{dC_2} - (L_1\beta_1 + L_2\beta_2)\frac{dx}{dC_2} = 0 \qquad \text{(VIII-A-11)}$$

and

$$\frac{dC_1}{dC_2} = \frac{L_1\gamma_1 + L_2\gamma_2}{L_1\alpha_1 + L_2\alpha_2} = -\frac{L_1\delta_1 + L_2\delta_2}{L_1\beta_1 + L_2\beta_2} = \xi \qquad \text{(VIII-A-12)}$$

Thus, we have

$$L_1\alpha_1\xi + L_2\alpha_2\xi = -(L_1\gamma_1 + L_2\gamma_2) \qquad \text{(VIII-A-13a)}$$
$$L_1\beta_1\xi + L_2\beta_2\xi = -(L_1\delta_1 + L_2\delta_2) \qquad \text{(VIII-A-13b)}$$

or

$$\begin{bmatrix} \alpha_1\xi + \gamma_1 & \alpha_2\xi + \gamma_2 \\ \beta_1\xi + \delta_1 & \beta_2\xi + \delta_2 \end{bmatrix} \begin{bmatrix} L_1 \\ L_2 \end{bmatrix} = 0 \qquad \text{(VIII-A-14)}$$

In order for this equation to have a nontrivial solution, it is necessary that the determinant of its coefficients be equal to zero. This leads to the equation

$$(\alpha_1\beta_2 - \alpha_2\beta_1)\xi^2 - [\beta_1\gamma_2 - \beta_2\gamma_1 + \alpha_2\delta_1 - \alpha_1\delta_2]\xi + \gamma_1\delta_2 - \gamma_2\delta_1 = 0 \qquad \text{(VIII-A-15)}$$

Actually, this equation is the same as equation VIII-30'. Introducing now the value of the coefficients of this equation in the ideal model of chromatography (eqns. VIII-25a and VIII-25b) and replacing ξ by dC_1/dC_2, we can rewrite eqn. VIII-A-15 as

$$\frac{\partial q_2}{\partial C_1}\left(\frac{dC_1}{dC_2}\right)^2 - \left(\frac{\partial q_1}{\partial C_1} - \frac{\partial q_2}{\partial C_2}\right)\frac{dC_1}{dC_2} - \frac{\partial q_1}{\partial C_2} = 0 \qquad \text{(VIII-A-16)}$$

Multiplying its two sides by dC_2/dC_1 gives

$$\frac{\partial q_1}{\partial C_1} + \frac{\partial q_1}{\partial C_2}\frac{dC_2}{dC_1} = \frac{\partial q_2}{\partial C_2} + \frac{\partial q_2}{\partial C_1}\frac{dC_1}{dC_2} \qquad \text{(VIII-A-17)}$$

or

$$\frac{Dq_1}{DC_1} = \frac{Dq_2}{DC_2} \qquad \text{(VIII-A-18)}$$

where

$$\frac{D}{DC_1} = \frac{\partial}{\partial C_1} + \frac{\partial}{\partial C_2}\frac{dC_2}{dC_1}$$
$$\frac{D}{DC_2} = \frac{\partial}{\partial C_2} + \frac{\partial}{\partial C_1}\frac{dC_1}{dC_2}$$

Equation VIII-A-16 is the explicit form of eqn. VIII-A-18. It is the differential equation which gives the relationship between the concentrations C_1 and C_2 of the two components in binary, nonlinear, ideal chromatography. This is an important result of the Hodograph transform. It gives the relationship that exists between C_1 and C_2 during the development of the chromatogram and it allows the management of the interference between the bands of the two compounds.

Comparing eqns. VIII-30, which gives σ, and VIII-A-15, which gives ξ, we can derive

$$\xi_{\pm} = -\frac{\alpha_1\gamma_2 - \alpha_2\gamma_1}{\alpha_1\beta_2 - \alpha_2\beta_1}\sigma_{\pm} + \frac{\beta_1\gamma_2 - \beta_2\gamma_1}{\alpha_1\beta_2 - \alpha_2\beta_1} \qquad \text{(VIII-A-19)}$$

This relationship is important because it relates $\xi_{\pm}$ and $\sigma_{\pm}$. Geometrically, the curves $\Sigma_{\pm}$ are the characteristic lines in the x,t-plane. They are determined by $\sigma_{\pm}$. The curves $\Gamma_{\pm}$ are the characteristic lines in the C_1,C_2 - plane and are determined by $\xi_{\pm}$. Therefore, equation VIII-A-19 also shows the correspondence between the characteristic lines $\Sigma_{\pm}$ and the characteristic lines $\Gamma_{\pm}$. Actually, the curves $\Gamma_{\pm}$ are obtained by the mapping of the characteristic lines $\Sigma_{\pm}$, as shown in Figure VIII-1. In the following discussion, we show that this correspondence is related to the correspondence between the Riemann invariants $R_{\pm}(C_1, C_2)$ and $\rho_{\pm}(C_1, C_2)$ which was mentioned earlier.

Actually, the two functions $\Gamma_{\pm}$ represent the relationships between C_1 and C_2 along the characteristic lines, so that they correspond to the Riemann invariant. Therefore, the Hodograph transform simplifies the calculation of the Riemann invariant. However, to calculate the functions $\Gamma_{\pm}(C_1, C_2)$, we need to integrate the differential equation VIII-A-16. In order to achieve this step, we first need to know the form of the relationships $q_1(C_1, C_2)$ and $q_2(C_1, C_2)$, $i.e.$ the equilibrium isotherms. The particular case of the Langmuir isotherm is discussed in section VII.

EXERCISES

1) Compare the values of the retention times of the two components of a binary mixture derived from the equations in this chapter with those obtained using a computer to calculate numerical solutions (see Chapter V).

2) In two-component ideal, nonlinear chromatography, the height of the breakthrough curve of the faster component is higher than the concentration of the injection pulse. Explain why, using the mechanism of nonlinear chromatography. Compare with the same situation in single-component, ideal, nonlinear chromatography.

3) Why is the simple-wave always related to a constant state zone.

4) Check the relationship between $\xi_\pm$ and $\sigma_\pm$ (eqn. VIII-A-19).

5) Show that they will increase when σ_+ changes along PE and σ_- along PB (Fig VIII-3).

6) Derive $\frac{dC_1}{dC_2}$ from eqns. VIII-27 and VIII-28, in the case of a Langmuir isotherm.

LITERATURE CITED

[1] G. Guiochon, S. Golshan-Shirazi and A. M. Katti, *"Fundamentals of Preparative and Nonlinear Chromatography"*, Academic Press, Boston, MA, 1994.

[2] S. C. Jacobson, S. Golshan-Shirazi, A. M. Katti, M. Czok, Z. Ma and G. Guiochon, *J. Chromatogr*, **484** (1989) 103.

[3] I. Quiñones, J. C. Ford and G. Guiochon, *Anal. Chem.*, **72** (2000) 1495.

[4] H.-K. Rhee, R. Aris and N.R. Amundson, *Trans. Roy. Soc. (London)*, **A267** (1970) 419.

[5] D. M. Ruthven, *"Principles of Adsorption and Adsorption Processes,"* Wiley, New York, NY, 1984.

[6] E. Glueckauf *Proc. Roy. Soc. (London)*, **A186** (1946) 35.

[7] F. Helfferich and G. Klein, *"Multicomponent Chromatography. A Theory of Interference"*, M. Dekker, New York, NY, 1970.

[8] R. Courant and D. Hilbert, *"Methods of Mathematical Physics,"* 1992.

[9] R. Aris and N. R. Amundson, *"Mathematical Methods in Chemical Engineering "*, Prentice- Hall, Englewood Cliffs, NY, 1973.

[10] H.-K. Rhee, R. Aris and N.R. Amundson, *"First-Order Partial Differential Equations. II Theory and Application of Hyperbolic Systems of Quasilinear Equations,"* Prentice Hall, Englewood Cliffs, NJ, 1989.

[11] M.D. LeVan and T. Vermeulen, *J. Phys. Chem.*, **85**, (1981) 3247.

[12] R. Bai and R. T. Yang, *J. Coll. Interf. Sci.*, **239** (2001) 296.

[13] R. Courant and K. O. Friedrichs, *"Supersonic Flow and Shock Waves"*, Wiley, New York, NY, 1948.

[14] G. B. Whitham, *"Linear and Nonlinear Waves "*, Wiley, New York, NY, 1974.

[15] A. H. Shapiro, *"The Dynamics and Thermodynamics of Compressible Fluid Flow"*, Ronald Press, New York, NY, 1953.

[16] S. Golshan-Shirazi and G. Guiochon, *J. Phys. Chem.*, **94** (1990) 495.

[17] Z. Ma, A.M. Katti, B. Lin and G. Guiochon, *J. Phys. Chem.*, **94** (1990) 6911.

[18] S. Golshan-Shirazi and G. Guiochon, *J. Phys. Chem.*, **93** (1989) 4143.

[19] A. M. Katti, Z. Ma and G. Guiochon, *AIChE J.*, **36** (1990) 1722.

CHAPTER IX

NUMERICAL ANALYSIS
OF CHROMATOGRAPHY PROBLEMS

As was made clear in the previous chapters, it is extremely difficult or entirely impossible to derive analytical solutions of the equations of chromatography in most cases. The complexity of the mathematical difficulties that need to be overcome is the most serious limitation encountered in fundamental investigations of chromatography. Except in the case of the ideal

TABLE IX.1

ANALYTICAL SOLUTIONS OF
CHROMATOGRAPHIC MODELS

Model	Single Component	Binary Mixture
Linear, Ideal	Trivial: Elution profile = Injection Profile	Trivial: Elution profile = Sum of the Injection Profiles
Linear, nonideal[a]		
Elution, Dirac	Eqns. IV-9, IV-10, IV-21	Sum of the solutions for the
Elution, wide rectangle	Eqn. IV-11	two single-component
Breakthrough	Eqns. IV-8	problems
Nonlinear, Ideal	For all isotherms:	Langmuir isotherm only
Elution, Dirac	Eqns. V-28a and V-57	see Section IX-VII
Elution, wide rectangle	Eqns. V-28a and V-30b (with $z = L$)	
Breakthrough	Eqn. V-16 or V-30b	Eqns.
Nonlinear, Nonideal		None
Parabolic isotherm[b]	Eqns. VII-7 to VII-10 or Eqns. VII-7 to VII-10b	
Thomas model[c]	See Section VII-II-1 Eqns.	

[a] Different solutions, depending on the phenomenon causing nonideal behavior: axial dispersion, mass transfer resistance, etc., and on the exact initial and boundary conditions.

[b] Assuming axial dispersion as the only cause for nonideal behavior.

[c] Assuming no axial dispersion.

model, for which such solutions do exist for most types of boundary conditions, we have to do with either approximate algebraic solutions or with numerical solutions. Even in linear chromatography, solving models that take the mass transfer resistances into account proves to be highly complicated. Finally, in all previous discussions of nonlinear chromatography, we have assumed a Langmuir single-component or a Langmuir competitive isotherm. This isotherm model is the simplest nonlinear isotherm and, in many applications, it does not properly represent the equilibrium data. More sophisticated models are often needed [1]. Except again for solutions of the ideal model [1], band profiles can be derived only through numerical calculations of the solutions of the chromatographic equations. Table IX-1 summarizes the algebraic solutions that are available for the different chromatographic problems. The list is rather short and there is little hope to see it ever becoming longer.

One of the major advantages of numerical solutions is the possibility, made highly practical by the progress of digital computing in the last fifteen years, of carrying out calculations or virtual experiments which are faster, cleaner, and more informative than actual ones. Not only does this method save considerable time and avoid wasting expensive and often toxic chemicals, it allows the detailed investigation of the influence of any single parameter while keeping all the others constant. It also allows an easy and creative combination of the theoretical analysis of chromatographic phenomena and the actual experimental aspects of chromatography. It has now become a necessary means for the prediction of the influence of the various experimental parameters on the band profiles, for research on new processes, for the optimization of experimental conditions and for the control of specific applications. To perform any numerical analysis of chromatography, it is first necessary to discretize the chromatography equation [2-5]. For the purposes of chromatography and chemical engineering investigations, the different discrete models correspond to plate models. For example, as shown by the analysis given later in this chapter, the relationship between the "artificial dispersion" (or numerical dispersion) produced in the discrete treatment and the "space length" of the integration increment is similar to the experimental relation between "longitudinal diffusion" and the "height equivalent to a theoretical plate". Under certain conditions, it is even possible to simulate the complicated parabolic chromatography equation (that includes axial dispersion) by using the discrete model of the simpler ideal chromatography that is expressed by a hyperbolic differential equation [6].

The main purpose of this book is a discussion of the mathematics of nonlinear chromatography. Therefore, it includes a discussion of the numerical analysis of the calculation methods which must be used to perform these virtual experiments. The same numerical analysis methods that can be used to calculate solutions of the chromatography equation under linear isotherm conditions can be used also for the calculation of solutions of this equation under nonlinear conditions, as long as the solution is continuous. But an important requirement for a calculation method in this latter case is that it also properly represents the discontinuity which, as we have shown in the two previous chapters, is a characteristic feature of the solutions of the nonlinear equation, at least in the ideal model. The difficulties encountered when this requirement is not, or rather is incompletely satisfied, are illustrated by the problems encountered in the early days of numerical calculations of solutions of the equations of chromatography under nonlinear conditions [23]. Furthermore, the stability conditions for discrete approximations are different in the linear and the nonlinear cases.

The general theory of nonlinear partial differential equations is still incomplete. There are no theorems regarding the existence of a solution, let alone the conditions required for the convergence of a discrete solution toward the true solution when the grid size tends toward 0. The determination of the stability of the various possible difference schemes is difficult. In numerical analysis, two general methods of approximation coexist, the finite elements methods and the finite difference methods [7-9]. Generally, the finite-difference schemes are "point approximations" while the finite-elements methods are "segment approximations." So, when the diffusion coefficient, D, is very small and, thus, a shock or a shock layer is present, the difference scheme is better. By contrast, when D is large, the finite-elements methods, for example the Galerkin method [10-13] and the orthogonal collocation method [14,15], are usually better [16-21].

As a matter of principle, we can say that good schemes for the calculation of numerical, hence approximate, solutions of mathematical models keep the physical characteristics of the mathematical model, provided that it converges. For example, the existence of "characteristic lines" (*i.e.*, lines along which the concentration wave propagates), the "mass conservation", and the "total diminishing variation" (the monotonous decrease of the shock amplitude during its migration) are important physical characteristics of chromatographic models. Algorithms or numerical calculation schemes have been developed that are based on one of these physical features. For example, the characteristic scheme emphasizes the velocity of propagation of the concentrations; the Lax–Wendroff conservation scheme emphasizes mass conservation. Yet, both schemes conserve the masses of the components involved and follow the actual velocities of the propagation of their concentrations,[1] at least to an approximation that depends on the scheme selected and on its implementation (*i.e.* on the values of the integration increments). To illustrate the high degree of fundamental equivalence of all the calculation schemes discussed here, we start from the observation that the fundamental mass conservation equation in chromatography is derived from the continuity principle of physics which is written

$$\frac{\partial \rho}{\partial t} = -J$$

where ρ is the density of the compound considered and J its flux. In chromatography, we have $\rho = C + Ff(C)$ and the flux is $J = u\partial C/\partial x$, hence the classical mass balance equation

$$\frac{\partial[C + Ff(C)]}{\partial t} + u\frac{\partial C}{\partial x} = 0$$

This equation can be rewritten without any change in its physical meaning as

$$[1 + F\frac{df(C)}{dC}]\frac{\partial C}{\partial t} + u\frac{\partial C}{\partial x} = 0$$

---

[1] This is true provided that they converge toward the true solution of the problem. But if they do not, they are useless and there is no point in discussing their properties here.

Mathematically, however, this equation is the starting point of the characteristic theory since it allows the derivation of the characteristic velocity

$$\frac{dx}{dt} = \frac{u}{1 + F\frac{df(C)}{dC}}$$

The discretization of these two identical equations and the derivation of calculation schemes of their numerical solutions should give the same result. For the first equation, this approach leads to the Lax–Wendroff conservation scheme. For the second equation, it leads to the characteristic scheme. It is obvious that the two approaches should give the same solution. The differences observed between the numerical solutions obtained following these different schemes will originate only from the different ways in which the calculation errors due to the finite sizes of the integration elements arise and propagate.

I. DISCRETIZATION OF THE CHROMATOGRAPHY EQUATION AND CHARACTERISTIC SCHEME

1 - Discretization of the Problem

To solve a partial differential equation using a finite difference method, it is first necessary to discretize it [2-5]. This means that the domain of space within which the equation is solved should be replaced by a regular mesh-like network or grid. For the chromatography equation, the space domain is $0 < x < L$ and the time domain, $t \geq 0$ *i.e.*, the integration domain is the right-up quarter of the x, t plane (see Figure IX-1).

Usually the grid replacing the continuous (*i.e.*, analog) space is made of equidistant parallel straight lines. Their distances are $\Delta x = h$ and $\Delta t = \tau$. These distances are the space and the time step size of the grid, respectively. So, the coordinates of the network nodes can be written

$$\begin{aligned} x = x_j = j\Delta x = jh, & \qquad j = 0, 1, 2, \cdots, \\ t = t_n = n\Delta t = n\tau & \qquad n = 0, 1, 2, \cdots, \end{aligned} \qquad \text{(IX-1)}$$

and these points can be denoted (x_j, t_n) or, more simply, (j, n).

The discretization of the chromatography equation can be performed based on this replacement of the continuous or analog space on which the equation is initially defined by this grid. If the finite difference method is used, the differentials in the chromatography equation are replaced by the

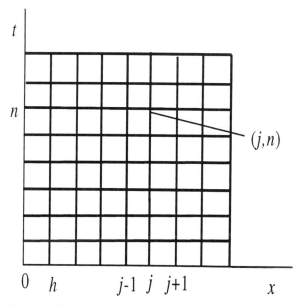

Figure IX-1.   The substitution of a grid to the continuous plane.

corresponding finite difference terms and a computation scheme is established. The solution of the finite difference equation must approach the solution of the corresponding differential equation as the net distances, h and τ, are indefinitely decreased and tend toward zero, *i.e.*, the numerical solution must converge toward the true solution. In the general case, there are no theorems demonstrating such a convergence, thus we have no general results indicating under which set of conditions practical convergence can be expected. A useful theorem exists under linear conditions. It is usually assumed that this result is also valid under a range of conditions covering most of those that are of practical interest in our field (see later).

In the following, we analyze the process of writing the finite difference scheme in the case of the ideal chromatography equation. For convenience sake, the ideal chromatography equation can be rewritten as

$$\frac{\partial C}{\partial x} + B \frac{\partial C}{\partial t} = 0 \tag{IX-2}$$

with

$$B = \frac{1 + F \, f'(C)}{u}$$

$$f(C) = \frac{aC}{1 + bC}$$

The classical initial and boundary conditions in elution chromatography are

$$C(x, 0) = 0$$

$$C(0, t) = \Psi(t)$$

There are many different ways to write the two finite difference quotients which replace the two differentials in eqn. IX-2 at the grid summits (j, n). For example, the space differential can be given by any of the following relationships:

$$\left(\frac{\partial C}{\partial x}\right)_j^n = \frac{C_{j+1}^n - C_j^n}{h} + O(h)$$

$$\left(\frac{\partial C}{\partial x}\right)_j^n = \frac{C_j^n - C_{j-1}^n}{h} + O(h)$$

$$\left(\frac{\partial C}{\partial x}\right)_j^n = \frac{C_{j+1}^n - C_{j-1}^n}{2h} + O(h^2)$$

$$\cdots$$

where the term $O(h^n)$ is the order of the error made in replacing the differential by the finite difference. This term indicates the power dependence of the truncation error on the increment h. This truncation error can be determined through a Taylor expansion of the difference term and is related to the nature of the higher order differentials.

Similarly, the time differential can be expressed in many different ways, as indicated by the following possible expressions

$$\left(\frac{\partial C}{\partial t}\right)_j^n = \frac{C_j^{n+1} - C_j^n}{\tau} + O(\tau)$$

$$\left(\frac{\partial C}{\partial t}\right)_j^n = \frac{C_j^n - C_j^{n-1}}{\tau} + O(\tau)$$

$$\left(\frac{\partial C}{\partial t}\right)_j^n = \frac{C_j^{n+1} - C_j^{n-1}}{2\tau} + O(\tau^2)$$

$$\cdots$$

2 - Main Properties of the Calculation Schemes

There are two types of calculation schemes, the **explicit** and the **implicit** schemes.

• The **explicit** schemes are those in which any term, C_j^n, involved in the scheme is a function of previously calculated similar terms. Thus, in an explicit scheme we have $C_j^n = f(C_j^{n-1}, C_{j-1}^n, C_{j-1}^{n-1}, \cdots)$.

• The **implicit** schemes are those in which at least one of the terms, C_j^n, that are involved in the scheme is a function of a yet unknown term, e.g., in $C_j^{n+1}, C_{j+1}^n, C_{j+1}^{n+1}, \cdots$. The procedure used to determine all the successive terms is complex. Its discussion is beyond the scope of this book. Interested readers should consult books dedicated to the numerical analysis of PDE problems [14].

The calculations required when using an explicit scheme are easier and faster than those made with an implicit scheme but the stability of the implicit schemes is better than the stability of the explicit schemes.

The previous two series of expressions show that each differential in the chromatography equation (eqn. IX-2) can be replaced by one of a number of different difference quotients. Thus, there are many different ways to replace the partial differential equation by a finite difference calculation scheme. These finite difference equations are all approximations of the differential equation. However, they are not equivalent. To be useful in practice, a difference scheme must possess the three following critical properties: **compatibility**, **stability**, and **convergence** (see next section). This means that:

• The solution of the difference equation must approach that of the corresponding differential equation (*i.e.*, the true solution) when the integration increments tend toward 0. This requires compatibility between the finite difference terms substituted to the different differentials and convergence of the algorithm.

• The calculation process must be stable, *i.e.* it must not diverge during the computation process. This requires the stability of the algorithm.

Thus, an acceptable difference scheme cannot be obtained by the arbitrary combination of two difference quotients picked from the two lists at the end of the previous section. All possible combinations are not equal. Some give far better results than others, as will be shown later in this chapter. The example of the ideal chromatography equation can be used to illustrate how the proper finite-difference schemes are selected, using the characteristic direction introduced in the previous chapters. This issue is discussed in the next section.

3 - Characteristic Scheme

Those features of the characteristic lines that were listed earlier and which are important in this context are that the propagation direction of the concentration wave in the (x, t)-plane is the characteristic direction, and that the characteristic line is the curve whose tangent has the characteristic direction in any of its points. The characteristic lines of the chromatography equation are determined by the following equations:

$$\frac{dC}{dx} = 0$$

$$\frac{dt}{dx} = B$$

These equations show that the concentration $C(x,t)$ remains constant along the characteristic direction, B. The characteristic line is SP in Figure IX-2, representing the local portion of the grid. The slope of its tangent is B. Let us assume that point P is a node of the grid. Then point S is not one of them.

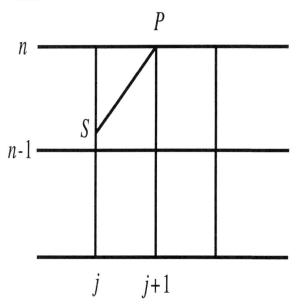

Figure IX-2 Representation of the characteristic line in the grid.

Let

$$C_S = C(x_S, t_S) = C(x_j, t_{n-a})$$
$$C_P = C(x_P, t_P) = C(x_{j+1}, t_n)$$

Since the concentration C remains constant along the characteristic lines, we have $C_S = C_P$, or

$$C(x_j, t_{n-a}) = C(x_{j+1}, t_n) \tag{IX-3}$$

t_{n-a} being the time coordinate of point S ($n - a$ is not an integer). Thus, $C_j^{n-a} = C_{j+1}^n$. Figure IX-2 shows that we have

$$t_n - t_{n-a} = hB = \frac{h(1 + Ff'(C))}{u}$$

Since t_{n-a} is not the time coordinate of a node of the grid, we need to interpolate $C(x_j, t_{n-a})$ using the value of C at the nearest nodes. With a linear interpolation, we have

$$C(x_j, t_{n-a}) = \frac{t_{n-a} - t_{n-1}}{t_n - t_{n-1}} C(x_j, t_n) + \frac{t_n - t_{n-a}}{t_n - t_{n-1}} C(x_j, t_{n-1})$$

$$= \left(1 - \frac{hB}{\tau}\right) C(x_j, t_n) + \frac{hB}{\tau} C(x_j, t_{n-1})$$

$$= C(x_j, t_n) - \frac{hB}{\tau} (C(x_j, t_n) - C(x_j, t_{n-1}))$$

Combining this result and eqn. IX-3 and letting C_j^n stand for $C(x_j, t_n)$ gives

$$C_j^n - \frac{hB}{\tau}(C_j^n - C_j^{n-1}) = C_{j+1}^n$$

or $\qquad \dfrac{C_{j+1}^n - C_j^n}{h} + B\dfrac{C_j^n - C_j^{n-1}}{\tau} = 0$ $\qquad$ (IX-4)

This is one of the possible characteristic schemes derived from the combinations of finite differences mentioned above. The results of the computations made with this scheme are better than most others, not only in the calculation of solutions of the model of ideal, linear chromatography but also in that of solutions of ideal, nonlinear chromatography. This scheme allows the calculation of the band profiles in the continuous or diffuse regions of the solution. It also allows the correct determination of the position and amplitude of the discontinuity or shock of the profile. The good results obtained with this scheme arise from it reflecting, in part, the physical characteristics of the process.

In fact, a finite difference scheme is a discrete mathematical model. It should coincide with the physical model of the chromatographic process. In fact, a basic principle of the derivation and improvement of finite difference schemes is that the scheme should correspond as closely as possible to the physical reality. For example, the conservation type scheme, which will be introduced later, reflects the conservation relation in the actual process; the TVD scheme reflects the monotonic character of the variation of the shock amplitude in the finite-difference calculation of the shock.

To solve the difference equation, we also need the discretized equivalent of the initial and boundary conditions. Those are:

$$C_j^0 = 0$$
$$C_0^n = \Psi^n(t) \qquad \text{for } j = 1, 2, \cdots, \text{and } n = 1, 2, \cdots$$

where $\Psi^n(t)$ is the result of the appropriate discretization of the continuous function $\Psi(t)$.

II. STABILITY, COMPATIBILITY, AND CONVERGENCE

1 - Stability

When we use the characteristic scheme of the linear equation of chromatography, the solution is obtained by iteration of the following computation

$$C_{j+1}^n = C_j^n - \frac{Bh}{\tau}(C_j^n - C_j^{n-1}) \qquad\qquad n = 0, 1, 2, \cdots, \qquad \text{(IX-5)}$$

Since this procedure involves a rather large amount of iterative computation, the stability of the computation process becomes an important problem [2-5]. Actually, the stability of the calculation is the result of its response to small perturbations. If an error is made during a step, if this error increases during further iterations, and if this error becomes eventually very large, the computation process is said to be unstable. If the error tends to decrease or, at least, if it remains small, the process is said to be stable. The stability of a given difference scheme may be either unconditional or conditional. In this latter case, the stability condition is a relationship between the time and space integration increments, h and τ.

We discuss now the stability analysis of the characteristic scheme of the ideal chromatography equation. Actually, this stability analysis is the analysis of the propagation of the error. In order to analyze the error propagation of a finite difference scheme, we introduce a perturbation. Suppose that the boundary condition is disturbed, *i.e.*

$$C(0, t) \to C(0, t) + \varepsilon(0, t)$$

where $\varepsilon(0, t)$ is small. Accordingly, the solution is also perturbed and

$$C(x, t) \to C(x, t) + \varepsilon(x, t)$$

Obviously, the new solution satisfies the same finite difference scheme and

$$\frac{(C + \varepsilon)_{j+1}^n - (C + \varepsilon)_j^n}{h} + B\frac{(C + \varepsilon)_j^n - (C + \varepsilon)_j^{n-1}}{\tau} = 0 \qquad \text{(IX-6)}$$

In linear chromatography, B is constant. Comparing eqns. IX-5 and IX-6 shows that we must have

$$\frac{\varepsilon_{j+1}^n - \varepsilon_j^n}{h} + B\frac{\varepsilon_j^n - \varepsilon_j^{n-1}}{\tau} = 0$$

The perturbation can be expanded into a Fourier series. Without any loss of generality, we can assume that the perturbation of the boundary condition is given by

$$\varepsilon_0^n = e^{I\omega n\tau} \tag{IX-7}$$

where $I = (-1)^{1/2}$ and that it is a sine wave with an angular frequency ω. When the computation is carried out along the X axis, the error will develop with x. Suppose that

$$\varepsilon_j^n = e^{\mu(\omega)h} e^{I\omega n\tau} \tag{IX-8}$$

where $\exp(\mu(\omega)h)$ is the amplification factor. The term $\mu(\omega)$ is a function of ω because the amplification factor may depend on the angular frequency. If the difference scheme is stable, the error will not diverge during the propagation process. This means that the amplification factor must be smaller than one. But the amplification factor depends on the difference equation used. Different difference equations will give different functions for the amplification factor, from which different stability conditions can be derived. Porting the expression IX-8 giving ε_j^n into eqn. IX-6 and letting $\lambda = \lambda(\omega) = \exp(\mu(\omega)h)$, we have

$$\frac{\lambda^{j+1} e^{I\omega n\tau} - \lambda^j e^{I\omega n\tau}}{h} + B \frac{\lambda^j e^{I\omega n\tau} - \lambda^j e^{I\omega(n-1)\tau}}{\tau} = 0$$

Eliminating $\lambda^j e^{I\omega n\tau}$, we have

$$\frac{\lambda - 1}{h} - B \frac{1 - e^{-I\omega\tau}}{\tau} = 0$$

or

$$\lambda = 1 - \frac{Bh}{\tau} \left(e^{-I\omega\tau} - 1\right)$$

Let

$$A = \frac{\tau}{Bh} = \frac{u\tau}{(1 + Fa)h} = \frac{u_z\tau}{h}$$

we have

$$\lambda(\omega) = 1 - A^{-1} + A^{-1} e^{-I\omega\tau} \tag{IX-9}$$

The value of $\lambda(\omega)$ given by eqn. IX-9 is a complex number. Obviously, the condition $|\lambda| < 1$ is required for the stability of the computation. To satisfy this condition, we must continue the analysis in the complex plane of λ. When the value of ω changes, the trace of $\lambda(\omega)$ in the complex plane is a

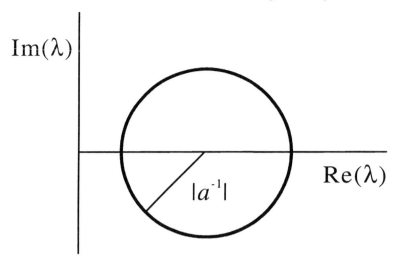

Figure IX-3. Complex plane

circle of radius $|A^{-1}|$ that is centered at the point of coordinates $(1 - A^{-1}, 0)$ (see Figure IX-3).

When $A^{-1} > 1$, the circle is larger than the unity circle. In this case, the absolute value of λ may be larger than 1. So, when $A^{-1} > 1$, the error ε_j^n will increase with increasing j; thus, under the corresponding set of calculation conditions, the difference scheme is unstable. But if the value of a satisfies the condition $0 < A^{-1} < 1$, the circle is smaller than the circle of radius unity and the error decreases exponentially with increasing j for any value of ω. In this case, the difference scheme is stable. Thus, the stability condition for the characteristic scheme is

$$0 < A^{-1} < 1 \qquad\qquad \text{or} \qquad A > 1 \qquad\qquad \text{(IX-10)}$$

The concrete expression of the above condition

$$\frac{u\tau}{(1 + Fa)h} > 1 \qquad\qquad \text{(IX-11)}$$

So the values of the integration increments h and τ must satisfy eqn. IX-11 in the computation of the solutions of the linear equation of ideal chromatography using the characteristic scheme. Equation IX-11 is the condition that is necessary for the successful completion of the numerical calculation. The number A in eqn. IX-10 is named the **Courant number** and the condition in eqn. IX-10 is named the **Courant–Friedrich–Lax** condition or CFL condition. This condition is opposite to the one generally used in the numerical calculations of solutions of hyperbolic equations. The reason is that the role of the variables x and t in chromatography is opposite to their role in the general investigations of this type of PDE, which is well known for its critical role in aerodynamics.

In nonlinear chromatography, the same method can also be used. In this case, we have

$$u_z = u_z(C) = \frac{u}{1 + F\frac{df}{dC}}$$

and a is a function of C. If the maximum value of $A^{-1}(C)$ satisfies the condition

$$\sup_{C} A^{-1}(C) < 1$$

it i.e., if the minimum value of A satisfies the condition

$$\inf_{C} A(C) > 1$$

and since $A(C) = u_z(C)\tau/h$, we have

$$\frac{\tau}{h} \inf_{C} u_z(C) > 1 \qquad (\text{IX-12})$$

In the Langmuir case, from eqn. IX-12, we have

$$\inf_{C} u_z(C) = \inf_{C} \frac{u}{1 + F\frac{df}{dC}} = \inf_{C} \frac{u}{1 + F\frac{a}{(1+bC)^2}} = \frac{u}{1 + Fa} = u_z(0) \quad (\text{IX-13})$$

Equations IX-12 and IX-13 demonstrate the important result that the stability condition is the same in the characteristic scheme whether we are dealing with linear or nonlinear chromatography.

2 - Compatibility

Whenever the ratio of the space to the time increment is selected according to the stability condition, the execution of the profile computation is stable. However, this does not guarantee the convergence of the numerical solution of the difference scheme toward the true solution of the partial differential equation. This convergence depends on the compatibility of the difference equation and the partial differential equation. The former must approach the latter for an infinitely small mesh size. Now we analyze the compatibility of the characteristic scheme. As mentioned above, if the difference scheme is compatible, the truncation error made in the calculation of solutions of the difference equation will approach zero when Δt and Δx both tend toward 0. The truncation error of the characteristic scheme of the ideal linear chromatography equation is

$$R_j^n = \left[\frac{C_{j+1}^n - C_j^n}{h} + B\frac{C_j^n - C_j^{n-1}}{\tau} \right] - \left[\left(\frac{\partial C}{\partial x}\right)_j^n + B\left(\frac{\partial C}{\partial t}\right)_j^n \right] \qquad (\text{IX-14})$$

Expanding $C_{j+1}^n = C(x+h, t_n)$ and $C_j^{n-1} = C(x, t-\tau)$ in Taylor series, we have

$$C_{j+1}^n = C_j^n + \left(\frac{\partial C}{\partial x}\right)_j^n h + \frac{1}{2}\left(\frac{\partial^2 C}{\partial x^2}\right)_j^n h^2 \tag{IX-15a}$$

$$C_j^{n-1} = C_j^n - \left(\frac{\partial C}{\partial \tau}\right)_j^n \tau + \frac{1}{2}\left(\frac{\partial^2 C}{\partial \tau^2}\right)_j^n \tau^2 \tag{IX-15b}$$

Combining eqns. IX-14 and IX-15, we have

$$R_j^n = \frac{h}{2}\left(\frac{\partial^2 C}{\partial x^2}\right)_j^n - \frac{B\tau}{2}\left(\frac{\partial^2 C}{\partial t^2}\right)_j^n = O(h+\tau)$$

The error made is of the first order with respect to the size of the integration increments. Therefore, for the characteristic scheme (eqn. IX-4) we have obviously

$$\lim_{h\to 0, \tau\to 0} R_j^n = 0$$

The characteristic scheme is compatible with the equation of ideal linear chromatography.

3 - Convergence

The compatibility of the difference scheme is a necessary but not a sufficient condition for the numerical solution to converge toward the true solution of the partial differential equation. The independent demonstration of the convergence of a finite difference scheme is generally difficult and complex. In the linear case, however, Lax demonstrated a theorem, the Lax equivalence theorem, which establishes that for a linear initial value (or for a periodic boundary value) problem, if the difference scheme is compatible, then it is also convergent, as long as this scheme is stable. The ideal linear chromatography equation derives from the classical equation of hydrodynamics by a simple exchange of the time and space variables. So, in general, the Lax equivalence theorem applies.

However, the Lax equivalence theorem does not apply to nonlinear problems. It has not yet been possible to extend any general results regarding the stability and the convergence of the numerical solutions of the linear problem to the stability and the convergence of the corresponding nonlinear problems. We have no other solutions than to assume that the results obtained for the stability and the convergence of the numerical solutions of the linear problem are valid also in the case of nonlinear problems. It turns out that, in practice, this assumption works and that these results are most usually valid.

III. ARTIFICIAL DISSIPATION
AND DIFFUSION COMPENSATION

The analysis made in the previous section showed that the difference equation is only an approximation of the partial differential equation [6]. This observation is true for any finite difference scheme. A Taylor expansion of the characteristic scheme contains higher order derivatives that are absent from the original equation of ideal chromatography. Therefore, the characteristic scheme of the ideal chromatography equation does not give exact numerical solutions. It gives solutions of a partial differential equation of a higher order than the original partial differential equation.

Porting the Taylor expansion (eqn. IX-15) of C_{j+1}^n and C_j^{n-1} into the left hand side of the equation of the characteristic scheme of the linear ideal chromatography equation (eqn. IX-4), we have

$$\frac{C_{j+1}^n - C_j^n}{h} + B\frac{C_j^n - C_j^{n-1}}{\tau} =$$
$$\left(\frac{\partial C}{\partial x}\right)_j^n + \frac{h}{2}\left(\frac{\partial^2 C}{\partial x^2}\right)_j^n + B\left(\frac{\partial C}{\partial t}\right)_j^n - \frac{B\tau}{2}\left(\frac{\partial^2 C}{\partial \tau^2}\right)_j^n$$

This equation means that the characteristic scheme (eqn. IX-4) gives a numerical solution of the following equation, at least with a second order accuracy

$$\frac{\partial C}{\partial x} + B\frac{\partial C}{\partial t} = \frac{B\tau}{2}\frac{\partial^2 C}{\partial t^2} - \frac{h}{2}\frac{\partial^2 C}{\partial x^2} \tag{IX- 16}$$

From the ideal chromatography equation (eqn. IX-2), we have

$$\frac{\partial C}{\partial t} = -\frac{1}{B}\frac{\partial C}{\partial x} = -u_z\frac{\partial C}{\partial x}$$

Under linear condition, $u_z = u/(1 + Fa)$ is constant (*i.e.*, independent of C, x and t), thus further differentiation gives

$$\frac{\partial^2 C}{\partial t^2} = u_z^2\frac{\partial^2 C}{\partial x^2}$$

Porting this expression into eqn. IX-16, gives

$$\frac{\partial C}{\partial x} + B\frac{\partial C}{\partial t} = \left[\frac{u_z}{2}\tau - \frac{h}{2}\right]\frac{\partial^2 C}{\partial x^2}$$

and, since $B = (1 + Fa)/u$, multiplying the two sides by u gives

$$(1 + Fa)\frac{\partial C}{\partial t} + u\frac{\partial C}{\partial x} = \frac{hu}{2}\left[\frac{u_z\tau}{h} - 1\right]\frac{\partial^2 C}{\partial x^2}$$

$$= \frac{hu}{2}(\mathbf{a} - 1)\frac{\partial^2 C}{\partial x^2} \qquad\qquad \text{(IX-17)}$$

The numerical parameter $\mathbf{a}$ is the Courant–Friedrichs–Levy number or, more simply, the Courant number [22]. This discussion demonstrates that following the characteristic scheme given by eqn. IX-4 leads to the calculation of solutions of eqn. IX-17, not of solutions of eqn. IX-4. Provided that $\mathbf{a} > 1$, the right hand side of eqn. IX-17 is a dispersive term with a dispersion coefficient $D_a = uh(\mathbf{a} - 1)/2$. This is a numerical coefficient that is a function of the integration increments. Only when this coefficient is equal to zero does the dispersion term also become zero (at the second order; there are obviously higher order terms in this case; the reader is encouraged to derive them). Since this dispersive term arises from the artificial treatment made to eqn. IX-4 in order to calculate its approximate numerical solutions, (i.e., from the replacement of the PDE by the finite-difference equation needed for numerical computation), this term is called an artificial or numerical dispersion.

When suitable values of h and a are selected and, thus, the relationship $D_a = uh(a - 1)/2$ allows the derivation of the value of D_a, this number behaves as an apparent dispersion coefficient, the numerical dispersion compensates almost exactly the physical dispersion (to the second order), and the characteristic scheme gives numerical solutions of the following equation

$$(1 + Fa)\frac{\partial C}{\partial t} + u\frac{\partial C}{\partial x} = D_a\frac{\partial^2 C}{\partial x^2} \qquad\qquad \text{(IX-18)}$$

This equation is the mass balance equation of linear, nonideal chromatography (see eqn. IV-2). A particularly convenient selection of h and τ is made by deriving a relationship between h and the height equivalent to a theoretical plate (HETP), H, in linear chromatography. We have

$$H = \frac{\sigma^2}{L} \qquad\qquad \text{(IX-19)}$$

where σ^2 is the variance of the Gaussian profile obtained as the approximate solution of the linear nonideal model.

Actually, in linear nonideal chromatography, the equation involves axial diffusion and is often written:

$$\frac{\partial C}{\partial t} + \frac{u}{1 + Fa}\frac{\partial C}{\partial x} = \frac{D_a}{1 + Fa}\frac{\partial^2 C}{\partial x^2} \qquad\qquad \text{(IX-20a)}$$

For the conventional Dirac boundary condition (*i.e.*, for $C(x,0) = C_0(x) = A\delta(x)$ for $-\infty < x < +\infty$), its solution is (see eqn. IV-9)

$$C(x,t) = \frac{A}{\sqrt{4\pi D_R t}} e^{-\frac{(x-u_R t)^2}{4D_R t}} \qquad \text{(IX-20b)}$$

where $D_R = D_a/(1 + Fa)$ and $u_R = u/(1 + Fa)$. The right hand side of eqn. IX-20 is a Gaussian function with a variance $\sigma^2 = 2D_R t$. When $x = L$, $t = t_R = L(1 + Fa)/u$ and the variance of the Gaussian profile is

$$\sigma^2 = 2D_R t_R = 2\frac{D}{1 + Fa}\frac{L}{u_z} = \frac{2DL}{u} \qquad \text{(IX-21)}$$

Porting eqn. IX-21 into eqn. IX-19, we have

$$H = \frac{2D}{u}$$

or

$$D = \frac{Hu}{2} \qquad \text{(IX-22)}$$

Comparing eqns. IX-17, IX-18, and IX-22, we see that if we take

$$\frac{hu}{2}(a-1) = \frac{Hu}{2} \qquad \text{or} \qquad h = \frac{H}{a-1}$$

then we have $D_a = D$, *i.e.*, the experimental diffusion is almost exactly compensated by the numerical dispersion. Since the stability condition is $a > 1$, we can take $a = 2$ and $h = H$, *i.e.*, we can take the space integration increment equal to the HETP, the time increment, $\delta\tau$ equal to $2H/u_z = 2H(1 + Fa)/u$ and then the computation result will just compensate for the experimental dispersion [6]. This means that the numerical solutions of the nonideal, linear chromatography process, which involves axial diffusion (IX-18), can be calculated using the characteristic scheme (eqn. IX-4) of the simpler ideal chromatography equation, *i.e.* the numerical solutions of the parabolic equation are calculated using a numerical scheme written to determine numerical solutions of the corresponding hyperbolic equation.

Finally, if we use the characteristic scheme with increments selected so that $a = 0$, there should be no numerical dispersion at the second order. Then, the numerical solution calculated is the solution of eqn. IX-2, at least to the second order. This is often a convenient way to calculate reasonably accurate solutions of the ideal model. To obtain a stable algorithm, however, the integration increments should be selected so as to have $A = 1.000001$. Caution must be exercised as incorrect results are sometimes obtained, particularly with convex downward isotherms.

IV. THE CHARACTERISTIC SCHEME OF THE IDEAL
TWO-COMPONENT CHROMATOGRAPHY EQUATION

The importance of the characteristic scheme arise from the fact that the numerical solutions which it affords are almost always in excellent agreement with the solutions obtained with more complex, more accurate but more time consuming methods (*e.g.*, with solutions calculated using the method of collocation on finite elements). Thus, we have every reason to believe that this scheme affords numerical solutions that agree well with the corresponding true solutions of the equation of nonlinear chromatography. So far, however, we have discussed only the application of this scheme to calculating the numerical solution of single-component problems. As we will show, excellent results are also obtained in two-component chromatography and, in the ideal case, not only in the continuous region of the solution but also in the discontinuous region [2-5].

When dealing with the characteristic scheme, we need to know the characteristic directions. The characteristic directions in ideal, nonlinear, two-component chromatography were studied in the previous chapter. From the characteristic directions, the characteristic form of the numerical equations can be determined. The characteristics directions are defined by the components L_1 and L_2 of its vector that we calculate now (see also eqn. VIII-10).

The ideal two-component equations are

$$\frac{\partial C_1}{\partial x} + \frac{1 + F\, f_{11}}{u}\frac{\partial C_1}{\partial t} + \frac{F\, f_{12}}{u}\frac{\partial C_2}{\partial t} = 0 \qquad \text{(IX-23a)}$$

$$\frac{\partial C_2}{\partial x} + \frac{1 + F\, f_{22}}{u}\frac{\partial C_2}{\partial t} + \frac{F\, f_{21}}{u}\frac{\partial C_1}{\partial t} = 0 \qquad \text{(IX-23b)}$$

where $f_{ij} = \partial f_i(C_1, C_2)/\partial C_j$ and f_i is the competitive equilibrium isotherm of component i. Combining eqns. IX-23a and IX-23b, *i.e.*, adding the products of the first equation and L_1 and of the second equation and L_2, where L_1 and L_2 are the unknowns to be determined (*i.e.*, the characteristic directions), we have

$$L_1\frac{\partial C_1}{\partial x} + \left[L_1\frac{1 + F\, f_{11}}{u} + L_2\frac{F\, f_{21}}{u}\right]\frac{\partial C_1}{\partial t} +$$

$$L_2\frac{\partial C_2}{\partial x} + \{L_1\frac{F\, f_{12}}{u} + L_2\frac{1 + F\, f_{22}}{u}\}\frac{\partial C_2}{\partial t} = 0$$

This combined equation can be rewritten as:

$$L_1\left[\frac{\partial C_1}{\partial x} + \sigma\frac{\partial C_1}{\partial t}\right] + L_2\left[\frac{\partial C_2}{\partial x} + \sigma\frac{\partial C_2}{\partial t}\right] = 0 \qquad \text{(IX-24a)}$$

where

$$\sigma = \frac{L_1 \frac{1+F\ f_{11}}{u} + L_2 \frac{F\ f_{21}}{u}}{L_1} = \frac{L_1 \frac{F f_{12}}{u} + L_2 \frac{1+F f_{22}}{u}}{L_2} \qquad \text{(IX-24b)}$$

σ is a characteristic direction and is related to the velocity of the characteristics, ρ (with $\rho = 1/\sigma$). Equation IX-24a is the characteristic form. The values of L_1 and L_2 are needed for the determination of the actual characteristic form. Equation IX-24b can be rewritten

$$\begin{bmatrix} \frac{1+Ff_{11}}{u} - \sigma & \frac{F f_{21}}{u} \\ \frac{F f_{12}}{u} & \frac{1+F f_{22}}{u} - \sigma \end{bmatrix} \begin{bmatrix} L_1 \\ L_2 \end{bmatrix} = 0 \qquad \text{(IX-24c)}$$

The values of σ are the roots of the determinant of the first matrix in eqn. IX-24c. There are two such roots, given by:

$$\sigma_\pm = \frac{1}{2u}(2 + F f_{11} + F f_{22})$$

$$\pm \frac{\sqrt{(2 + F f_{11} + F f_{22})^2 - 4[(1 + F f_{11})(1 + F f_{22}) - F^2 f_{12} f_{21}]}}{2u}$$

Putting $\sigma_\pm$ into eqn. IX-24c, we obtain

$$\left(\frac{L_1}{L_2}\right)_\pm = \frac{F f_{21}}{u\sigma_\pm - (1 + F f_{11})} \qquad \text{(IX-25)}$$

Combining eqns. IX-24a and IX-25, we have

$$\left(\frac{\partial C_1}{\partial x} + \sigma_+ \frac{\partial C_1}{\partial t}\right) + \frac{u\sigma_+ - (1 + F f_{11})}{F f_{21}}\left(\frac{\partial C_2}{\partial x} + \sigma_+ \frac{\partial C_2}{\partial t}\right) = 0 \qquad \text{(IX-26a)}$$

$$\left(\frac{\partial C_1}{\partial x} + \sigma_- \frac{\partial C_1}{\partial t}\right) + \frac{u\sigma_- - (1 + F f_{11})}{F f_{21}}\left(\frac{\partial C_2}{\partial x} + \sigma_- \frac{\partial C_2}{\partial t}\right) = 0 \qquad \text{(IX-26b)}$$

Equations IX-26a and IX-26b are the characteristic forms of eqns. IX-23a and IX-23b, respectively.

Replacing the derivatives in eqns. IX-26a and IX-26b with the corresponding finite difference quotients, we can derive the characteristic difference scheme for the calculation of numerical solutions of the system of eqns. IX-23a and IX-23b

$$\left[\frac{C_{1,j+1}^n - C_{1,j}^n}{h} + \sigma_{+j}^n \frac{\Delta_1 C_{1,j}^n}{\tau}\right] + \left[\frac{u\sigma_+ - (1 + F f_{11})}{F f_{21}}\right]_j^n$$

$$\times \left[\frac{C_{2,j+1}^n - C_{2,j}^n}{h} + \sigma_{+j}^n \frac{\Delta_1 C_{2,j}^n}{\tau}\right] = 0 \quad \text{(IX-27a)}$$

and

$$\left[\frac{C_{1,j+1}^n - C_{1,j}^n}{h} + \sigma_{-j}^n \frac{\Delta_2 C_{1,j}^n}{\tau}\right] + \left[\frac{u\sigma_- - (1 + Ff_{11})}{Ff_{21}}\right]_j^n$$

$$\times \left[\frac{C_{2,j+1}^n - C_{2,j}^n}{h} + \sigma_{-j}^n \frac{\Delta_2 C_{2,j}^n}{\tau}\right] = 0 \quad \text{(IX-27b)}$$

where

$$\Delta_1 C_j^n = \begin{cases} C_j^n - C_j^{n-1}, & \text{if} \quad \sigma_{+j}^n \geq 0 \\ C_j^{n+1} - C_j^n, & \text{if} \quad \sigma_{+j}^n < 0 \end{cases}$$

$$\Delta_2 C_j^n = \begin{cases} C_j^n - C_j^{n-1}, & \text{if} \quad \sigma_{-j}^n \geq 0 \\ C_j^{n+1} - C_j^n, & \text{if} \quad \sigma_{-j}^n < 0 \end{cases}$$

In the previous chapter, we showed that, when both isotherms follow competitive Langmuir behavior, the values of $\sigma_\pm$ are both larger than zero. So, under these conditions $\Delta_i C_j^n$ in eqns. IX-27a and IX-27b are both equal to $C_j^n - C_j^{n+1}$. With the linear approximation, the approximate stability condition for the scheme described above is

$$\max_n \left(\sigma_j^n \frac{h}{\tau}\right) \leq 1$$

This finite difference scheme is called the characteristic scheme because it is constructed on eqns. IX-23a,b and emphasizes the conservation of the directions of the characteristics. Other schemes are possible; they are discussed in the next three sections. Each scheme is a different approximation and has a different precision.

Figures IX-4a and IX-4b show the results obtained in two different cases, the first one under ideal conditions, the second under nonlinear conditions. These results are discussed in more detail in the last section of this chapter (X - Comparison of The Accuracy of The Different Numerical Schemes).

V. THE CONSERVATION TYPE DIFFERENCE SCHEME OF THE IDEAL CHROMATOGRAPHY EQUATION

The essential property of the chromatography equation is that it is a mass conservation law. Therefore, whether a discrete or a continuous mathematical scheme is used, it should partake of the conservation character of the mass balance equation. Lax *et al.* [17-19] pointed out that a difference

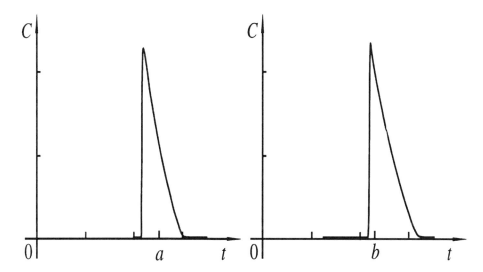

Figure IX-4.   Band Profiles Obtained with the Characteristic Scheme. Left, Ideal, Nonlinear Case. Right, Nonideal, Nonlinear Case. Points a or b indicate the exact position of the peak maximum derived from the ideal model.

scheme used to calculate numerical solutions of conservation type differential equations should be conservative too. The conservation law of ideal chromatography is written

$$\frac{\partial(C + Ff)}{\partial t} + u\frac{\partial C}{\partial x} = 0 \tag{IX-28}$$

where $f = f(C)$ is the adsorption or partition equilibrium isotherm. Equation IX-28 corresponds to

$$\frac{\partial \rho}{\partial t} + div\mathbf{J} = 0$$

where $\rho = C + Ff$ and $\mathbf{J}$ is the mass flux of the corresponding compound with $\mathbf{J} = \mathbf{u}C$. In the one-dimensional case (since we assume here that the chromatographic column is radially homogeneous), $\mathbf{J} = J_x\mathbf{i}^\circ = u_x C\mathbf{i}^\circ$, $\mathbf{i}^\circ$ being the unity vector along the OX direction, and

$$div\mathbf{J} = \frac{\partial J_x}{\partial x} = u\frac{\partial C}{\partial x}$$

The Lax–Friedrichs conservation type difference equation corresponding to eqn. IX-28, is

$$\frac{(C_j^{n+1} - C_j^{n-1}) + F[f(C_j^{n+1}) - f(C_j^{n-1})]}{2\tau} + u\frac{C_{j+1}^n - \frac{1}{2}(C_{j+1}^n - C_{j-1}^n)}{h} = 0$$

$$\tag{IX-29}$$

This scheme is different from the characteristic scheme which is

$$[1 + Ff'(C_j^n)]\frac{C_j^n - C_j^{n-1}}{\tau} + u\frac{C_{j+1}^n - C_j^n}{h} = 0$$

or

$$\frac{C_j^n - C_j^{n-1}}{\tau} + \frac{u}{1 + Ff'(C_j^n)}\frac{C_{j+1}^n - C_j^n}{h} = 0$$

In the conservation type difference equation, the change of time at the jth plate is from $(n-1)\tau$ to $(n+1)\tau$. The change in concentration is $[C + Ff'(C)]^{n+1} - [C + Ff'(C)]^{n-1}$. But in the characteristic scheme the corresponding concentration change is $[1 + Ff'(C_j^n)](C_j^{n+1} - C_j^n)$. Of course, when $\tau = 0$, these two variations are the same. However, when $\tau \neq 0$, they are different. During the computation, the conservation type scheme is able not only to keep the conservation character of the equation but also the convergence character of the calculation.

The Lax–Wendroff conservation type scheme for the single-component ideal chromatography equation is noted the L–W scheme in the following. We now discuss this scheme, first in the simpler single-component case, then in the two-component case.

1 - Single Component Case

Expanding $C(x+h, t)$ in a Taylor series, we have in the single-component case

$$C(x + h, t) = C(x, t) + h\frac{\partial}{\partial x}C(x, t) + \frac{h^2}{2}\frac{\partial^2}{\partial x^2}C(x, t) + O(h^2) \qquad \text{(IX-30)}$$

From the ideal chromatography equation (eqn. IX-28), we have

$$\frac{\partial C}{\partial x} = -\frac{\partial(C + Ff)/u}{\partial t} = -\frac{1 + Ff'}{u}\frac{\partial C}{\partial t}$$

so

$$\begin{aligned}
\frac{\partial^2 C}{\partial x^2} &= -\frac{\partial}{\partial x}\frac{\partial(C + Ff)/u}{\partial t} = -\frac{\partial}{\partial t}\left[\frac{\partial}{\partial x}\frac{C + Ff}{u}\right] \\
&= -\frac{\partial}{\partial t}\left[\left(\frac{1 + Ff'}{u}\right)\frac{\partial C}{\partial x}\right] \\
&= \frac{\partial}{\partial t}\left[\left(\frac{1 + Ff'}{u}\right)^2\frac{\partial C}{\partial t}\right]
\end{aligned}$$

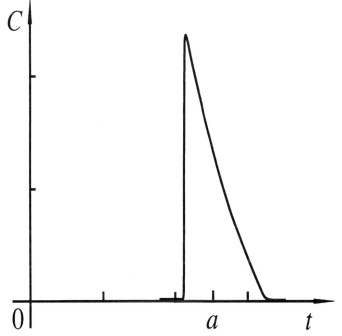

Figure IX-5. **Band Profile Obtained with the Lax–Wendroff Scheme. Point** **a indicates the exact position of the peak maximum derived from the ideal** **model.**

Porting these two results into eqn. IX-30 gives

$$C(x+h,t) = C(x,t) - h\frac{\partial}{\partial t}\left(\frac{C+Ff}{u}\right) + \frac{h^2}{2}\frac{\partial}{\partial t}\left[\frac{1+Ff'}{u}\frac{\partial}{\partial t}\frac{C+Ff}{u}\right] + O(h^2)$$

Replacing the derivatives in the equation above with the center difference quotient and letting $R = h/\tau$, we have

$$C_{j+1}^n = C_j^n - \frac{R}{2u}(C_j^{n+1} + Ff_j^{n+1} - C_j^{n-1} - Ff_j^{n-1})$$

$$+ \frac{R^2}{2u^2}\{(1+Ff')_j^{(n+1)/2})[(C_j^{n+1} + Ff_j^{n+1}) - (C_j^n + Ff_j^n)]$$

$$- (1+Ff')_j^{(n-1)/2}[(C_j^n + Ff_j^n) - (C_j^{n-1} + Ff_j^{n-1})]\} \quad \text{(IX-31)}$$

where

$$f_j^n = f(C_j^n) \qquad C_j^{(n+1)/2} = \frac{C_j^{n+1} + C_j^n}{2} \qquad C_j^{(n-1)/2} = (C_j^n + C_j^{(n-1)})/2$$

$$\text{(IX-32)}$$

The equation above is the Lax–Wendroff conservation type scheme for the ideal chromatography equation, eqn. IX-28. This difference equation ensures mass conservation.

Figure IX-5 shows the result with the Lax–Wendroff scheme under non-ideal, nonlinear conditions. This result is discussed in more detail in the last section of this chapter (X - Comparison of The Accuracy of The Different Numerical Schemes).

2 - Two-Component Case

In multicomponent (N components), ideal chromatography, the vectorial form of the conservation law is

$$\frac{\partial}{\partial t}\frac{\mathbf{C}+F\mathbf{f}}{u} + \frac{\partial \mathbf{C}}{\partial x} = 0 \tag{IX-33}$$

where $\mathbf{C}+F\mathbf{f}$ and $\mathbf{C}$ are two vectors in the N-dimension space, with components $C_1 + Ff_1, C_2 + Ff_2, \cdots, C_N + Ff_N$ and $C_1, C_2, \cdots, C_N$, respectively. The equilibrium isotherms are functions of the concentrations of all the components, $f_i = f_i(C_1, C_2, \cdots, C_N)$.

The corresponding Lax–Wendroff scheme is

$$\mathbf{C}^n_{j+1} = \mathbf{C}^n_j - \frac{R}{2u}[(\mathbf{C}+F\mathbf{f})^{n+1}_j - (\mathbf{C}+F\mathbf{f})^{n-1}_j] + \frac{R^2}{2u^2}$$
$$\{A^{(n+1)/2}_j[(\mathbf{C}+F\mathbf{f})^{n+1}_j - (\mathbf{C}+F\mathbf{f})^{n-1}_j] - A^{(n-1)/2}_j[(\mathbf{C}+F\mathbf{f})^n_j - (\mathbf{C}+F\mathbf{f})^{n-1}_j]\} \tag{IX-34a}$$

where $f^{(n\pm1)/2}_j(C) = f(C^{(n\pm1)/2}_j)$, $C^{(n+1)/2}_j = (C^{n+1}_j + C^n_j)/2$, and $C^{(n-1)/2}_j = (C^n_j + C^{(n-1)}_j)/2$, and $A^{(n\pm1)/2}_j$ are two matrices, since eqn. IX-34a is a vectorial equation. These matrices are derived from the following one

$$A = \begin{vmatrix} \dfrac{\partial(C_1+Ff_1)/u}{\partial C_1} & \dfrac{\partial(C_1+Ff_1)/u}{\partial C_2} & \cdots & \dfrac{\partial(C_1+Ff_1)/u}{\partial C_N} \\ \dfrac{\partial(C_2+Ff_2)/u}{\partial C_1} & \dfrac{\partial(C_2+Ff_2)/u}{\partial C_2} & \cdots & \dfrac{\partial(C_2+Ff_2)/u}{\partial C_N} \\ \cdots & \cdots & \cdots & \\ \dfrac{\partial(C_N+Ff_N)/u}{\partial C_1} & \dfrac{\partial(C_N+Ff_N)/u}{\partial C_2} & \cdots & \dfrac{\partial(C_N+Ff_N)/u}{\partial C_N} \end{vmatrix} \tag{IX-34b}$$

or, in abbreviation

$$A = \frac{\partial\left(\dfrac{C_1+Ff_1}{u}, \dfrac{C_2+Ff_2}{u}, \cdots \dfrac{C_N+Ff_N}{u}\right)}{\partial(C_1, C_2, \cdots, C_N)}$$

The matrices $A^{(n\pm1)/2}_j$ are the matrix A with $t = (n \pm 1/2)\tau$ and $x = jh$ (i.e., $C = C^{(n\pm1)/2}_j$ and $f = f^{(n\pm1)/2}_j$).

In the two-component case, we have

$$C_{1,j+1}^n = C_{1,j}^n - \frac{R}{2u}[(C_1 + Ff_1)_j^{n+1} - (C_1 + Ff_1)_j^{n-1}] + \frac{R^2}{2u^2}$$
$$\{(1 + Ff_{11})_j^{n+\frac{1}{2}}[(C_1 + Ff_1)_j^{n+1}) - (C_1 + Ff_1)_j^n)] +$$
$$+ Ff_{12,j}^{n+\frac{1}{2}}[(C_2 + Ff_2)_j^{n+1} - (C_2 + Ff_2)_j^n] -$$
$$- (1 + Ff_{11})_j^{n-\frac{1}{2}}[(C_1 + Ff_1)_j^n - (C_1 + Ff_1)_j^{n-1}] -$$
$$- Ff_{12,j}^{n-\frac{1}{2}}[(C_2 + Ff_2)_j^n - (C_2 + Ff_2)_j^{n-1}]\} \quad \text{(IX-35a)}$$

$$C_{2,j+1}^n = C_{2,j}^n - \frac{R}{2u}[(C_2 + Ff_2)_j^{n+1} - (C_2 + Ff_2)_j^{n-1}] + \frac{R^2}{2u^2}$$
$$\{(1 + Ff_{22})_j^{n+\frac{1}{2}}[(C_2 + Ff_2)_j^{n+1}) - (C_2 + Ff_2)_j^n)] +$$
$$+ Ff_{21,j}^{n+\frac{1}{2}})[(C_1 + Ff_1)_j^{n+1} - (C_1 + Ff_1)_j^n] -$$
$$- (1 + Ff_{22})_j^{n-\frac{1}{2}}[(C_2 + Ff_2)_j^n - (C_2 + Ff_2)_j^{n-1}] -$$
$$- Ff_{21,j}^{n-\frac{1}{2}}[(C_1 + Ff_1)_j^n - (C_1 + Ff_1)_j^{n-1}]\} \quad \text{(IX-35b)}$$

where $f_{ij} = \partial f_i/\partial C_j$, $i,j = 1,2$ is the matrix element of A. To simplify the calculations, Richtmayer gave a two-step form to the L-W scheme [2]. The first step of this method is to calculate

$$\mathbf{C}_{(j+1)/2}^{(n+1)/2} = \frac{1}{2}(\mathbf{C}_j^{n+1} + \mathbf{C}_j^n) - \frac{R}{2u}[(\mathbf{C} + F\mathbf{f})_j^{n+1} - (\mathbf{C} + F\mathbf{f})_j^n]$$

and the second step is

$$\mathbf{C}_{j+1}^n = \mathbf{C}_j^n - \frac{R}{u}[(\mathbf{C} + F\mathbf{f})_{(j+1)/2}^{(n+1)/2} - (\mathbf{C} + F\mathbf{f})_{(j+1)/2}^{(n-1)/2}]$$

where $\mathbf{f}_{(j+1)/2}^{(n\pm1)/2} = \mathbf{f}(\mathbf{C}_{(j+1)/2}^{(n\pm1)/2})$. The first step of this method is called the predictor formula, but the method can be used only if the two steps are combined.

a - Convergence of the L-W Scheme

The stability condition of the L-W scheme is

$$R\max|\lambda_i| \leq 1 \qquad\qquad i = 1, 2, \cdots, N \qquad\qquad \text{(IX-36a)}$$

where the parameters λ_i are the eigen values of matrix A, *i.e.*, the roots of the following equation:

$$
\begin{vmatrix}
\frac{\partial(C_1+Ff_1)/u}{\partial C_1} - \lambda & \frac{\partial(C_1+Ff_1)/u}{\partial C_2} & \cdots & \frac{\partial(C_1+Ff_1)/u}{\partial C_N} \\
\frac{\partial(C_2+Ff_2)/u}{\partial C_1} & \frac{\partial(C_2+Ff_2)/u}{\partial C_2} - \lambda & \cdots & \frac{\partial(C_2+Ff_2)/u}{\partial C_N} \\
\cdots & \cdots & \cdots & \\
\frac{\partial(C_N+Ff_N)/u}{\partial C_1} & \frac{\partial(C_N+Ff_N)/u}{\partial C_2} & \cdots & \frac{\partial(C_N+Ff_N)/u}{\partial C_N} - \lambda
\end{vmatrix} = 0
$$

$$\text{(IX-36b)}$$

In single-component chromatography, $\lambda = \frac{1+Ff'}{u}$. When the isotherm is given by the Langmuir equation, $f = aC/(1+BC)$, then $f' = a/(1+bC)^2$, $\lambda = [1+Fa/(1+bC)^2]/u$ and $\max|\lambda| = (1+Fa)/u$. Then, from eqn. IX-36a, the stability condition is

$$\frac{h}{u\tau}(1+Fa) \leq 1$$

$$\frac{h}{\tau} \leq \frac{u}{1+Fa} \qquad \text{(IX-36c)}$$

This means that the stability condition demands that the difference velocity, *i.e.*, h/τ, be smaller than the characteristic velocity. This is the same condition as in eqn. IX-11.

b - Truncation error in the L–W Scheme

We now analyze the truncation error of the L–W scheme when it is applied to the mass balance equation of a single component in linear ideal chromatography:

$$\frac{\partial C}{\partial x} + B\frac{\partial C}{\partial t} = 0$$

with $B = (1+Fa)/u$ (see eqn. IX-28). Replacing C_j^n with $C(x+jh, t+n\tau)$, the L–W scheme corresponding to the linear equation above can be written as

$$
\frac{C(x+h,t) - C(x,t)}{h} + B\frac{C(x,t+\tau) - C(x,t-\tau)}{2\tau}
$$
$$
- \frac{B^2 h}{2}\frac{C(x,t+\tau) - 2C(x,t) + C(x,t-\tau)}{\tau^2} = 0 \qquad \text{(IX-37)}
$$

From the Taylor expansion, we derive

$$C(x+h,t) = C(x,t) + h\frac{\partial C}{\partial x} + \frac{h^2}{2}\frac{\partial^2 C}{\partial x^2} + \frac{h^3}{6}\frac{\partial^3 C}{\partial x^3} + O(h^3)$$

$$C(x,t+\tau) = C(x,t) + \tau\frac{\partial C}{\partial t} + \frac{\tau^2}{2}\frac{\partial^2 C}{\partial t^2} + \frac{\tau^3}{6}\frac{\partial^3 C}{\partial t^3} + O(h^3)$$

$$C(x,t-\tau) = C(x,t) - \tau\frac{\partial C}{\partial t} + \frac{\tau^2}{2}\frac{\partial^2 C}{\partial t^2} - \frac{\tau^3}{6}\frac{\partial^3 C}{\partial t^3} + O(h^3)$$

Porting these expansions into eqn. IX-37 gives

$$\frac{\partial C}{\partial x} + B\frac{\partial C}{\partial t} = -\frac{h}{2}\frac{\partial^2 C}{\partial x^2} - \frac{h^2}{6}\frac{\partial^3 C}{\partial x^3} + \frac{B^2 h}{2}\frac{\partial^2 C}{\partial t^2} - \frac{B\tau^2}{6}\frac{\partial^3 C}{\partial t^3} \qquad \text{(IX-38a)}$$

If we rewrite eqn. IX-28 and differentiate with respect to x, we obtain:

$$\frac{\partial C}{\partial x} = -B\frac{\partial C}{\partial t} \qquad \text{(IX-39a)}$$

$$\frac{\partial^2 C}{\partial x^2} = -B\frac{\partial}{\partial x}\frac{\partial C}{\partial t} = -B\frac{\partial}{\partial t}\frac{\partial C}{\partial x} = B^2\frac{\partial^2 C}{\partial t^2} \qquad \text{(IX-39b)}$$

$$\frac{\partial^3 C}{\partial x^3} = -B^3\frac{\partial^3 C}{\partial t^3} \qquad \text{(IX-39c)}$$

Combining these equations with eqn. IX-38a gives

$$\frac{\partial C}{\partial x} + B\frac{\partial C}{\partial t} = -\frac{1}{6}(h^2 - \frac{\tau^2}{B^2})\frac{\partial^3 C}{\partial x^3} \qquad \text{(IX-38b)}$$

This equation shows that the precision of the L-W scheme is of the order of h^2 and τ^2. The L-W scheme gives true numerical solutions of eqn. IX-38b but not the exact solutions of eqns. IX-28 or IX-33 (in the two-component case). The right hand side of eqn. IX-38b is the main term of the truncation error. Obviously, if the integration elements are chosen so that $h = \tau/B$, the error will be only of the fourth order in the linear case. In the nonlinear case, B is a function of C, eqns. IX-39a-c are no longer valid nor is eqn. IX-38b. The truncation error is larger, depends on C, and is complex to calculate.

More generally, for eqn. IX-28, if the equation containing the main term of the truncation error of a difference scheme is

$$\frac{\partial C}{\partial x} + B\frac{\partial C}{\partial t} = D_a\frac{\partial^2 C}{\partial x^2}$$

the scheme introduces an artificial or numerical diffusion which depends on the calculation increments, h and τ.

By contrast, if the equation containing the truncation error of the difference scheme is

$$\frac{\partial C}{\partial x} + B\frac{\partial C}{\partial t} = \gamma\frac{\partial^3 C}{\partial x^3}$$

the scheme introduces an oscillation error or pseudo-oscillations that are also of numerical origin. In the L-W scheme for calculation of solutions of eqn. IX-28, $\gamma = -(h^2 - \tau^2/B^2)/6$. The oscillation error affects first the calculation of the concentration profile, then that of the shock velocity, and thus that of the retention time. So, in the L-W scheme, although mass conservation is held, an oscillation error appears. This pseudo-oscillation may be important and may give large lobes with a negative concentration (it is the algebraic area that is conserved). So, it is much more difficult to account for these oscillations or to use them to advantage than the numerical diffusion term (see previous section on the Characteristic Scheme). To eliminate this pseudo-oscillation, some other restrictions must be introduced to improve the difference scheme. One of these improved schemes is the TVD scheme which will be introduced now.

VI. THE TVD SCHEME OF THE IDEAL CHROMATOGRAPHY EQUATION

Recently, the Total Variation Diminishing scheme (TVD) became one of the focal points in the study of difference-based calculation schemes of the hyperbolic conservation equations [20,21,23]. The reason for this is that the TVD scheme is able to restrain the pseudo-oscillations of the L-W scheme and to give a higher precision and a higher resolution for the shock solution [20]. It ensures that the solution is monotonous, which is an essential characteristic of the shock solution. So the TDV scheme will appropriately reflect the characters of the nonlinear convection and diffusion processes.

The conservation equation of ideal chromatography is now written

$$\frac{\partial \phi(C)}{\partial t} + \frac{\partial C}{\partial x} = 0$$

$$\phi(C) = \frac{C + Ff(C)}{u}$$

$$f(C) = \frac{aC}{1 + bC}$$

$$\text{(IX-40)}$$

The solution of the Riemann problem for this equation is monotonous with respect to the variable x. This means concretely that the height of the band decreases monotonously along the x axis, from the column inlet to its exit. This monotonous decay corresponds to a diminishing total variation in mathematics. The total variation of $C(x,t)$ in the whole period, $TV[C(x)]$, is defined as

$$TV[C(x)] = \int_0^\infty \left| \frac{\partial}{\partial t} C(x,t) \right| dt \qquad \text{(IX-41a)}$$

The name "total variation diminishing scheme" arises from the observation that, if we have two positions, x_1 and x_2 such that $x_2 > x_1$, we have always

$$TV[C(x_2)] \leq TV[C(x_1)] \qquad \text{(IX-42a)}$$

In a finite difference scheme, the total variation of the concentration can be denoted as

$$TV[C] = \sum_{n-1}^{\infty} |\Delta_{(n+1)/2}C| \tag{IX-41b}$$

where $\Delta_{(n+1)/2}C = C^{n+1} - C^n$ and the subscript $n = 1, 2, \cdots$ corresponds to the discrete values of the time, $t_1, t_2, \cdots$. So, the summation over n in eqn. IX-41b corresponds to the numerical integration of eqn. IX-41a over time. Equation IX-42a states that the total variation of C along the x axis diminishes constantly. In the difference scheme, this can be expressed by writing that

$$TV[C_{j+1}] \leq TV[C_j] \tag{IX-42b}$$

where $j = 1, 2, \cdots$ correspond to $x_1, x_2, \cdots$. The difference scheme which satisfies this equation is named the total variation diminishing scheme or TVD scheme. Harten *et al* [21] discussed the TVD and proved that it possesses the property of maintaining a monotonous calculation process. These authors introduced the TVD function, g^{TVD}, and other switch functions to maintain monotonous the calculation and to restrict the pseudo-oscillations. This method is excellent for the accurate determination of retention times in ideal nonlinear chromatography [23].

1 - The TDV Scheme for Nonlinear Single-Component Chromatography

The TDV scheme for ideal nonlinear chromatography (eqn. IX-40) is written

$$C_{j+1}^n = C_j^n - R \left(H_j^{(n+1)/2} - H_j^{(n-1)/2} \right) \tag{IX-43a}$$

where

$$H_j^{(n+1)/2} = \frac{1}{2} \left(\phi_j^n + \phi_j^{n+1} - \Phi_j^{(n+1)/2} \right) \tag{IX- 43b}$$

and

$$\Phi_j^{(n+1)/2} = \left[Q \left(a_j^{(n+1)/2} \right) - R \left(a_j^{(n+1)/2} \right)^2 \right] g_j^{(n+1)/2} -$$
$$- Q \left(a_j^{(n+1)/2} \right) \Delta_j^{(n+1)/2} C \tag{IX-43c}$$

with $R = h/\tau$, h and τ being the space and the time increments, respectively, and

$$a_j^{(n+1)/2} = \begin{cases} \left(\phi_j^{n+1} - \phi_j^n \right) / \Delta_j^{(n+1)/2}C, & \Delta_j^{(n+1)/2}C \neq 0 \\ \left(\dfrac{\partial \phi}{\partial C} \right)_j^{(n+1)/2} & \Delta_j^{(n+1)/2}C = 0 \end{cases} \tag{IX-43d}$$

where

$$\Delta_j^{(n+1)/2} C = C_j^{n+1} - C_j^n \qquad \text{(IX-43e)}$$

and

$$Q(x) = \begin{cases} |x|, & |x| \geq \varepsilon \\ \frac{1}{2\varepsilon}(x^2 + \varepsilon^2), & |x| < \varepsilon \end{cases} \qquad \text{(IX-43f)}$$

where ε is a small quantity, depending on the required precision of the result,

$$g_j^{(n+1)/2} = \min \bmod \left(\Delta_j^{(n-1)/2} C, \Delta_j^{(n+1)/2} C, \Delta_j^{(n+3)/2} C \right) \qquad \text{(IX-43g)}$$

where

$$\min \bmod(\alpha, \beta.\gamma) = \begin{cases} sign(\alpha) \min(|\alpha|, |\beta|, |\gamma|), \text{if } \alpha, \beta, \gamma \text{ have the same sign} \\ 0, \text{if } \alpha, \beta, \gamma \text{ have different signs} \end{cases}$$

and

$$sign(\alpha) = \begin{cases} 1, & \alpha > 0 \\ 0, & \alpha = 0 \\ -1, & \alpha < 0 \end{cases}$$

2 - The TDV Scheme for Nonlinear Two-component Chromatography

The system of nonlinear mass balance equations of chromatography for two components is

$$\frac{\partial C_1}{\partial x} + \frac{\partial \phi_1}{\partial t} = 0 \qquad \text{(IX-44a)}$$

$$\frac{\partial C_2}{\partial x} + \frac{\partial \phi_2}{\partial t} = 0 \qquad \text{(IX-44b)}$$

with

$$\phi_1 = [C_1 + F f_1(C_1, C_2)] / u, \qquad f_1(C_1, C_2) = \frac{a_1 C_1}{1 + b_1 C_1 + b_2 C_2}$$

$$\phi_2 = [C_2 + F f_2(C_1, C_2)] / u, \qquad f_2(C_1, C_2) = \frac{a_2 C_2}{1 + b_1 C_1 + b_2 C_2}$$

The vector form of the set of eqns. IX-44a and IX-44b is

$$\frac{\partial \mathbf{C}}{\partial x} + \frac{\partial \bar{\phi}}{\partial t} = 0 \qquad \text{(IX-44c)}$$

where

$$\mathbf{C} = \begin{bmatrix} C_1 \\ C_2 \end{bmatrix}, \qquad \tilde{\phi} = \begin{bmatrix} \phi_1 \\ \phi_2 \end{bmatrix}$$

The TDV scheme for eqn. IX-44c is analogous to the one for the single component case.

$$\mathbf{C}_{j+1}^n = \mathbf{C}_j^n - \frac{\Delta x}{\Delta t} \left(\tilde{\phi}_j^{(n+1)/2} - \tilde{\phi}_j^{(n-1)/2} \right)$$

$$\tilde{\phi}_j^{(n+1)/2} = \frac{1}{2} \left(\tilde{\phi}_j^n + \tilde{\phi}_j^{n+1} - \tilde{\phi}_j^{(n+1)/2} \right)$$

$$\Phi_j^{(n+1)/2} = Q(\lambda_{(n+1)/2}) \left[1 - S \left(r_{(n+1)/2}^-, r_{(n+1)/2}^+ \right) \right] \Delta_{(n+1)/2} \mathbf{C}_j$$

where $\Delta_{(n+1)/2} \mathbf{C}_j = \mathbf{C}_j^{n+1} - \mathbf{C}_j^n$, and

$$r_{(n+1)/2}^- = \frac{\left(\Delta_{(n-1)/2} \mathbf{C}_j, \ \Delta_{(n+1)/2} \mathbf{C}_j \right)}{\left(\Delta_{(n+1)/2} \mathbf{C}_j, \ \Delta_{(n+1)/2} \mathbf{C}_j \right)}$$

$$r_{(n+1)/2}^+ = \frac{\left(\Delta_{(n+3)/2} \mathbf{C}_j, \ \Delta_{(n+1)/2} \mathbf{C}_j \right)}{\left(\Delta_{(n+1)/2} \mathbf{C}_j, \ \Delta_{(n+1)/2} \mathbf{C}_j \right)}$$

where $(a, b) = a_1 b_1 + a_2 b_2$, and

$$S \left(r_{(n+1)/2}^-, r_{(n+1)/2}^+ \right) = \frac{r_{(n+1)/2}^- + |r_{(n+1)/2}^-|}{1 + r_{(n+1)/2}^-} + \frac{r_{(n+1)/2}^+ + r_{(n+1)/2}^+}{1 + r_{(n+1)/2}^+} - 1$$

$$Q(\lambda_{(n+1)/2}) = \begin{cases} |\lambda_{(n+1)/2}|, & \text{if } \lambda_{(n+1)/2} \geq \varepsilon \\ \frac{\lambda_{(n+1)/2}^2 + \varepsilon^2}{2\varepsilon}, & \text{if } \lambda_{(n+1)/2} < \varepsilon \end{cases}$$

$$\lambda_{(n+1)/2} = \max_k |\lambda_{(n+1)/2}^k|, \quad k = 1, 2$$

where $\lambda_{(n+1)/2}^k$ is the eigenvalue of the matrix $A = \partial(\phi_1, \phi_2)/\partial(C_1, C_2)$, i.e. the root of the following equation:

$$\begin{vmatrix} \frac{\partial \phi_1}{\partial C_1} - \lambda & \frac{\partial \phi_1}{\partial C_2} \\ \frac{\partial \phi_2}{\partial C_1} & \frac{\partial \phi_2}{\partial C_2} - \lambda \end{vmatrix} = 0$$

In some respects, the TVD scheme could be considered as an improvement of the Lax–Wendroff conservation scheme. In the TVD scheme, the term $\Phi_j^{(n+1)/2}$ contains a restriction series that was introduced to decrease the pseudo-oscillations. This increases the accuracy in the position of the concentration shock. The discussion of the stability condition of the TVD scheme

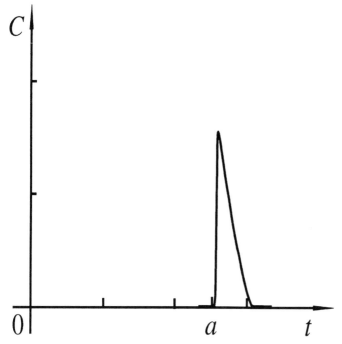

Figure IX-6.   Band Profile Obtained with the TVD Scheme. Point *a* indicates the exact position of the peak maximum derived from the ideal model.

is too complex to be within the scope of this book. However, it is worth mentioning that the same condition as the one used for the Lax–Wendroff conservation scheme is generally used and allows the improved accuracy just reported. Figure IX-6 shows the result obtained in under nonideal, nonlinear conditions. These results are discussed in more detail in the last section of this chapter (X - Comparison of the Accuracy of the Different Numerical Schemes).

We have used the improved TDV scheme to solve the single-component and a system of two-component equations of ideal, nonlinear chromatography. The results showed one concentration jump in the single-component case, two such jump in the two-component case. These discontinuities correspond to the shocks predicted by the respective exact mathematical solutions (see Chapter V and VIII, respectively). The error observed on the prediction of the retention time was less than the experimental error of measurement in the single-component case. In the two-component case, this error was observable but still very small.

VII. NUMERICAL ANALYSIS OF
THE EQUILIBRIUM DISPERSIVE
MODEL OF CHROMATOGRAPHY

The equation of the equilibrium–dispersive model of chromatography is

$$[1 + Ff'(C)]\frac{\partial C}{\partial t} + u\frac{\partial C}{\partial x} = D\frac{\partial^2 C}{\partial x^2} \qquad \text{(IX-45)}$$

It was pointed out earlier, in Section III, that longitudinal diffusion can be simulated by numerical diffusion, at least in the linear case, provided that the calculation increments of time and space are chosen in an appropriate fashion so that the coefficient of axial dispersion, D_a, in eqn. IX-45 is equal to the coefficient of artificial dispersion [8-10]. This means that if the net ratio of the time and space increments is selected appropriately, solutions of the equation of the equilibrium–dispersive model of chromatography can be calculated using the proper finite difference scheme of the equation of the ideal model of chromatography. This compensation, however, is restricted by rather strict conditions. Accordingly, it is still often necessary to develop numerical schemes that take explicitly into account the effect of axial diffusion or dispersion (and in which the effects of numerical dispersion are small enough to be negligible).

The difference scheme of the second order term in eqn. IX-45 (RHS term of this equation) is written

$$\left(\frac{\partial^2 C}{\partial x^2}\right)_j^n = \frac{C_{j+1}^n - 2C_j^n + C_{j-1}^n}{h^2}$$

The difference scheme for eqn. IX-45 is the combination of the difference scheme for the diffusion term (above) and of the difference scheme for the equation of ideal chromatography. Thus, several kinds of difference schemes have been introduced for the calculation of numerical solutions of equation IX-45, similar to those that were used for the calculation of numerical solutions of the equation of the ideal model.

1. The Left Bias Explicit Scheme

In the linear case, $f(C) = aC$ and the difference scheme is:

$$\frac{1 + Fa}{\tau}\left(C_j^{n+1} - C_j^n\right) + \frac{u}{h}\left(C_j^n - C_{j-1}^n\right) - \frac{D}{h^2}\left(C_{j+1}^n - 2C_j^n + C_{j-1}^n\right) = 0$$

This numerical scheme is named the left bias explicit scheme. Its precision is $E = O(\tau + h)$, *i.e.* its precision is of the first order. Its stability condition is

$$2D + uh \geq 0$$

$$\tau \leq \min \left(\frac{h^2(1 + Fa)}{2D + uh}, \ \frac{(2D + uh)(1 + Fa)}{u^2} \right)$$

This scheme is very simple. When $D = 0$, it reduces to the characteristic scheme of the convection process. Conversely, the results obtained with this scheme are not very accurate.

Figure IX-7 shows the results obtained with this scheme under nonideal, nonlinear conditions. These results are discussed in more detail in the last section of this chapter (X - Comparison of the Accuracy of the Different Numerical Schemes).

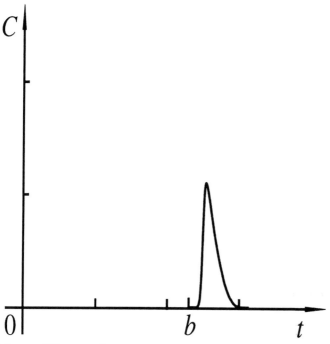

Figure IX-7.   **Band Profile Obtained with the Left Bias Explicit Scheme. Point *b* indicates the exact position of the peak maximum derived from the ideal model.**

2. The Lax–Wendroff Scheme

A two-step form of the Lax–Wendroff conservation scheme is often used [18,19]. In the linear case, the two-step form can be reduced into a one-step form. In this case, the conservation scheme for eqn. IX-45 is

$$\frac{1 + Fa}{\tau} \left(C_j^{n+1} - C_j^n\right) + \frac{u}{2h} \left(C_{j+1}^n - C_{j-1}^n\right)$$
$$- \left(\frac{u^2\tau}{2(1 + Fa)} + D\right) \frac{C_{j+1}^n - 2C_j^n + C_{j-1}^n}{h^2} = 0 \quad \text{(IX-46)}$$

In fact, in the linear case, the conservation scheme of the equation of the equilibrium–dispersive model of chromatography can be written as:

$$\frac{\partial C}{\partial t} + \frac{\partial \Phi}{\partial x} = D_R \frac{\partial^2 C}{\partial x^2}$$

with $\Phi = uC/(1 + Fa)$ and $D_R = D/(1 + Fa)$. The corresponding Lax–Wendroff two-step scheme of the above equation is

$$C_{(j+1)/2}^{(n+1)/2} = \frac{1}{2} \left(C_{j+1}^n + C_j^n\right) - \frac{\tau}{2h} \left(\Phi_{j+1}^n - \Phi_j^n\right)$$

$$C_j^{n+1} = C_j^n - \frac{\tau}{h} \left(\Phi_{(j+1)/2}^{(n+1)/2} - \Phi_{(j-1)/2}^{(n+1)/2}\right) + \frac{\tau D_R \left(C_{j+1}^n - 2C_j^n + C_{j-1}^n\right)}{h^2}$$
$$\text{(IX-47)}$$

where $\Phi_j^n = \Phi(C_j^n)$. In the linear case,

$$\Phi_j^n = \frac{uC_j^n}{1 + Fa} \tag{IX-48a}$$

$$C_{(j+1)/2}^{(n+1)/2} = \frac{1}{2}(C_{j+1}^n + C_j^n) - \frac{u\tau}{2h(1 + Fa)}(C_{j+1}^n - C_j^n) \tag{IX-48b}$$

$$C_{(j-1)/2}^{(n+1)/2} = \frac{1}{2}(C_j^n + C_{j-1}^n) - \frac{u\tau}{2h(1 + Fa)}(C_j^n - C_{j-1}^n)$$

Combining eqns. IX-48a and IX-48b with eqn. IX-47 gives

$$C_j^{n+1} = C_j^n - \frac{u\tau}{2h(1 + Fa)}(C_{j+1}^n - C_{j-1}^n) + \frac{u^2\tau^2(C_{j+1}^n - 2C_j^n + C_{j-1}^n)}{2(1 + Fa)^2 h^2}$$
$$+ \frac{D\tau(C_{j+1}^n - 2C_j^n + C_{j-1}^n)}{h^2(1 + Fa)}$$

This equation is the same as eqn. IX-46, already written at the beginning of this subsection. The term $u^2\tau/[2(1 + Fa)]$ is a compensation term which offsets the artificial dispersion introduced by the rounding errors in the numerical calculations. This term improves the precision. In the linear case, the precision of this scheme is $E = O(\tau^2 + h^2)$ and the stability condition is

$$\frac{u^2 \tau^2}{(1 + Fa)^2} + \frac{2D\tau}{1 + Fa} \leq h^2$$

The results of this two-step scheme are still excellent when applied under nonlinear conditions to the convection–diffusion equation in hydrodynamics. However, in this latter case, the nonlinear behavior is introduced by a space-derivative term. By contrast, in nonlinear, nonideal chromatography, the nonlinear behavior is introduced by a time-derivative term. The latter is not equivalent to the former when diffusion is involved. Thus, the nonlinear chromatography problem involving diffusion is different from the general nonlinear convection-diffusion problem, and the Lax–Wendroff two-step scheme cannot be applied directly in chromatography.

3. The jump point scheme

As mentioned above, the stability condition depends not only on the convection factor, $u\tau/h$, but also on the diffusion factor, $D\tau/h^2$. Generally, when diffusion is involved, the stability condition is more complicated and may include several inequations. From the physical point of view, diffusion is different from the migration of a concentration wave. The propagation of a wave is a propagation process of the solute, so the results of the explicit scheme (see Section I-2) are still good, but the diffusion of the solute concentration in any point will be affected not only by the earlier concentration at the point considered but also by the concentrations obtained at the same time, in points that are close to the one considered (diffusion is controlled by the concentration gradient, see Chapter II, Section III). So, the results of the following scheme, called the implicit scheme,

$$C_j^{n+1} = H(C_j^n, C_{j+1}^n, C_{j-1}^n, C_j^{n+1}, C_{j-1}^{n+1}, C_{j+1}^{n+1})$$

will be better than those given by other schemes. The reason is that, in the implicit scheme, the value of C_j^{n+1} depends not only on C_j^n (*i.e.* the previous value of the concentration) but it also depends on the values of C_{j-1}^{n+1} and C_{j+1}^{n+1}), which are the simultaneous values of the concentration, in the nearest points. However, in the implicit scheme, it is necessary to solve algebraic equations at each step of the calculation to obtain the numerical solution. This is more complex and causes longer CPU time than explicit schemes.

The jump points scheme possesses the advantages of both the explicit and the implicit schemes. Because it is an explicit scheme, the numerical calculations that it requires are simpler than those which must be made with the implicit scheme. Yet, because an implicit scheme has been introduced, the stability of the algorithm is better than that of an explicit scheme. The concrete form of the jump points scheme is as follows

$$\frac{1+Fa}{\tau}\left(C_j^{n+1} - C_j^n\right) + \frac{u}{2h}\left(C_{j+1}^n - C_{j-1}^n\right) - \frac{D}{h^2}(C_{j+1}^n - 2C_j^n + C_{j-1}^n) = 0$$

$$\text{if}\quad n+j = 2p+1$$

$$\frac{1+Fa}{\tau}\left(C_j^{n+1} - C_j^n\right) + \frac{u}{2h}\left(C_{j+1}^{n+1} - C_{j-1}^{n+1}\right) -$$
$$- \frac{D}{h^2}(C_{j+1}^{n+1} - 2C_j^{n+1} + C_{j-1}^{n+1}) = 0 \qquad \text{if}\quad n+j = 2p$$

where p is any integer.

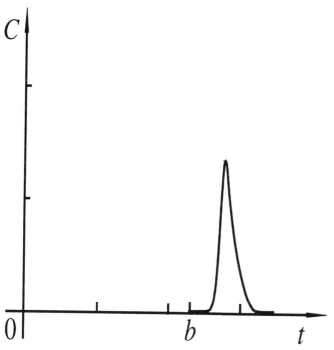

Figure IX-8.   Band Profile Obtained with the Jump Point Scheme. Point b indicates the exact position of the peak maximum derived from the ideal model.

In this scheme, the odd and even terms are interchanged. However, the whole scheme is still explicit, hence it is simpler to implement than an implicit scheme. Yet, because this scheme contains the implicit concept, the stability condition is also simple. The results of the jump points are better in linear chromatography. In nonlinear chromatography, the situation is far more complex and cannot be discussed in detail here. Like all the other schemes, the jump point scheme is applied to nonlinear cases, using the same approach as in the linear case. Figure IX-8 shows the results obtained with the Jump Point Scheme under nonideal, nonlinear conditions. These results are discussed in more detail in the last section of this chapter (X - Comparison of the Accuracy of the Different Numerical Schemes).

4. The Predictor–Corrector Scheme

The general equation of the nonlinear equilibrium–dispersive model is eqn. IX-45, written as follows

$$[1 + Ff'(C)]\frac{\partial C}{\partial t} + u\frac{\partial C_1}{\partial x} = D\frac{\partial^2 C}{\partial x^2}$$

Its predictor–corrector is composed of two equations. The predictor is

$$\frac{D}{h^2}\delta^2 C_j^{(n+1)/2} = [1 + Ff'(C_j^n)]\frac{2\left(C_j^{(n+1)/2} - C_j^n\right)}{\tau} + \frac{u}{2h}(C_{j+1}^n - C_{j-1}^n)$$

$$(\text{IX-49})$$

with $\delta^2 C_j^{(n+1)/2} = (C_{j+1}^{(n+1)/2} - 2C_j^{(n+1)/2} + C_{j-1}^{(n+1)/2})$ and $j = 1, 2, \cdots, n = 0, 1, 2, \cdots$. The corrector is given by

$$\frac{D}{2h^2}\delta^2(C_j^n + C_j^{n+1}) = \left[1 + Ff'\left(C_j^{(n+1)/2}\right)\right]\frac{C_j^{n+1} - C_j^n}{\tau}$$

$$+ \frac{u}{4h}(C_{j+1}^n - C_{j-1}^n + C_{j+1}^{n+1} - C_{j-1}^{n+1}) \quad (\text{IX-50})$$

Both the predictor and the corrector are linear equations. The predicted value, $C_j^{(n+1)/2}$, is first derived from eqn. IX-49 using the "pursuit" method. Then the corrected value, C_j^{n+1} is obtained using the same pursuit method, after introducing $C_j^{(n+1)/2}$ into eqn. IX-50.

The great advantage of the Predictor–Corrector Scheme is that both steps in this scheme *i.e.*, eqns. IX-49 and IX-50, are linear. The nonlinear terms are $f'(C_j^n)$ in eqn. IX- 49 and $f'(C_j^n)f'(C_j^{(n+1)/2})$ in eqn. IX-50. These nonlinear terms are known, so the nonlinear problem becomes easier to solve. Figure IX-9 shows the results obtained with the Predictor Corrector scheme under

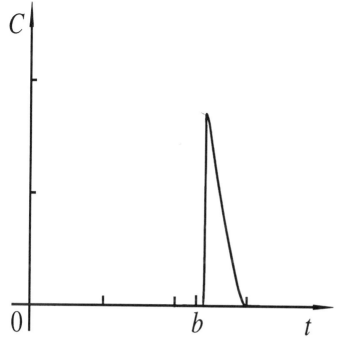

Figure IX-9.   **Band Profile Obtained with the Predictor Corrector Scheme.**
Point b **indicates the exact position of the peak maximum derived from the**
ideal model.

nonideal, nonlinear conditions. These results are discussed in more detail
in the last section of this chapter (X - Comparison of the Accuracy of the
Different Numerical Schemes).

5. The Difference Scheme of Convective Diffusion when Convection is Dominant

In chromatography, and especially in liquid chromatography, the axial
dispersion coefficient is very small, so the convection term is always the dominant one. When one of the general numerical analysis methods is used to
solve a parabolic problem in such a case, (*i.e.*, a problem in which diffusion is
significant but not dominant), artificial dispersion and pseudo-oscillations of
numerical origin are important and they are a serious nuisance. In chromatographic problems, convection is dominant and the basic hyperbolic character
of the mass-balance equation is kept. This means that we can still consider
concentration waves and use their properties. These waves propagate along
the characteristic lines of the equation of the corresponding ideal model, in
which diffusion is not involved. Accordingly, the finite difference scheme
used to calculate numerical solutions of nonideal chromatographic problems
should reflect the hyperbolic character mentioned above.

For the sake of convenience, the equation of the linear equilibrium–dispersive model of chromatography can be rewritten as

$$\frac{\partial C}{\partial t} + u_z \frac{\partial C}{\partial x} = \frac{D}{1 + Fa} \frac{\partial^2 C}{\partial x^2}$$

When $D = 0$, the characteristic velocity is $u_z = u/(1 + Fa)$ and the characteristic lines are straight lines from $(j - \beta, n)$ to $(j, n + 1)$. Since the concentration remains constant along a characteristic line, we have:

$$C_j^{n+1} = C_{j-\beta}^n$$

where $\beta = u_z \tau / h$. We extend this result to the case in which $D \neq 0$ but in which also convection remains the dominant contribution to the variations of the solute concentration with time and position in space and in which we can still consider that a concentration wave $C(x, t)$ propagates along the characteristic lines. Then, we construct a finite difference scheme keeping the same convection terms as for the ideal model and deriving the diffusion term from $D\tau/[(1 + Fa)h^2]$. This scheme is:

$$C_j^{n+1} = C_{j-\beta}^n + \frac{D\tau}{(1 + Fa)h^2} \left(C_{j+1}^n - 2C_j^n + C_{j-1}^n \right)$$

The term $C_{j-\beta}^n$ in this equation can be expressed as a function of C_j^n and C_{j-1}^n by interpolation. This method can also be extended to the calculation of solutions of nonlinear problems. In nonlinear chromatography, where diffusion is most usually very small, the results of this method are in excellent agreement with the experimental band profiles.

VIII. THE GALERKIN METHOD

Difference schemes are "point approximation" methods. Consequently, they provide the proper approach to calculate the positions and the amplitudes of discontinuities or shocks. Accordingly, the best results in the calculation of numerical solutions of the equations of ideal, nonlinear chromatography are obtained when using a finite difference scheme. When numerical solutions of the equilibrium–dispersive model are required, however, diffusion is involved and the mass balance equation becomes parabolic. Then, the finite difference schemes are generally no longer the best approximation methods available. In this case, finite element methods, which are segment approximations, are in general more appropriate to calculate the continuous and smooth solutions of the convection–diffusion equations.

The Galerkin method introduced in this section is one of the finite element methods, which are segment approximations based on the use of test functions [11-14]. The basic idea of the Galerkin method is similar to the "virtual displacement principle" in solid dynamics. In the Galerkin method, it is not necessary to transform the partial differential equation into a variation problem of the functional. Therefore, this method is simpler than most other finite element methods. The basic ideas of the Galerkin method can be expressed as follows.

Suppose a partial differential equation written as:

$$\mathcal{L}w = f \qquad\qquad (IX\text{-}51)$$

in which the operator $\mathcal{L} = \mathcal{L}(\partial/\partial x, \partial^2/\partial x^2, \partial/\partial t)$ is the differential operator, and f is an arbitrary function. Suppose also that $\{\phi_k(x)\}, k = 1, 2, \cdots$ is a complete set of test functions, $\phi_k(x)$, that is a set of orthogonal functions that can be used as a basis for the expansion of the function that is the solution of our problem. The Galerkin method gives an approximate solution which has the following form

$$w_h = \sum_{k=0}^{N} A_k(t)\phi_k(x) \qquad\qquad (IX\text{-}52)$$

where the $A_k(t)$ are restricted and determined by the Galerkin principle which demands that

$$\int_0^L (\mathcal{L}w_h - f)\phi_i(x)dx = 0 \qquad\qquad i = 0, 1, 2, \cdots, N \qquad (IX\text{- }53)$$

where L is the column length. Thus, we have a set of $N + 1$ equations where 0 and L are the coordinates of the boundary points. If we consider the functions ϕ_i as "virtual displacements", these equations (and Galerkin principle) correspond to the "virtual displacement principle". Combining these last two equations gives

$$\sum_{k=0}^{N} \int_0^L [\mathcal{L}A_k(t)\phi_k(x) - f]\phi_i(x)dx = 0 \qquad\qquad i = 0, 1, 2, \cdots, N$$

$$(IX\text{-}54)$$

with $i = 0, 1, 2, \cdots, N$.

An approximate solution for w is derived from the $\{A_k(t)\}$ and from eqn. IX-52. Introducing the boundary condition into this equation, allows the derivation from eqn. IX-54 of $N + 1$ ordinary differential equations involving the functions $A_k(t)$. Introducing then the initial conditions allows the determination of $\{A_k(t)\}$.

Now, we will derive numerical solutions of the chromatography equation with diffusion using the Galerkin method. This equation is written

$$\frac{\partial(C + Ff)}{\partial t} + u\frac{\partial C}{\partial x} = D\frac{\partial^2 C}{\partial x^2} \qquad \text{(IX-55)}$$

The initial and boundary condition are

$$\left\{ \begin{array}{c} C(x,0) = 0 \\ \left(uC - D\frac{\partial C}{\partial x}\right)_{x=0} = \Phi(t), \quad \left(\frac{\partial C}{\partial x}\right)_{x=L} = 0 \end{array} \right. \qquad \text{(IX-56)}$$

Expanding the function $C(x,t)$ on the set of functions $\{\phi_j(x)\}$, we have

$$C(x,t) = \sum_{i=0}^{N} A_j(t)\phi_j(x) \qquad \text{(IX-57)}$$

From the Galerkin principle,

$$\int_0^L \left[(C + Ff)_t + uC_x - DC_{xx} \right] \phi_i(x)dx = 0$$

where

$$C_t \equiv \frac{\partial C}{\partial t}, \qquad C_x \equiv \frac{\partial C}{\partial x}, \qquad C_{xx} \equiv \frac{\partial^2 C}{\partial x^2}$$

The partial integration of the Galerkin equation gives

$$\int_0^L \left[\frac{\partial}{\partial t(C + Ff)}\phi_i - \left(uC\frac{\partial \phi_i}{\partial x} - D\frac{\partial C}{\partial x}\frac{\partial \phi_i}{\partial x}\right) \right] dx =$$
$$- \phi_i(uC - D\frac{\partial C}{\partial x})|_0^L \qquad \text{(IX-58a)}$$

$$\int_0^L \left[\frac{\partial}{\partial t(C + Ff)}\phi_i - uC\frac{\partial \phi_i}{\partial x} + D\frac{\partial C}{\partial x}\frac{\partial \phi_i}{\partial x} \right] dx =$$
$$- \Phi(t)\phi_i(0) - uC(L,t)\phi_i(L) \qquad \text{(IX-58b)}$$

with $k = 0, 1, 2, \cdots, N$. We introduce now a pseudo-boundary, $L_\infty > L$, such that near this point the concentration wave has not yet arrived and, thus, $C = 0$. Equation IX-58b can be rewritten

$$\int_0^{L_\infty} \left[\frac{\partial}{\partial t(C + Ff)}\phi_i - \left(uC\frac{\partial \phi_i}{\partial x} + D\frac{\partial C}{\partial x}\frac{\partial \phi_i}{\partial x}\right) \right] dx = -\phi_i(0)\Phi(t) \qquad \text{(IX-58c)}$$

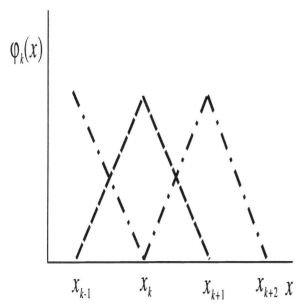

Figure IX-10   Calculation of Numerical Solutions of the Equilibrium–dispersive Model. Plot of the Galerkin Functions, $\phi_i(x)$, versus x.

If the set of functions $\phi_i(x)$ is the following set of linear polynomials (other sets of functions could do, but polynomials make the simplest set, the derivation is easiest to understand, and the results are quite satisfactory).

$$\phi_k = \begin{cases} \frac{x - x_{k-1}}{x_k - x_{k-1}}, & x_{k-1} \leq x \leq x_k \\ \frac{x_{k+1} - x}{x_{k+1} - x_k}, & x_k \leq x < x_{k+1} \\ 0, & x \leq x_{k-1} \quad \text{or} \quad x_{k+1} < x \end{cases} \qquad (\text{IX-59})$$

where $k = 0, 1, 2, \cdots, N$, $x_0 = 0$, and $x_N = L$. A graph of the variation of ϕ_i as a function of x is shown in Figure IX-10.

From eqn. IX-59 and Figure IX-10, we see that

$$\begin{cases} \phi_k(0) = \delta_{k0} \\ \phi_k(L) = \delta_{kN} \end{cases} \qquad (\text{IX-60})$$

where $\delta_{ki} = 0$ if $i \neq k$ and $\delta_{ki} = 1$ if $i = k$. It is clear from eqn. IX-59 that the functions in the set $\{\phi_k(x)\}$ are nearly orthogonal, $i.e.$ that

$$\int_0^L \phi_k(x)\phi_l(x) \neq 0 \quad \text{if and only if} \quad l = k - 1, \ k, \ \text{or} \ k + 1 \qquad (\text{IX-61})$$

So, the integration of eqn. IX-58b can be simplified sufficiently to allow a relatively easy numerical calculation. Combining eqns. IX-57, IX-58b, IX-59 and IX-60 and considering that $C(L,t) = A_N(t)\phi_N(L)$ and $\phi_N(L) = 1$, we have

$$\frac{d}{dt}\int_0^L \left(\sum_{i=0}^N A_i(t)\phi_i(x) + F\frac{a\sum_{i=0}^N A_i(t)\phi_i(x)}{1 + b\sum_{i=0}^N A_i(t)\phi_i(x)}\right)\phi_k(x)dx$$

$$-\int_0^L \left(u\sum_{i=0}^N A_i(t)\phi_i(x)\right)\frac{d\phi_i(x)}{dx}dx + \int_0^L D\left(\sum_{i=0}^N A_i(t)\frac{d\phi_i(x)}{dx}\right)dx$$

$$= \phi(t)\delta_{k0} - uA_N(t)\delta_{kN} \qquad \text{with} \quad k = 0,1,2,\cdots,N \quad \text{(IX-62)}$$

Equation IX-62 is an ordinary differential equation of the type

$$\frac{dP_k}{dt} + Q_k = \Phi(t)\delta_{k0} - uA_N(t)\delta_{kN} \qquad\qquad \text{(IX-63a)}$$

where

$$P_k = \int_0^L \left(\sum_{i=0}^N A_i(t)\phi_i(x) + F\frac{a\sum_{i=0}^N A_i(t)\phi_i(x)}{1 + b\sum_{i=0}^N A_i(t)\phi_i(x)}\right)\phi_i(x)dx \qquad \text{(IX-64a)}$$

$$Q_k = -\int_0^L \left(u\sum_{i=0}^N A_i(t)\phi_i(x)\right)\frac{d\phi_k(x)}{dx}dx + \int_0^L D\left(\sum_{i=0}^N A_i(t)\frac{d\phi_k(x)}{dx}\right)dx \quad \text{(IX-65a)}$$

Considering the regions of space within which each of the different functions $\phi_i(x)$ is defined and the near-orthogonality of the functions in the set $\{\phi_i(x)\}$, we obtain the following results

a - When $k = 0$:

$$\frac{dP_0}{dt} + Q_0 = \Phi(t) \qquad\qquad \text{(IX-63b)}$$

$$P_0 = \int_{x_0}^{x_1} \left(A_0\phi_0 + A_1\phi_1 + Fa\frac{A_0\phi_0 + A_1\phi_1}{1 + b(A_0\phi_0 + A_1\phi_1)}\right)\phi_0 dx \qquad \text{(IX-64b)}$$

and

$$Q_0 = -u\int_{x_0}^{x_1}(A_0\phi_0 + A_1\phi_1)\frac{d\phi_0(x)}{dx}dx +$$

$$+ D\int_{x_0}^{x_1}\left(A_0\frac{d\phi_0}{dx} + A_1\frac{d\phi_1}{dx}\right)\frac{d\phi_0(x)}{dx}dx \quad \text{(IX-65b)}$$

b - When $0 < k < N$, we have

$$\frac{dP_k}{dt} + Q_k = 0 \qquad \text{(IX-63c)}$$

with

$$P_k = \int_{x_{k-1}}^{x_k} \left(A_{k-1}\phi_{k-1} + A_k\phi_k + Fa\frac{A_{k-1}\phi_{k-1} + A_k\phi_k}{1 + b(A_{k-1}\phi_{k-1} + A_k\phi_k)} \right) \phi_k dx +$$
$$\int_{x_k}^{x_{k+1}} \left(A_k\phi_k + A_{k+1}\phi_{k+1} + Fa\frac{A_k\phi_k + A_{k+1}\phi_{k+1}}{1 + b(A_k\phi_k + A_{k+1}\phi_{k+1})} \right) \phi_k dx \qquad \text{(IX-64c)}$$

$$Q_k = -u \int_{x_{k-1}}^{x_k} (A_{k-1}\phi_{k-1} + A_k\phi_k) \frac{d\phi_k(x)}{dx} dx$$
$$+ D \int_{x_{k-1}}^{x_k} \left(A_{k-1}\frac{d\phi_{k-1}}{dx} + A_k\frac{d\phi_k}{dx} \right) \frac{d\phi_k(x)}{dx} dx$$
$$- u \int_{x_k}^{x_{k+1}} \left(A_k\frac{d\phi_k}{dx} + A_{k+1}\frac{d\phi_{k+1}}{dx} \right) \frac{d\phi_k(x)}{dx} dx$$
$$+ D \int_{x_k}^{x_{k+1}} \left(A_k\frac{d\phi_k}{dx} + A_{k+1}\frac{d\phi_{k+1}}{dx} \right) \frac{d\phi_k(x)}{dx} dx \qquad \text{(IX-65c)}$$

c - Finally, if $k = N$, we have

$$\frac{dP_N}{dt} + Q_N = -uA_N \qquad \text{(IX-63d)}$$

with

$$P_N = \int_{x_{N-1}}^{x_N} \left(A_{N-1}\phi_{N-1} + A_N\phi_N + Fa\frac{A_{N-1}\phi_{N-1} + A_N\phi_N}{1 + b(A_{N-1}\phi_{N-1} + A_N\phi_N)} \right) \phi_N dx$$
$$\text{(IX-64d)}$$

$$Q_N = -u \int_{x_{N-1}}^{x_N} (A_{N-1}\phi_{N-1} + A_N\phi_N) \frac{d\phi_N(x)}{dx} dx$$
$$+ D \int_{x_{N-1}}^{x_N} \left(A_{N-1}\frac{d\phi_{N-1}}{dx} + A_N\frac{d\phi_N}{dx} \right) \frac{d\phi_N(x)}{dx} dx \qquad \text{(IX-65d)}$$

In all cases, the functions P_k and Q_k are obtained by introducing the expressions of the functions $\phi_k(x)$ in eqn. IX-59 into eqns. IX-63a to IX-63d. The equations obtained are nonlinear, ordinary differential equations.

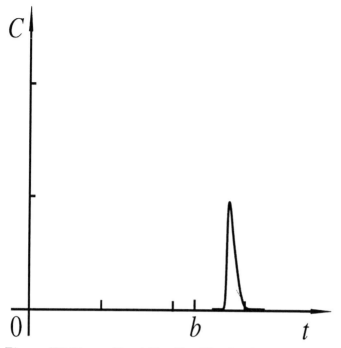

Figure IX-11. Band Profile Obtained with the Galerkin Method. Point b indicates the exact position of the peak maximum derived from the ideal model.

They can be solved using a difference scheme. In such a scheme, the eqns. IX-63a to IX-63d can be expressed as the following nonlinear equations:

$$H(A_{k-1}^n, A_{k-1}^{n-1}, A_k^{n-1}, A_k^n, A_{k+1}^n, A_{k+1}^{n+1}, \tau, h) = 0 \qquad \text{(IX-66)}$$

In this equation, $n\tau$ corresponds to the entire time during which the chromatogram is considered, τ and h are the time and space increments in the digitized (x, t)-space. The initial condition in eqn. IX-56 is $C(x, 0) = 0$. For any arbitrary value of k, we have

$$A_k(0) = 0$$

In the difference scheme, this corresponds to

$$A_k^0 = 0$$

The problem of deriving numerical solutions of the equation of chromatography has now been reduced to the calculation of the numerical solutions of a large set of nonlinear algebraic equations. In practice, the Newton iteration method is used to solve these equations. The solution of the corresponding linear equation is taken as the "initial testing value" in the iteration calculation of the solution of the nonlinear equation.

The Galerkin and other finite element methods give better results than the methods based on the use of finite difference schemes in accounting for the influence of axial dispersion on the band profiles because they are segment approximations while the finite difference methods are point approximations [12]. Figure IX-11 shows the results obtained with this scheme under non-ideal, nonlinear conditions. These results are discussed in more detail in the last section of this chapter (X - Comparison of the Accuracy of the Different Numerical Schemes).

IX. ORTHOGONAL COLLOCATION METHOD

Orthogonal collocation belongs to the group of methods called the Methods of Weighted Residual (MWR) [15]. It is similar to the Galerkin method because it actually determines a polynomial approximation of the solution, *i.e.*, it searches for an approximation of the solution expressed by

$$C(x,t) = \sum_{i=1}^{N} A_i \phi_i \tag{IX-67}$$

where the i segments of the solution are each obtained as the product of a function ϕ_i and a coefficient A_i and both sets A_i and ϕ_i are determined separately. The main difference with the Galerkin method is in the nature of the constraint condition applied to the residuals, $R_N(\mathbf{A}, x)$, when determining the coefficients A_i. In the collocation method, this constraint condition is

$$R_N(\mathbf{A}, x) = 0 \tag{IX-68a}$$

where $R_N = C(x,t) - \sum_{i=1}^{N} A_i \phi_i$. In the Galerkin method the corresponding condition is

$$\int R_N(\mathbf{A}, x) \phi_k dx = 0 \tag{IX-68b}$$

As to the difference between the orthogonal collocation method and the general collocation method, it is only in the choice of the collocation points. In the orthogonal collocation method, the collocation points are chosen at the roots (zero points) of the orthogonal polynomials. In the orthogonal collocation method, the interpolation polynomials may be orthogonal or may belong to another type of polynomials (for example, the Lagrange interpolation polynomials, which are not orthogonal, are used sometimes).

1. The orthogonal collocation method [15]

Let $x_1, x_2, \cdots, x_{n+1}$ be the roots of an $n+1$ degree Legendre polynomial, on the interval $[0, 1]$. We denote as $y(x_i)$ the value of the unknown function $y(x)$ at the position x_i. The points $x_i, i = 1, 2, \cdots, n+1$ are taken as the node points and the Lagrange interpolation polynomials are used to construct the function $y(x)$. Thus, we have

$$y(x) = L_n(x) = \sum_{i=1}^{n+1} l_i(x) y(x_i) \qquad \text{(IX-69)}$$

where the interpolation basis function is

$$l_i(x) = \frac{\Pi(x)}{(x - x_i)\Pi'(x_i)} \qquad \text{(IX-70)}$$

with $\Pi(x) = (x - x_1)(x - x_2) \cdots (x - x_{n+1})$. Obviously, $l_i(x_j) = \delta_{ij}$. So, eqn. IX-69 is an identity at the node points.

When the unknown function, $y(x)$, is approximated with an interpolation polynomial, the derivative of $y(x)$ can be written as follows

$$\frac{d}{dx} y(x) = \sum_{i=1}^{n+1} \frac{dl_i(x)}{dx} y(x_i) \qquad \text{(IX-71)}$$

From eqn. IX-71, it is easy to see that when $x = x_i$,

$$\left[\frac{dl_i(x)}{dx} \right]_{x=x_i} = \frac{1}{2} \frac{\Pi^{(2)}(x_i)}{\Pi^{(1)}(x_i)} \qquad \text{(IX-72a)}$$

$$\left[\frac{d^2 l_i(x)}{dx^2} \right]_{x=x_i} = \frac{1}{3} \frac{\Pi^{(3)}(x_i)}{\Pi^{(1)}(x_i)} \qquad \text{(IX-72b)}$$

and when $x_j \neq x_i$,

$$\left[\frac{dl_i(x)}{dx} \right]_{x_j} = \frac{1}{x_j - x_i} \frac{\Pi^{(1)}(x_i)}{\Pi^{(1)}(x_i)} \qquad \text{(IX-72c)}$$

$$\left[\frac{d^2 l_i(x)}{dx^2} \right]_{x_j} = 2 l_i^{(1)}(x_j) \left[l_i^{(1)}(x_i) - \frac{1}{x_j - x_i} \right] \qquad \text{(IX-72d)}$$

where

$$l_i^{(k)}(x) = \frac{d^k l_i(x)}{dx^k} \qquad\qquad \Pi^{(k)}(x) = \frac{d^k \Pi(x)}{dx^k}$$

Combining eqns. IX-72a to IX-72d, we have

$$\left[\frac{dy(x)}{dx}\right]_{x_j} = \sum \left[\frac{dl_i(x)}{dx}\right]_{x_j} y(x_i) = \sum l_{ji}^{(1)} y_i \qquad \text{(IX-73a)}$$

or $y_j^{(1)} = \sum_{i=1}^{n+1} l_{ji}^{(1)} y_i$, and

$$\left[\frac{d^2y(x)}{dx^2}\right]_{x_j} = \sum \left[\frac{d^2l_i(x)}{dx^2}\right]_{x_j} y(x_i) = \sum l_{ji}^{(2)} y_i \qquad \text{(IX-73b)}$$

or $y_j^{(2)} = \sum_{i=1}^{n+1} l_{ji}^{(2)} y_i$. In matrix form, we have

$$\begin{aligned}\vec{y}^{\,(1)} &= [A]\vec{y} \\ \vec{y}^{\,(2)} &= [B]\vec{y}\end{aligned} \qquad \text{(IX-73c)}$$

where $[A] = [j_{ji}^{(1)}]$ and $[B] = [l_{ji}^{(2)}]$ or

$$\vec{y} = (y(x_1), y(x_2), \cdots, y(x_{n+1}))^T$$

$$\vec{y}^{\,(1)} = \left(\left(\frac{dy}{dx}\right)_{x_1}, \left(\frac{dy}{dx}\right)_{x_2}, \cdots, \left(\frac{dy}{dx}\right)_{x_{n+1}}\right)^T$$

$$\vec{y}^{\,(2)} = \left(\left(\frac{d^2y}{dx^2}\right)_{x_1}, \left(\frac{d^2y}{dx^2}\right)_{x_2}, \cdots, \left(\frac{d^2y}{dx^2}\right)_{x_{n+1}}\right)^T$$

So, in the orthogonal collocation method, the derivative term of the ordinary differential equation can be expressed as an algebraic form, $[A]\vec{y}$ or $[B]\vec{y}$. Thus, the ordinary differential equation can be replaced by a suitable set of algebraic equations.

2. Application of the collocation method in chromatography

We consider now the equations of the transport–dispersive model of chromatography [16], which are written as

$$\frac{\partial C}{\partial t} + F\frac{\partial q}{\partial t} + u\frac{\partial C}{\partial x} = D\frac{\partial^2 C}{\partial x^2} \qquad \text{(IX-74a)}$$

$$\frac{\partial q}{\partial t} = -k[q - f(C)] \qquad \text{(IX-74b)}$$

In order to calculate numerical solutions of this system, *i.e.*, to calculate band profiles, a splitting method is used. In the arbitrary interval $[n\Delta t - (n+1)\Delta t]$, the numerical calculation of the solution is split into two successive steps with

$$\frac{\partial C}{\partial t} = -u\frac{\partial C}{\partial x} + D\frac{\partial^2 C}{\partial x^2} \qquad n\Delta t \le t < (n+\frac{1}{2})\Delta t \qquad \text{(IX-75a)}$$

during the first half-step and

$$\frac{\partial C}{\partial t} = -F\frac{\partial q}{\partial t} \qquad\qquad (n+\frac{1}{2})\Delta t \le t < (n+1)\Delta t \qquad\text{(IX-75b)}$$

during the second half-step. Actually, the splitting method is a fictitious process which is used only for the purpose of computation. It means that each time step is split into two half parts. During the first half part ($n\Delta t - (n+1/2)\Delta t$), the time differential $\frac{\partial C_i}{\partial t}$ is calculated as the sum of the contributions of the convection and the diffusion processes. During the second half part ($(n+1/2)\Delta t - (n+1)\Delta t$), it is calculated from the mass-transfer kinetics. The details of this method can be found elsewhere [3].

Combining eqns. IX-74b and IX-75 gives

$$\frac{\partial q}{\partial t} = -k[q - f(C)] \qquad\qquad\qquad\qquad\text{(IX-76a)}$$

$$\frac{\partial C}{\partial t} = -F\frac{\partial q}{\partial t} \qquad (n+1/2)\Delta t \le t < (n+1)\Delta t$$
$$\text{(IX-76b)}$$

This method transform the initial partial differential equation, eqn. IX-74a into two equations. The first one (eqn. IX-75a) is a second order linear partial differential equation while the second one (eqn. IX-76a) is a nonlinear ordinary differential equation.

In order to carry out the orthogonal collocation, the domain $[0, L]$ is divided into a number $N_h = L/H$ of finite elements, $i = 1, 2, \cdots, L/H$ (L, column length, H, length of the finite element). On each element, the nodes are collocated, the roots of the N_pth degree orthogonal polynomial (displaced-Legendre polynomials) being taken as the node points. A Lagrange polynomial is used to construct the interpolated function. At the jth node, on the jth element, the derivatives are transformed into

$$\frac{\partial C}{\partial x}\Big|_{C=C_{j,i}} = \frac{1}{\Delta x}\sum_{k=1}^{N_p+2} A_{j,k}C_{k,j} = \frac{1}{\Delta x}[A][C] \qquad\text{(IX-77a)}$$

$$\frac{\partial^2 C}{\partial x^2}\Big|_{C=C_{j,i}} = \frac{1}{(\Delta x)^2}\sum_{k=1}^{N_p+2} B_{j,k}C_{k,j} = \frac{1}{(\Delta x)^2}[B][C] \qquad\text{(IX-77b)}$$

So, eqn. IX-75a is transformed into

$$\frac{\partial C}{\partial t}\Big|_{C=C_{j,i}} = \frac{D}{\Delta x^2}\sum_{k=1}^{N_p+2} B_{j,k}C_{j,k} - \frac{u}{\Delta x}\sum_{k=1}^{N_p+1} A_{j,k}C_{j,k} \qquad\text{(IX-78)}$$

where $C_{j,i}$ is the concentration at the jth collocation point in the ith element, and $A_{i,j} = l_i^{(x_j)}$ and $B_{j,k}$ are the elements of the matrices $[A]$ and $[B]$ which are the values of the roots of the orthogonal polynomials. At the junctions of two successive finite elements, the continuity condition of $C(x)$ and of its derivative are respectively:

$$\left[C_{N_p+2}\right]_j = [C_1]_{j+1} \tag{IX-79a}$$

$$\left[\sum_{k=1}^{N_p+2,k} A_{N_p+2}C_k\right]_j = \left[\sum_{k=1}^{N_p+2} A_{1,k}C_k\right]_{j+1} \qquad \text{for } j = 1, 2. \cdots, N_p + 2 \tag{IX-79b}$$

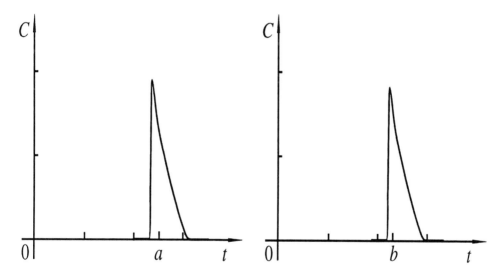

Figure IX-12.   **Band Profiles Calculated with the Orthogonal Collocation Method.   Left, Ideal, Nonlinear Case.   Right, Nonideal, Nonlinear Case. Points a or b indicate the exact position of the peak maximum derived from the ideal model.**

The position of the nodes in the calculation of a band profile by orthogonal collocation on finite elements can be determined based on the use of the displaced-Legendre polynomial and of a Lagrange interpolation. Further details on the determination of the matrices $[A]$ and $[B]$, on the relationships between the precision required for the numerical solution that will be obtained and the calculation conditions (simple or complex calculations, short or long calculation time), and on the choice of the number of node points (*e.g.*, two-node points, three-node points, etc.) can be found in treatises of numerical analysis [14,15,24-26]. The fourth order Runge–Kutta scheme is used for the calculation of solutions of the first order ordinary differential equations IX-78 and IX-79 and the Gill scheme is used for the solution of eqn. IX-75b. Figures IX-12a and IX-12b show the results obtained with this calculation method under ideal and nonideal, nonlinear conditions, respectively. These results are discussed in more detail in the next section (X - Comparison of the Accuracy of the Different Numerical Schemes).

X. COMPARISON OF THE ACCURACY OF THE DIFFERENT NUMERICAL SCHEMES

The earlier discussions of the different calculation schemes presented in this chapter were illustrated with the numerical solution obtained with each of them in the specific case of the elution band of a single compound following the Langmuir isotherm. Now, we compare these different results. We compare them also to the theoretical results that are available *i.e.*, to the exact solution of the ideal model (see Chapter V). In this case, we can assess precisely the errors made with the calculation scheme. Accordingly, we can derive a figure of merit for a calculation scheme from the difference between the exact solution of the model and its numerical solution with that scheme. Unfortunately, in the case of the nonideal model, there is no similar possibility of assessing the accuracy of the different calculation schemes.

TABLE IX-2

SCHEME ACCURACY IN IDEAL NONLINEAR CHROMATOGRAPHY[†]

Scheme	bC_0	L_f (%)	t_R^t (min)	t_R^c (min)	Δt_R	A^t	A^c	ΔA[‡]
Characteristic	0.02	0.20	11.41	11.66	0.25	0.020	0.020	0.000
Characteristic	0.05	0.50	10.92	11.12	0.20	0.050	0.057	0.007
Characteristic	0.69	6.8	7.72	6.94	0.78	0.050	0.074	0.024
L–W Conserv.	0.02	0.20	11.41	11.66	0.25	0.020	0.020	0.000
L–W Conserv.	0.05	0.50	10.92	11.12	0.20	0.050	0.057	0.007
L–W Conserv.	0.69	6.8	7.72	6.69	1.03	0.050	0.084	0.034

[†] In these calculations, the isotherm is $q = \frac{29.2C}{1+bC}$, with $b = 1$ in rows 1, 2, 4 and 5, and $b = 13.77$ in rows 3 and 6; the phase ratio is $F = 0.275$; the injection parameters are $t_p = 60$ sec and $C_0 = 0.05452$ mg/ml; the column length is $L = 100$ mm and the mobile phase velocity, $u = 80$ mm/min.

[‡] Peak areas in [min mg ml^{-1}]. The exponents t and c mean the true value and the calculated value, respectively.

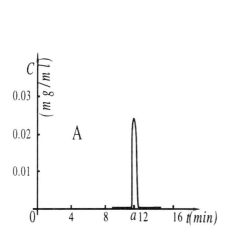

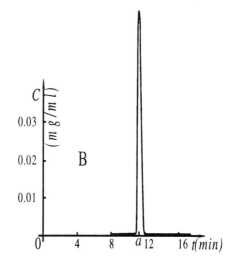

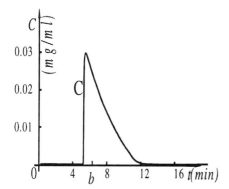

Figure IX-13   Band Profiles Calculated with the Characteristic Scheme.
Top Left, $bC_0 = 0.02$. Top Right, $bC_0 = 0.05$. Bottom Left, $bC_0 = 0.69$. See
Table IX-2. Points a and b show the exact location of the peak maximum.

First, we discuss the ideal case. Calculations were made using four different algorithms and determining eight band profiles corresponding to three sets of experimental conditions. The main features of the calculation results are reported in Tables IX-2 and IX-3. The two algorithms used in this case are the characteristic scheme (Figure IX-4a) and the Lax–Wendroff conservation scheme (Figure IX-5). The results obtained with these algorithms are compared in Figures IX-13 (characteristic scheme) and IX-14 (Lax–Wendroff conservation scheme). Since the experimental conditions correspond to the ideal model ($D = 0$), the two algorithms should give the exact result in the linear case, except for the consequences of the truncation error (see Figures IX-13a, IX-13b, IX-14a, and IX-14b). Under quasi-linear conditions ($L_f =$

TABLE IX-3

SCHEME ACCURACY IN IDEAL
NONLINEAR CHROMATOGRAPHY[†]

Scheme	bC_0	L_f (%)	t_R^t	t_R^c	Δt_R	A^t	A^c	ΔA
TVD	0.69	1.7	10.897	10.965	0.068	0.015	0.015	0.000
Orthogonal Col.	0.69	1.7	10.897	10.95	0.053	0.015	0.015	0.000

[†] In these calculations, the isotherm is $q = \frac{29.2C}{1+13.77C}$; the phase ratio is $F = 0.275$; the injection parameters are $t_p = 18$ sec and $C_0 = 0.050$ mg/ml; the column length is $L = 150$ mm and the mobile phase velocity, $u = 100$ mm/min. The units are the same as in Table IX-2.

0.2 %), the agreement is excellent for the peak area and the error is only 2 % for the retention time of the peak maximum. Under slightly overloaded conditions ($L_f = 0.5$ %), the error on the retention remains the same while a slight error on the peak area appears.

As explained in earlier discussions, however, numerical errors occur under nonlinear conditions because the higher order terms arising in the expansion of the mass balance equation are neglected in the numerical schemes of integration. While these high-order terms are zero under linear conditions, they are no longer negligible under nonlinear conditions. They increase with an increasing degree of nonlinear behavior of the isotherm, hence they increase with increasing sample size. The calculations under nonlinear conditions were performed with the characteristic scheme (Figures IX-4a), the Lax–Wendroff conservation scheme (Figure IX-5), the TVD scheme (Figure IX-6), and the orthogonal collocation method (Figure IX-12a). The calculated band profiles are shown in Figures IX-13 (characteristic scheme), IX-14 (Lax–Wendroff conservation scheme), IX-15a (TVD scheme), and IX-15b (orthogonal collocation scheme). In Tables IX-2 and IX-3, the results of the calculations (retention time of the shock and band area) are compared with the theoretical values of the retention time (t_R^t).

For an overloaded profile with $L_f = 6.9\%$, the errors made with the first two calculation schemes become important. The errors made on the retention time are 10 and 13 % with the characteristic and Lax–Wendroff schemes, respectively, while the errors made on the area are of the order of a third, which is barely acceptable (Table IX-2). As expected the TVD scheme and the collocation method are markedly more accurate since the relative error made on the retention time calculated with these two schemes is of the order of 5% for an overloaded band ($L_f = 1.7\%$). The peak areas calculated with these two schemes are very close to the true value, the error

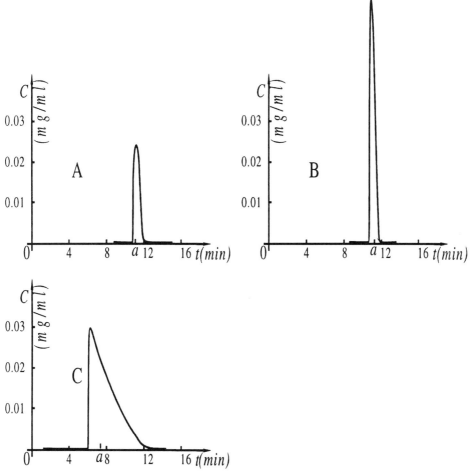

Figure IX-14   Band Profiles Calculated with the Lax–Wendroff Scheme. Top Left, $bC_0 = 0.02$. Top Right, $bC_0 = 0.05$. Bottom Left, $bC_0 = 0.69$. See Table IX-2. Point a shows the exact location of the peak maximum.

made being negligible in practice. Note that, as we show later, the exact calculation algorithm and the computer used may have a significant bearing on the numerical accuracy of the results. At any rate, a simple check on the accuracy of the calculation consists in determining by a simple numerical integration the area of the profiles calculated. They should be equal to the area injected within better than 1% or so, otherwise the results are not acceptable for accurate modeling of band profiles.

Finally, the same approach was used to compare the results of band profiles calculations performed with six different schemes that were all applied to the same nonideal, nonlinear problem. These schemes are the characteristic scheme (Figure IX-4b), the left-bias explicit scheme (Figure IX-7), the jump

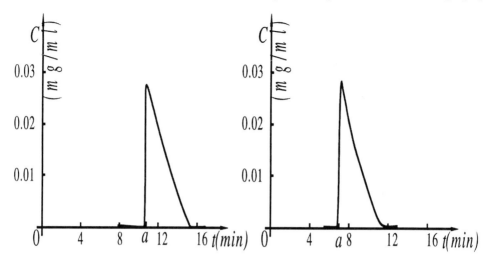

Figure IX-15.   Band Profiles Calculated with the TVD Scheme (left) and with Orthogonal Collocation (right). See Table IX-2. Point a shows the exact location of the peak maximum.

TABLE IX-4

SCHEME ACCURACY IN NONIDEAL NONLINEAR CHROMATOGRAPHY[†]

Scheme	L_f (%)	t_R^t	t_R^c	Δt_R	A^t	A^c	ΔA
Orthogonal Col	2.57	6.960	6.990	0.030	0.0150	0.0150	0.0000
Characteristic	2.57	6.960	6.970	0.010	0.0150	0.0160	0.0010
Predictor Corr.	2.57	6.960	7.130	0.170	0.0150	0.0134	0.0016
Left Bias Expl.	2.57	6.960	7.299	0.339	0.0150	0.0160	0.0010
Jump Point	2.57	6.960	7.730	0.770	0.0150	0.0150	0.0000
Galerkin	2.57	6.960	7.160	0.200	0.0150	0.0147	0.0003

[†] In these calculations, the isotherm is $q = \frac{29.2C}{1+13.77C}$; the phase ratio is $F = 0.275$; the injection parameters are $t_p = 18$ sec and $C_0 = 0.050$ mg/ml; the axial dispersion coefficient is $D = 0.0001$ cm^2 sec^{-1}; the column length is $L = 100$ mm and the mobile phase velocity, $u = 100$ mm/min. The units are the same as in Table IX-2.

The exponents t and c mean the true value and the calculated value, respectively.

point scheme (Figure IX-8), the predictor corrector scheme (Figure IX-9), the Galerkin method (Figure IX-11), and the orthogonal collocation method (Figure IX-12b). The profiles obtained with these six different schemes are compared in Figures IX-16, IX-17 and IX-18. Note that in nonideal chromatography, the retention time of the shock layer center is equal to that of the shock in the ideal model. This observation provides the only available test of the absolute accuracy of the calculation schemes. For this reason, the time of elution of the shock in the ideal model is shown as the point a in all these figures. The error made in estimating the correct retention time of the band from the retention time of its shock layer should be of the same order as the thickness of the shock layer. The error made when comparing the peak height and the value of the maximum concentration of the peak given by the ideal model is markedly larger. Table IX-4 compares the values of the retention times of the peak maximum and of the maximum concentration of the different bands calculated using the schemes listed above.

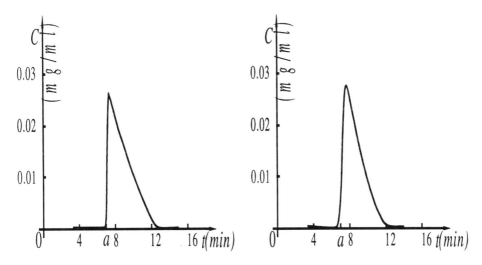

Figure IX-16.   **Band Profiles Calculated with the Characteristic Scheme (left) and with the Left Bias Explicit Scheme (right). See Table IX-2. Point a shows the exact location of the peak maximum.**

The data in Table IX-4 show that, rather surprisingly, the characteristic scheme gives the best estimate of the retention time, with an error of approximately 0.1%. The orthogonal collocation method is a close second, with a 0.4% error. The other schemes give larger errors, between 2.5 and 10%. Regarding mass conservation, i.e., the loss of peak area during the calculations, the orthogonal collocation scheme is, as expected, the best calculation method. The jump point scheme also gives excellent mass conservation but the large error made on the retention time suggests that orthogonal collocation should be preferred. It is remarkable, however, how the characteristic scheme gives excellent values of the retention times, values that are surprisingly more accurate than those given by all the other schemes. They are even three times more accurate than orthogonal collocation. This probably explains why this scheme has often been the one preferred by chromatographers [27].

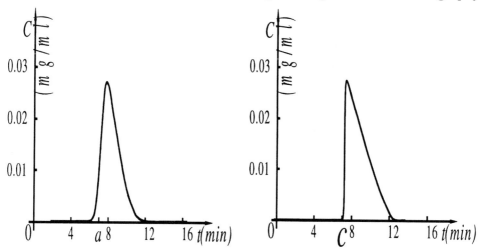

Figure IX-17   Band Profiles Calculated with the Jump Point Scheme (left) and with the Predictor Corrector Scheme (right). See Table IX-2. Point a shows the exact location of the peak maximum.

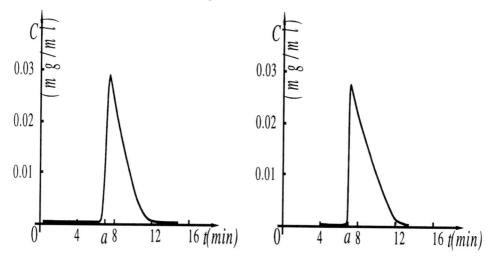

Figure IX-18   Band Profiles Calculated with the Galerkin Method (left) and with the Orthogonal Collocation Method (right). See Table IX-2. Point a shows the exact location of the peak maximum.

The comparison of the six profiles shown in Figures IX-16 to IX-18 shows also some surprising differences between the profiles obtained for this moderately overloaded band. For example, the profile of the peak calculated with the jump point scheme (Figure IX-17a) is far more dispersed than all the others. It seems almost Gaussian. The profiles of the peaks calculated with the left bias explicit scheme (Figure IX-16b) and the Galerkin method are also more dispersed than the four other ones, although to a lesser degree. These three profiles are also those that give the largest error on the retention time of the peak. The four other profiles are very similar. This observation confirms the selection of the algorithm used by most separation scientists for the calculation of band profiles in liquid chromatography, the characteristic scheme when a high accuracy is not required, orthogonal collocation on finite elements when it is. The former method, however, is much faster than the second and not much less accurate, which justifies its frequent use.

TABLE IX-5

ACCURACY OF AREA RECOVERY IN
NONIDEAL NONLINEAR CHROMATOGRAPHY

Area injected	Area Calculated with the ED Model[a]	Area Calculated with the ED Model[b]	Area Calculated with the POR Model
2.4000	2.3999	2.3999	2.3995
4.0000	3.9999	3.9997	3.9996
8.0000	7.9999	7.9999	7.9989
26.4000	26.3999	26.3984	26.3925
55.2000	55.2000	55.2013	55.1818
88.0000	87.9999	87.9961	88.9887
158.4000	158.400	158.3915	158.3880
368.0000	367.9999	368.0041	368.0016
1104.0000	1103.9999	1104.0169	1103.9851

[a] Characteristic Forward Backward Scheme [1].

[b] Orthogonal Collocation on Finite Elements.

We present in Table IX-5 another series of results in an attempt to assess the accuracy of the peak area recovery, *i.e.*, a comparison between the area of the rectangular band injected into the column, and the area of the corresponding profile, calculated with two different models, the equilibrium dispersive model (ED) and the POR model [28]. The latter is a simple version of the general rate model and uses two mass balance equations, one for the fluid percolating through the column bed, the other for the stagnant mobile phase. The calculations made in this last case use the method of orthogonal collocation on finite elements. For the implementation of the equilibrium dispersive model, we used two different programs, a conventional characteristic scheme, often used in the literature [1], and the orthogonal collocation method. The relative error or relative area lost during the numerical calculations is of the order of 1×10^{-4} or lower, which is more than satisfactory when band profiles are calculated for the purpose of modeling separations. The numerical results presented in Table IX-5 were obtained during the modeling of the band profiles of 3-phenyl 1-propanol, 4-tert-butylphenol, butylbenzene, and butyl benzoate on a classical packed column and on a monolithic column, using methanol/water RP-HPLC conditions [29]. The agreement between calculated and experimental profiles was excellent in all cases, although the isotherms measured for the four compounds were quite different. The differences between the calculated and experimental profiles are essentially explained by the difficulties that are still encountered in properly modeling the mass transfer kinetics when its rate is not very fast.

TABLE IX-6

ACCURACY OF AREA CALCULATIONS
IN NONIDEAL, NONLINEAR CHROMATOGRAPHY

Amount injected[1]	Area measured[2]	Area Calculated[3]	Area Calculated[4]
0.361	0.340	0.3611	0.3610
0.722	0.715	0.7223	0.7220
1.444	1.388	1.4440	1.4440
2.888	2,744	2.8881	2.8880
6.317	6.171	6.3176	6.3176
10.83	10.83	10.830	10.830
18.05	18.01	18.051	18.050

[1] Amounts actually injected into the column (mg)

[2] Amounts corresponding to the band profile recorded. curve).

[3] Areas of the profiles calculated with the equilibrium dispersive model.

[4] Areas of the profiles calculated with the POR model.

Similar results were obtained in another recent study, using the same models but the enantiomers of 1-Indanol [30]. A selection of results, covering a wide range of sample sizes is reported in Table IX-6. The injected sample sizes (column 1) were derived from the duration of the injection, the feed concentration and the flow rate (the injection was carried out using the pumping system of the chromatograph). The measured sample size was derived from the recorded band profiles and the calibration curve of the detector (which was not linear in the whole range of sample sizes used in this work). Finally the calculated band areas were obtained by integration of the profiles calculated with the equilibrium dispersive model (column 3) and the POR model (column 4). The agreements between the four numbers in each row is excellent, the main source of error in this work being the determination of the exact amount of sample eluted from the column, *i.e.* the quantitative analysis. This agreement suggests that significant calculations errors can be made when the integration increments are selected merely to satisfy the general criteria of stability and convergence. A more careful choice of these increments should allow a drastic reduction of the calculation errors.

XI. COMPARISON BETWEEN EXPERIMENTAL
AND CALCULATED BAND PROFILES

The literature contains many examples in which band profiles calculated with one or several different models and/or algorithms are compared with experimental band profiles (*e.g.*, 1,26,27]). These comparisons are valid only when the band profiles are calculated from isotherm data that have been accurately measured and properly modeled. The last operation is as critical as the first one. The use of too approximate an isotherm model (*e.g.*, the use of the Langmuir model when it does not really apply, which is far too often) may lead to important disagreements between experimental and calculated band profiles (even when accurate isotherm data had been acquired). More importantly, this results in serious errors in the prediction of band profiles, which is the main reason why isotherm data are measured and modeled and why band profiles are calculated: to determine accurate estimates of the best experimental conditions for optimum production costs.

We prefer to give here the comparison between two sets of experimental band profiles and the profiles calculated with the same isotherm model but using three widely different models of chromatography, hence different methods to account for the effect of the mass transfer kinetics on the band profiles. We have used three models of increasing complexity in this comparison (Figures IX-19 and IX-20). These models are the ideal model, the equilibrium dispersive model, and the POR model. The ideal model (Chapter V) neglects the influence of the mass transfer resistances that are assumed to be zero. It accounts only for the influence of the thermodynamics of phase equilibrium (Figures IX-19a and IX-20a). The equilibrium dispersive model assumes that the rate of the mass transfer kinetics is fast, so its influence is a mere correction (Figures IX- 19b and IX-20b). This correction is made by considering the contribution of the mass transfer resistances as a dispersive effect and by lumping it with the contributions of axial and eddy diffusion [31]. The apparent dispersion coefficient is proportional to the column HETP under linear conditions, which assumes that the rate of the mass transfer kinetics is independent of the sample concentration (see Chapter VII). Finally, the POR model (Figures IX-19c ans IX-20c) takes into account the influence of the mass transfer kinetics when it is much slower and can easily consider the influence of several contributions [28,30]. By contrast with the other two models, the POR model uses two mass balance equations, one in the liquid phase percolating through the column bed, the other one in the stagnant mobile phase in the pores of the stationary phase.

In Figure IX-19, we compare the experimental band profile (symbols) of a large sample of 4-tert-butyl phenol with the profiles calculated using

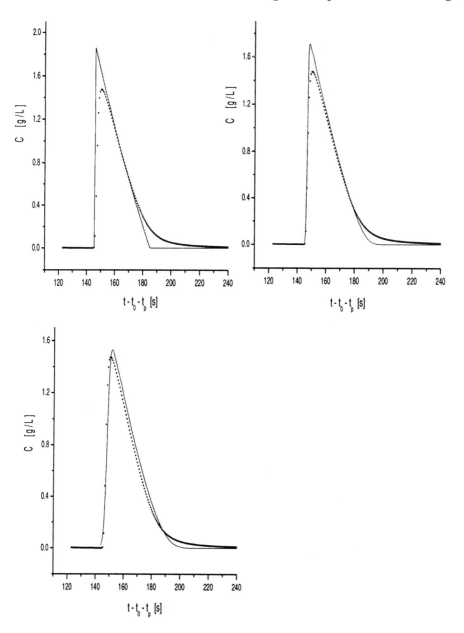

Figure IX-19. Comparison between Experimental and Calculated Band Profile. 4-tert-butyl phenol on 10cm long monolithic C_{18} silica, with CH_3OH / H_2O as the mobile phase (flow velocity, 0.100 cm/sec). Langmuir isotherm [29]. Top Left, Experimental profile and solution of the ideal model. Top Right, Experimental profile and profile calculated with the equilibrium-dispersive model. Column efficiency, 4000 theoretical plates. Bottom Left, Experimental profile and profile calculated with the POR model.

three different models of chromatography. The experimental profile was obtained with a monolithic silica column bonded with C_{18} and eluted with methanol/water as the mobile phase (Chromolith Performance from Merck). The equilibrium isotherm of 4-tert-butyl phenol under these experimental conditions fits best to the Langmuir model [29]. The band profiles have the corresponding Langmuirian shape, with a front shock layer. Figure IX-19a (top, left) shows that the profile predicted by the ideal model is in very good agreement with the experimental profile. The experimental shock layer is very steep, which should be expected with a column efficiency of 4000 theoretical plates and the calculated and experimental retention times are in excellent agreement. Obviously, the tail of the experimental profile is longer than that predicted by the ideal model, since this model does not include any diffusion. The agreement is better between the experimental profile and the profile calculated with the equilibrium dispersive model (Figure IX-19b, top right). The improvement compared with the profile calculated with the ideal model is important only in the tail of the peak, although the retention times are now in excellent agreement and the peak heights are closer. The profile predicted by the POR model (Figure IX-19c, bottom) brings further improvements in terms of peak height and tail profile.

In Figure IX-20, we compare the experimental band profile (symbols) of a large sample of butyl benzoate with the profiles calculated using the same three models of chromatography. The experimental profile was obtained with the same monolithic silica column as used for the tert-butyl phenol just discussed, under similar experimental conditions. The equilibrium isotherm is best accounted for by a modified BET model [29]. This isotherm has an S-shape profile. Accordingly, the band profile has two concentration shocks, one on the rear of the band, the other one on its front. This second shock appears only when the sample size is large enough and the maximum concentration of the elution profile exceeds the concentration of the inflection point of the isotherm [1]. As seen in Figure IX-20a (top left), there is a substantial agreement between the experimental profile and the solution of the ideal model. Obviously, the concentration shocks are eroded and turn into shock layers. In Figures IX-20b and IX-20c, the experimental profile is compared to the profiles calculated with the equilibrium dispersive (top, right) and the POR (bottom) models, respectively. The agreement is better, although the equilibrium dispersive model seems to account better for the front profile of the band and the POR model for the rear profile.

As shown in these two examples, simple models of chromatography account well for the high concentration band profiles of low molecular weight compounds that are neutral or weakly acidic and have a high rate of mass transfer in the chromatographic system used, at least provided that the column used is long enough and has a sufficiently high efficiency. Sophisticated

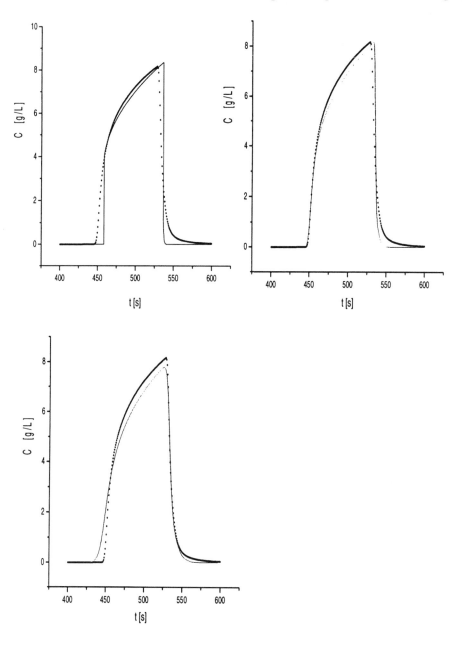

Figure IX-20.   Comparison between Experimental and Calculated Band Profile. Butyl benzoate on a 10 cm long, monolithic C_{18} silica, with CH_3OH / H_2O as the mobile phase. BET isotherm [29]. Top Left, Experimental profile and solution of the ideal model. Top Right, Experimental profile and profile calculated with the equilibrium-dispersive model. Column efficiency, 4000 theoretical plates. Bottom Left, Experimental profile and profile calculated with the POR model.

models such as the POR models or more complex versions of the generated model of chromatography are useful only when the kinetics of mass transfer is slow. There are many discussions and illustrations of the validity of this statement in the literature [32-34].

EXERCISES

1) Give the value of the truncation error and the stability condition of the finite difference equation:

$$\frac{C_{j+1}^n - C_j^n}{2h} + B\frac{C_j^n - C_j^{n-1}}{\tau} = 0$$

for the differential equation

$$\frac{\partial C}{\partial x} + B\frac{\partial C}{\partial t} = 0$$

2) Give the matrices $[A]$ and $[B]$ in the orthogonal collocation method based on the displaced-Legendre polynomials.

3) Analyze the effects of concentration and volume overload on the band profile of a compound using different programs to calculate numerical solutions of the mass balance of chromatography.

4) Calculate the band profiles of two compounds in nonlinear chromatography with a competitive Langmuir isotherm. Compare the effects of increasing the amount of one compound on the profile of the other one.

5) Compare the profiles of the front shock layer of an elution band when calculated with the characteristic scheme and the TVD scheme.

LITERATURE CITED

[1] G. Guiochon, S. Golshan-Shirazi and A. M. Katti, *"Fundamentals of Preparative and Nonlinear Chromatography "*, Academic Press, Boston, MA, 1994.

[2] R. D. Richtmayer and K. W. Morton, *"Difference Methods for Initial Value Problems,"* Interscience, New York, NY, 2nd edn., 1967.

[3] L. Lapidus and G. F. Pinder, *"Numerical Solution of Partial Differential Equations in Science and Engineering,"* Wiley, New York, NY, 1982, pp. 255-260.

[4] G. E. Forsythe and W. R. Wasow, *"Finite Difference Methods for Partial Differential Equations"*, Wiley, New York, NY, 1960.

[5] A. R. Mitachell ande D. F. Griffiths, *"The Finite Difference Method in Partial Differential Equations"*, Wiley, New York, NY, 1980.

[6] B. Lin and G. Guiochon, *Separat. Sci. Technol.*, **24** (1988) 31.

[7] L. Jacob, P. Valentin and G. Guiochon, *Chromatographia*, **4** (1971) 6.

[8] R. W. MacCormack, *SIAM - AMS Proc.*, **11** (1978) 130.

[9] J. R. Douglas and T. F. Russel, *SIAM J. Numer. Anal.*, **19** (1982) 871.

[10] V. R, Casulli, R. T. Cheng, and Burgurelli, *Numer. Meth. Nonlinear Probl.*, **2** (1984) 962.

[11] G. Fairweather, *"Finite Element Galerkin Methods for Differential Equations"*, M. Dekker, New York, NY, 1978.

[12] K. Wang and B. Lin, *Chin. J. Comput. Phys.*, **9** (1992) 330.

[13] C. Deboor amd B. Schwartz, *SIAM J. Numer. Anal.*, **10** (1973) 582.

[14] M. E. Davis, *"Numerical Methods and Modeling for Chemical Engineers"*, Wiley, New York, NY, 1984.

[15] J. Villadsen, M. L. Michelsen, *"Solutions of Differential Equation Models by Polynomial Approximation,"* Prentice-Hall, New York, NY, 1978. ex [104]

[16] Z. Ma and G. Guiochon, *Comput. and Chem. Eng.*, **15** (1991) 415.

[17] P. D. Lax, *Comm. Pure Appl. Math.*, **7** (1954) 159.

[18] P. D. Lax and B. Wendroff, *CPAM*, **13** (1980) 217.

[19] P. D. Lax and B. Wendroff, *"Difference Schemes with High Order of Accuracy for Solving Hyperbolic Equations*, NYU Rep. 9757, Courant Inst. Math. Sci., New York, NY, 1962.

[20] A. Harten, *Math. Comput.*, **36** (1978) 363.

[21] A. Harten, *J. Comput. Phys.*, **49** (1983) 357.

[22] R. Courant, K. Friedrichs and H. Lewy, *Math. Ann.*, **100** (1928) 32.

[23] B. Lin, J. Wang and S. Yang, *Chin. J. Comput. Phys.*, **12** (1995) 172.

[24] J. Baker, *"Finite Element Computational Fluid Mechanics,"* McGraw- Hill, New York, NY, 1983.

[25] G. Szego, *"Orthogonal Polynomials,"* American Mathematical Society, Colloquium Publications, **23** (1959).

[26] P. Henrici, *"Elements of Numerical Analysis"*, Wiley, New York, NY, 1964.

[27] G. Guiochon, *J. Chromatogr. A*, **965** (2002) 129.

[28] K. Kaczmarski, *Comput. Chem. Eng.*, **20** (1996) 49.

[29] F. Gritti, W. Piatkowski and G. Guiochon, *J. Chromatogr. A*, (2002) In Press.

[30] D. Zhou, K. Kaczmarski, A. Cavazzini, X. Liu and G. Guiochon, *J. Chromatogr. A*, (2002) In Press.

[31] J. C. Giddings, *"Dynamics of Chromatography,"* M. Dekker, New York, NY, 1965.

[32] K. Kaczmarski, D. Antos, *J. Chromatogr. A*, **756** (1996) 73.

[33] K. Kaczmarski, D. Antos, H. Sajonz, P. Sajonz and G. Guiochon, *J. Chromatogr. A*, **925** (2001) 1.

[34] K. Kaczmarski, A. Cavazzini, P. Szabelski, D. Zhou, X. Liu and G. Guiochon, *J. Chromatogr. A*, **962** (2002) 57.

SUBJECT INDEX